Modern Europe

Place, Culture and Identity

Edited by

BRIAN GRAHAM

Professor of Human Geography, University of Ulster
at Coleraine, Northern Ireland

A member of the Hodder Headline Group
LONDON • SYDNEY • AUCKLAND

First published in Great Britain in 1998 by Arnold,
a member of the Hodder Headline Group
338 Euston Road, London NW1 3BH

http://www.arnoldpublishers.com

British Library Cataloguing in Publication Data
A catalogue record for this book is available from the British Library

Library of Congress Cataloging-in-Publication Data
A catalog record for this book is available from the Library of Congress

Production Editor: Wendy Rooke
Production Controller: Sarah Kett
Cover designer: Terry Griffiths

ISBN 0 340 67698 1 (pb)

Typeset by Saxon Graphics Ltd, Derby
Printed and bound in Great Britain by J. W. Arrowsmith Ltd, Bristol

Contents

List of figures   vii

List of tables   ix

List of contributors   x

Preface   xi

List of abbreviations – Europe of the acronyms   xiv

Introduction   Modern Europe: fractures and faults
BRIAN GRAHAM

Europe and identity   1
Themes of the book   2
Europe as idea and place   3
Structure of the book   9
References   14

Part I: Europe and Europeanness

1   **The past in Europe's present: diversity, identity and the construction of place**
BRIAN GRAHAM

Introduction: contested geographies   19
The past in the present; the present in the past   21
Integration and diversity   29
The renegotiation and contestation of space and place   35
The renegotiation of the nation-state: towards a European
     identity and heritage   41
Conclusion   45
References   46

Part II: The past in the present; the present and the past

2   Economies in space and time: economic geographies of
    development and underdevelopment and historical
    geographies of modernization
    MICHAEL DUNFORD

    Introduction: economic disparities and pathways to modernization
        and development   53
    Time and space: understanding historical systems   60
    Secular trends and the transition to capitalism   64
    Geographies of modernization and of the transition to capitalism   66
    Modern industrial development   72
    Catching up and falling behind: uneven development   77
    Modernization and emigration   82
    Conclusion   85
    References   86

3   War and the shaping of Europe
    MICHAEL HEFFERNAN

    Introduction: geography and war   89
    War and the idea of Europe   91
    Europe built on peace?   95
    Europe built on war?   100
    Conclusion: war, memory and identity in Europe   113
    References   115

4   'The chickens of Versailles': the new Central and Eastern Europe
    DENIS J. B. SHAW

    Introduction   121
    Definitions of Europe: historic dimensions   122
    Imperialism and the state   128
    Economic dimensions   131
    Peoples and places   136
    Conclusion   139
    References   140

Part III: The nature of European integration and the consequences of diversity

5   The political geography of European integration
    RICHARD GRANT

    Introduction   145
    Evolution of European institutions and their roles in
        policy-making   148

The EU as an institutional network 151

Policy-making in the EU 153

Decision-making in the Council of Ministers 155

Reform in the EU 157

Conclusion: towards the creation of public legitimacy? 160

References 162

6 Convergence, cohesion and regionalism: contradictory trends in the new Europe

MARK HART

Introduction 164

Contemporary political and economic challenges for the EU 165

Regional disparities 170

The convergence debate 174

Cohesion and regionalism 178

Conclusion 182

References 183

7 Room to talk in a house of faith: on language and religion

COLIN H. WILLIAMS

Introduction 186

Religion, language and diversity: the socio-cultural renegotiation of Europe 189

The nature of civil rights: inclusion versus exclusion 192

The politics of equal respect in multicultural societies 196

Language policy and conflict in Europe 199

Linguistic hegemony 202

Conclusion 206

References 207

Part IV: Identity and the renegotiation of the meanings of European place

8 European landscape and identity

JOHN AGNEW

Introduction 213

Stories of landscape and identity, simple and complex 213

Italian unification and the problem of a landscape ideal 217

Florence and the Tuscan landscape ideal 219

Rome and the Roman landscape ideal 224

The difficulty of realizing a singular landscape ideal in Italy 230

Conclusion 232

References 233

9 The question of heritage in European cultural conflict
JOHN E. TUNBRIDGE

Introduction: culture, nationalism and heritage in Europe   236
European minorities old and new   239
Misplaced peoples, misplaced heritages: the case of
   'Lost Germany'   243
German heritage: misplaced ideology as cultural conflict   249
Cultural and heritage conflict resolution: a postmodern Europe?   255
Conclusion   258
References   259

10 The conserved European city as cultural symbol: the meaning of the text
G. J. ASHWORTH

Introduction: the city as text   261
Who is the producer of the conserved European urban
   landscape as cultural symbol?   264
What culture is symbolized by the conserved European city?   266
For whom is this European culture symbolized?   279
Conclusion: the conserved European city as the expression of
   power, culture and identity   283
References   283

11 The European countryside: contested space
HUGH D. CLOUT

Introduction: the perspective of time and space   287
The dynamics of rural change   291
New policies to meet new imperatives   295
The planned retreat   297
Standing back from the present   300
The quest for diversification   302
Conclusion: the future of the rural past   304
References   307

Epilogue   Europe's geographies: diversity and integration
BRIAN GRAHAM

References   316

Index   318

Figures

I.1   Physical geography of Europe
I.2   The extent of the Roman Empire, *c.* AD 280
I.3   The changing geographies of Germany, 1871–1990
I.4   The countries of Europe
1.1   The medieval pilgrimage routes to Santiago de Compostela
1.2   The Christian *Reconquista* of Iberia
1.3   *Santiago Matamoros*
1.4   *Sanctus Jacobus*
1.5   The fragmented political geography of France, *c.* 1200
1.6   The County of Toulouse, *c.* 1200
2.1   Europe's vital axis
2.2   Europe, *c.* 814: the Carolingian Empire; the Byzantine Empire (610–1453); Kievan Rus (880–1054); the Emirate of Córdoba
2.3   The Great Schism between Latin and Orthodox Christianity, 1054
2.4   Rise of the Ottoman Empire, 1300–1683
2.5   Western part of the Mongol Empire (1206–1696), *c.* 1300: the Khanate of the Golden Horde
2.6   Kondratieff cycles and the secular trend
2.7   Europe's major industrial areas in 1815
2.8   Europe's major industrial areas in 1875
3.1   The expansion of France, 1648–1715
3.2   Revolutionary and Napoleonic Europe
3.3   Napoleon's invasion of Russia, 1812–13
3.4   Impact of World War I on Europe's political geography
3.5   The Spanish Civil War: the collapse of Republican Spain
3.6   Maximum extent of German occupation, November 1942
3.7   The destruction of European Jewry during World War II
3.8   Estimated percentage population loss during World War II
3.9   British and Commonwealth World War I cemeteries, Western Front

4.1   Central and Eastern Europe, 1921–89
4.2   Central and Eastern Europe, *c.* 1815
4.3   Central and Eastern Europe, *c.* 1914
5.1   The countries of the European Union and its various enlargements
5.2   Decision-making in the European Union
6.1   European Union: GDP per capita by region, 1993
6.2   European Union: unemployment rates by region, 1995
7.1   Ethnic divisions of Europe in the early nineteenth century
7.2   Models of ethnic integration
8.1   Unification of Italy
8.2   Silvestro Lega: *Paese con contadini* (Landscape with peasants), *c.* 1871.
8.3   Rafaello Sernesi: *Radura nel bosco* (Forest glade), *c.* 1862–3.
8.4   Telemaco Signorini: *Sul greto d'Arno* (On the bed of the river Arno), *c.* 1863–5
8.5   Rome
8.6   Excavations of the Roman Forum at the turn of the twentieth century
8.7   The Via dell'Impero (now the Via dei Fori Imperiali) shortly after its opening in 1933
9.1   The changing political geography of North-Central and North-Eastern Europe during the twentieth century
9.2   Gdańsk: St Mary's Church (Gothic red brick) and the statue of Neptune in the central square
9.3   Gdańsk: the old city and recently reconstructed harbourside warehouses located in the area still ruined from World War II
9.4   Kaliningrad: reconstructed cathedral on former city-centre 'island' in the Pregolya river
9.5   Kaliningrad: eastern gate of the old city (Königs Tor) with trees growing from ruins
9.6   Weimar
9.7   Berlin
10.1 Paris
10.2 Barcelona
10.3 Historic cities in the Netherlands
10.4 Groningen: late nineteenth-century boulevards, built upon dismantled fortifications, interrupt the late medieval plan
11.1 Rural landscapes of Europe
11.2 National parks, nature reserves and environmentally disrupted areas in Poland
11.3 Medieval expansion of German settlement into East-Central Europe
11.4 Tourism routes and rural landscapes in the Calvados area of Normandy

Tables

I.1 Population of European countries, 1997
2.1 Concepts of historical time and geographical space (geohistorical time–space)
2.2 Population changes in Europe, 1500–1800
2.3 Agricultural employment and output in the European periphery, 1860 and 1910
2.4 Labour productivity in the cotton industry
2.5 Comparative economic development of European countries, 1820–1992
2.6 Trends in international inequality in Europe and the New World, 1820–1992
5.1 European Union enlargement: criteria for membership and applicant states as of 1997
5.2 Present and possible future allocations of institutional powers among EU Member States

Contributors

Professor John Agnew, Department of Geography, University of California, Los Angeles, 405 Hilgard Avenue, Los Angeles, California 90095–1524, USA

Professor Gregory Ashworth, Department of Physical Planning and Demography, University of Groningen, PO Box 800, 9700 Av Groningen, The Netherlands

Professor Hugh Clout, Department of Geography, University College London, 26 Bedford Way, London WC1H 0AP, UK

Professor Michael Dunford, School of European Studies, University of Sussex, Falmer, Brighton BN1 9QN, UK

Professor Brian Graham, School of Environmental Studies, University of Ulster, Coleraine BT52 1SA, UK

Dr Richard Grant, Associate Professor, Department of Geography, Syracuse University, 144 Eggers Hall, Syracuse, New York 13244–1090, USA

Dr Mark Hart, School of Public Policy, Economics and Law, University of Ulster, Jordanstown, Newtownabbey BT37 0QB, UK

Dr Michael Heffernan, Department of Geography, University of Loughborough, Loughborough, Leicestershire LE11 3TU, UK

Dr Denis J. B. Shaw, School of Geography, The University of Birmingham, Edgbaston, Birmingham B15 2TT, UK

Dr John E. Tunbridge, Associate Professor, Department of Geography, Carleton University, B-349 Loeb Building, 1125 Colonel By Drive, Ottawa, Canada K1S 5B6

Professor Colin H. Williams, Department of Welsh, University of Wales, Cardiff, PO Box 910, Cardiff CF1 3XW, UK

Preface

The general objective of this book is to examine the ways in which the contemporary heterogeneity of Europe's political, cultural and economic geography is implicated in often conflicting constructs of identity that, in turn, reflect a complex fusion of contemporary and historical processes and the contested meanings placed upon them. The book transcends the conventional geographical distribution of labour between the study of the past and that of the present. In so doing, the text eschews a conventional historical narrative approach, while still intent on addressing the time dimension largely ignored in many other texts on Europe's contemporary geography. To say that in its concern with the endless renegotiation of the meanings of place, the book visualizes Europe within either past-to-present or present-to-past terms would be to over-simplify. Rather – drawing on a wide array of examples and case studies – it explores the ways in which a meshing of past and present factors, policies and ambitions accounts for Europe's characteristic cultural diversity, which apparently stands at odds with the policies of political and economic integration being orchestrated through the European Union.

   The recent and potential future enlargement of the Union, the ratification of the integrationist Treaty of Maastricht, and the collapse of Communism in Eastern Europe are among factors stimulating optimistic scenarios of a 'New Europe', underwritten by a common European identity that might transcend the array of myriad local, regional and national identities which constitute a primary legacy of the past. However, the widespread – if variable – resistance to integration, the collapse of Yugoslavia through secession and civil war, the numerous conflicts within and between the republics of the former Soviet Union, and the persistence of nationalistic antagonisms in West as well as East are only the most telling examples of the enduring importance throughout Europe of overlapping layers of identity largely defined by selective interpretations of past events. It is also difficult to ignore the malign legacy of war and atrocity, which has always been one of the critical processes shaping the political and cultural map of Europe.

Nor is the contribution and relevance of the past restricted solely to issues of contested cultural identity. The patterns of uneven economic development that constitute one of the more potent destabilizing forces in contemporary European society are more than simply the products of capitalist modernization and post-Fordist economic change. They too are related to long-term transformations in the control and organization of economic production that date back to the Middle Ages, to Europe's internal and external history of imperialism, and to the historical emergence of identifiable cores and peripheries in patterns of wealth and power. Again, the concentration of contemporary wealth and power in urbanized regions and the repercussions of these patterns for various expressions of identity can be explained only by recourse to historical processes of urbanization, population mobility and agricultural change and modernization.

Thus, at the core of the book is the apparent paradox between political and economic integration, and the enduring relevance of social, cultural, economic and political diversity, derived directly from the continent's past. This broadly ranging content reflects my long-standing belief that the political, economic and cultural realms cannot be regarded as separate entities. All impact on the construction of simplifying synecdoches that relate the self to like-minded people in an array of identities. Again, the book's concern with the ramifications of the continuous renegotiation of the meanings of European place reflects the perspective that the past is a social construct, framed in contemporary circumstances and heavily implicated in processes of validation and legitimation. Although the tenor of the book is firmly in favour of a European dimension to identity, it does not underestimate the difficulty of creating such a construct in multicultural societies riven by many contested depictions of Otherness. It is clear that the absence of a readily accepted Europeanness, within the complex layerings of identity that already exist in Europe, remains one of the most intransigent barriers to the project of integration upon which much of the continent is embarked. While Europe is not yet the European Union, increasingly it will be as plans for enlargement become more transparent. But that latter process will serve only to complicate further the difficult and often irreconcilable tensions and patterns of inclusion and exclusion that emanate from Europe's contested pasts and presents.

Despite its general commitment to – and acceptance of – a European dimension to identity, the book embraces a variety of perspectives and concerns. The contributors do not necessarily subscribe to an agreed political agenda. There is, however, a consensus that Europe can be viewed only through narratives of diversity, inclusivity, hybridity and fluidity, contexts – whether political, economic or cultural – which regard multiculturalism and fragmented, multi-layered identities as essentially positive forces. No one advocates here that European identity be defined as a minimal common denominator to which all can hopefully relate, nor can Europe's distasteful past simply be discarded. I found editing the book to be a profoundly illuminating and thought-provoking process, and would like to express my gratitude to all the contributors, who met their various deadlines with unfailing good humour. My thanks also to Kilian McDaid of the University of Ulster, Hazel Lintott of the Geography Laboratory, University of Sussex, and Peter Robinson of the Department of Geography, Loughborough University, for their invaluable help in preparing the maps and diagrams. Above all, I would like to acknowledge my gratitude to Laura McKelvie at Arnold, for her initial faith in the project and for much subsequent enthusiastic and ever-encouraging support.

Brian Graham
Belfast, October 1997

Abbreviations – Europe of the acronyms

Europe – past and particularly present – has generated a mass of acronyms. Those employed in this book are listed here and are also defined at first point of usage in the text.

BSE	bovine spongiform encephalopathy
CAP	Common Agricultural Policy
CEC	Commission of the European Communities
CIS	Commonwealth of Independent States
CMEA	Council for Mutual Economic Assistance (also Comecon)
COR	Committee of the Regions
Coreper	Committee of Permanent Representatives (EU)
EC	European Community
ECJ	European Court of Justice
ECSC	European Coal and Steel Community
EEC	European Economic Community
EMU	European Monetary Union
ERDF	European Regional Development Fund
ERM	Exchange Rate Mechanism
ESA	Environmentally Sensitive Area
ESF	European Social Fund
ETA	An acronym in Basque for 'Homeland and Freedom'
EU	European Union
GATT	General Agreement on Tariffs and Trade
GDP	Gross Domestic Product
GDR	German Democratic Republic
GIS	Geographical Information Systems
GOR	Government Offices in the Regions (UK)
MEP	Member of the European Parliament
NATO	North Atlantic Treaty Organization
PEU	Pan-European Union
PPS	Purchasing Power Standards
SEM	Single European Market
USSR	Union of Soviet Socialist Republics

| Introduction |

Modern Europe: fractures and faults

BRIAN GRAHAM

Europe and identity

This book is concerned with the interactions between the European past and present and the ways in which these have created a complex geographical patterning of social and economic diversity. In examining the implications of this heterogeneity for the contestation of identity and the meaning of place throughout contemporary Europe, the central issue of the book is this. The political and economic integration of the continent is proceeding inexorably under the aegis of the European Union (EU), the spatial extent of which is set to expand in the near future as various countries in Central, Eastern and Southern Europe are admitted to membership. The development of the Single European Market (SEM), underpinned by its infrastructure of Trans-European Networks, can be viewed as an essentially centripetal force of integration, imposed on past political and economic heterogeneity. None the less, Europe remains characterized by manifest cultural diversity, which results in a complex fragmentation of identity and allegiance, largely shaped by contested readings of the past. How – or even can – these apparent contradictions be reconciled?

Despite its importance for state and society, identity remains a highly ambiguous concept. In general terms, it incorporates values, beliefs and aspirations, which are used to construct simplifying structures of sameness that identify the self with like-minded people. Identity is a multi-faceted phenomenon that embraces a range of human attributes, including language, religion, ethnicity, nationalism and shared interpretations of the past (Guibernau 1996), and constructs these into discourses of inclusion and exclusion – who qualifies and who does not (Said 1978, 1993). Central to the entire concept of identity, therefore, is the idea of the Other: groups with competing – and often conflicting – beliefs, values and aspirations. The attributes of Otherness are thus fundamental to representations of identity, which are constructed in counter-distinction to them. As Douglas (1997, pp. 151–2) argues,

> the function of identity lies in providing the basis for making choices and facilitating relationships with others while positively reinforcing these choices. . . . In emphasising sameness, group membership provides the basis for supportive social interaction, coherence and consensus. As identity is expressed and experienced through communal membership, awareness will develop of the Other. . . . Recognition of Otherness will help reinforce self-identity, but may also lead to distrust, avoidance, exclusion and distancing from groups so-defined.

These axes of belonging and rejection are not necessarily neatly reconciled in individual and group identity, nor – as Northern Ireland and Bosnia only too graphically illustrate – are they inevitably defined by orderly spatial divisions. We must resist also the idea that a sense of common identity comes into being because of an integration and homogenization of disparate cultures. As Colley (1992, p. 6) contends, Great Britain, for example,

> did not emerge by way of a blending of the different regional or older national cultures contained within its boundaries . . . the Welsh, the Scottish and the English remain in many ways distinct peoples in cultural terms, just as all three countries continue to be conspicuously subdivided into different regions.

In this statement can be isolated what essentially constitutes the central motif of the discussions that follow in this book. Identity is not a discrete social construction that is territorially defined in coherent spatial entities; rather, identities – and their defining criteria – overlap in complex ways and geographical scales. Perhaps best visualized as a multiplicity of superimposed layerings, identity has potentially conflicting supranational, national, regional and local expressions, in turn fractured by other manifestations of sameness – religion, language, high culture – that are not necessarily defined in terms of those same spatial divisions. Inevitably, therefore, identity is socially and geographically diverse rather than neatly bundled, with nationalism, perhaps its most potent expression, existing to simplify such heterogeneity into simplifying representations – synecdoches – of sameness.

Themes of the book

Within the general context of the contested and ambiguous nature of identity, which is a characteristic feature of Europe past and present, the arguments in this book embrace three broad themes. While these are addressed separately here, they should not be regarded as discrete theses. Rather the themes reflect varying – but intermeshing – perspectives upon the central European paradox of diversity and integration.

- First, it is argued that the contemporary heterogeneity of Europe's social, cultural, political and economic geography can be understood only through an understanding of the ways in which a complex fusion of present and past processes has created overlapping layers of place, which are not necessarily reconcilable. On one hand, these can be allied to the centrifugal forces of political integration, while simultaneously legitimating the centripetal tendencies of nationalism, regionalism or localism.
- Second, the discussion addresses the fundamental geographical diversity – operating at a variety of scales – which underlies the veneer of Europe's evolving politico-economic integration. Even that latter set of processes is itself circumscribed by patterns of uneven development that again reflect the fusion of past and present processes.
- Third, the book examines the ways in which the meanings of Europe and its places are – and always have been – continuously renegotiated, reformulated or remade in response to the demands of ever-changing constructs of society and economy and their concomitant power relationships. In so doing it addresses how a politically integrated, yet culturally and economically diverse, Europe might accommodate its heterogeneity of people and their mosaic of often conflicting identities. This theme can be visualized in two ways. The first is represented by the actual physical remaking of space through war and negotiation, the fluid and constantly altering political mosaic of past and present Europe. In some contrast, the second invokes the continuous representational remaking or renegotiation of place and identity, manifested particularly through nationalism and its contestation.

Europe as idea and place

Both as an idea and a place, Europe is ill-defined. The idea of Europe emerged only in the late seventeenth and early eighteenth centuries, part and parcel of an early modern secularization that sought to replace Judaeo-Christianity as the continent's common cultural focus (Wilson and van den Dussen 1993; see Chapter 3). That idea remains contested to the present day as Europeans seek a shared resolution to conflicting forces of globalization, nationalism and regionalism. As a place, much of Europe can be readily circumscribed by its Atlantic and Mediterranean littorals. For more than five centuries, however, 'the cardinal problem in defining Europe has centred on the inclusion or exclusion of Russia' (Davies 1996, p. 10) (not to mention Turkey). As Russian power declined in the late twentieth century and the former Soviet Union – the USSR – broke up into its constituent republics (creating the Commonwealth of Independent States – the CIS), and the EU has expanded – in part as a conscious counter-balance to the global power of the United States – the epicentre of Europe has shifted eastward. Inevitably, the former Communist countries of Central and Eastern Europe seek to emulate the wealthy Member States of the EU to the west, and several will soon be admitted within the Union's fortress frontiers of capitalist privilege. Attenuated Russia continues – as perhaps it always has done – to look in different directions. In some respects, it has oriented itself towards the West, not least because of the inward investment necessary for its financial and social reconstruction, whereas in others it seems increasingly to face eastwards towards Siberia, China and Japan.

In terms of physical extent, Europe is surprisingly small. London and Moscow are only 2400 km apart, little more than half the distance between New York and San Francisco, while the continent's wealthiest region stretches no more than 1000 km between London and Milan (roughly comparable, say, to the distance between New York and Chicago). None the less, Europe is far more fragmented in physical, political and cultural terms than is North America, and salient geographical changes occur over remarkably small distances. Europe is itself a peninsula, and its geography is dominated by a central continental core extended by peninsulas (Iberia, Italy, Scandinavia, the Balkans) and islands (Ireland, Britain, Corsica, Sardinia, Sicily) (Figure I.1). The Mediterranean Basin in the south, together with its subsidiary seas the Adriatic and Aegean, the Black Sea in the east, and the North and Baltic Seas in the north all contribute to this fragmentation, although, equally, they have also functioned historically as routeways that generated trade, movement and unity. Thus Fernand Braudel (1972), one of Europe's most distinguished historians, envisaged the Mediterranean world, for example, as a unified and coherent region, in which the almost timeless realities of physical environment and human geography transcended cultural and political fragmentation.

While its internal seas can unite as well as separate, the broad parameters of Europe's relief are more divisive. This is dominated by the high lands that stretch all of 10 000 km from Galicia in north-west Spain to the Bosporus, the entrance to the Black Sea and the south-eastern corner of Europe. The Cordillera Cantabrica of northern Spain is succeeded by the Pyrenees and the Massif Central, which – although not particularly high – effectively divides Germanic-influenced northern France from its Latin Mediterranean counterpart, the Midi. Although now tamed by road and rail tunnels, the Alps historically separated Mediterranean Europe – the continent's traditional heartland – from the less developed regions to north and west. The long and slow transfer of Europe's locus of power away from the Mediterranean across the Alps to the newly emergent trading and industrial cities of North-West Europe was one of the defining features of the Middle Ages (Graham 1997). East again, the Alps are succeeded by the sweeping arc of the Carpathians and the fractured topography of the Balkans. To the north and east lie the steppelands of Eurasia across which the frontiers of a 'tidal Europe' have ebbed and flowed (Parker 1968; Taylor 1996; see Chapter 4).

This swath of mountain and upland determined the historic pattern of communications in Europe and continues to do so into the present day.

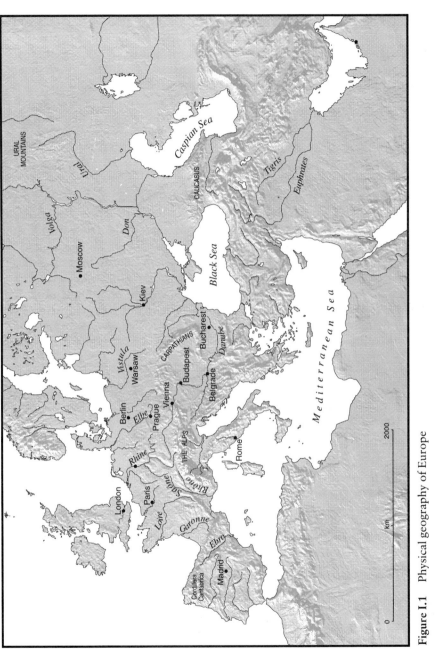

Figure I.1 Physical geography of Europe

North–south movement is still comparatively difficult and requires expensive infrastructure such as the Alpine tunnels. Even today, only the Rhône–Saône and Aude–Garonne valleys, the historic routeways that served Rome's legions, allow relatively easy terrestrial access between Mediterranean and Atlantic Europe. In Eastern Europe, the Danube has always acted as an axis of east–west movement, but its course is somewhat tangential to Europe's principal axes of transformation. The other salient feature of Europe's topography is the great northern plain – albeit divided by a succession of major rivers – that stretches from Russia to France, and across which so much of the continent's brutal and near-continuous history of warfare has occurred. Whether mountain or plain, a far more finely differentiated mosaic of physical geography lies beneath these broad strokes, one vital factor contributing to the intensely heterogeneous regionality of contemporary Europe.

Another is to be found in the political, economic and cultural fault lines that historically have transcended this physical diversity. Their significance lies in the ways in which they have created convoluted and complexly varying trajectories of economic and cultural change (see Chapter 2). For instance, much of Europe's social history over the past two millennia reflects divisions that can be traced to the influences of the Roman Empire. Between the first century BC and *c.* AD 350, that part of Europe west of the Rhine, south of Hadrian's Wall and south and east of the Danube was part of an empire, linked by military power, trade, roads and towns, that also embraced the entire Mediterranean littoral (Figure I.2). By the fourth century AD, it was

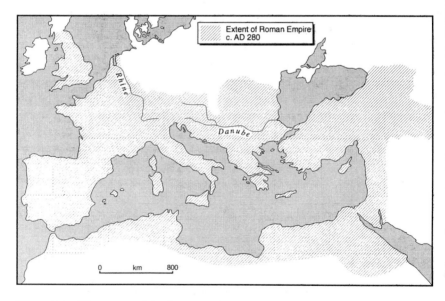

Figure I.2   The extent of the Roman Empire, *c.* AD 280
Source: Based on *The Times Atlas of World History* (1986, p. 89)

also predominantly Christian. Beyond its frontiers, urbanization and political organization developed much later; the first towns in Ireland, for example, date only to the ninth and tenth centuries. In the eleventh century, the schism between Latin Christianity in the west and its Orthodox counterpart in the east introduced a north–south cultural boundary that stretched from the Baltic to the eastern Mediterranean (see Figure 2.3). Meanwhile, in Iberia and along the fluid south-eastern and eastern borders of Europe, Christianity – in whatever guise – faced and often fought its Islamic rival. Another absolute fracturing was the post-World War II division of Europe between East and West, two power blocs separated by the Iron Curtain, one of the most systematically fortified boundaries in human history. Today, the boundary of the EU defines Europe's most significant fault line. Interwoven with each other and the underlying physical mosaic, these – and numerous other – cleavages help define the cultural, economic and political diversity that is modern Europe.

A geopolitical map of remarkable recency is superimposed upon this heterogeneity. The essence of the European state lies in its territoriality, and the imposition of sovereignty over a particular bounded space. The present boundaries of almost all European states, most of which reflect attempts to equate cultural nations with political territories in nation-states, are products of the nineteenth and twentieth centuries and owe much to war or its aftermath (see Chapter 3). Germany, for example, finally unified only in 1871, has altered dramatically in extent several times during the intervening period (Figure I.3). After World War I, it lost territory to the newly established Polish state, only to reclaim this (and more) following Hitler's invasion of Czechoslovakia and Poland in 1938–9. Following World War II, a much shrunken Germany was divided by the victors – the United States, Britain, France and the USSR – between the capitalist Federal Republic (West) and the Communist German Democratic Republic (East). The country's present boundaries were established as recently as October 1990, when West and East were reunited.

In total, around 800 million people (including 148 million in the Russian Federation and 63 million in Turkey, both with extensive Asian territories) live in almost 50 states in Europe, nearly half (370 million) being accounted for by the present 15 Member States of the EU (Figure I.4; Table I.1). (As a comparison, the United States has a population of just over 270 million.) These peoples are differentiated by a multiplicity of languages, mostly subdivided into numerous regional dialects, and are characterized by varied religious and ethnic allegiances. Moreover, in keeping with the fluid nature of the definition of Europe, it is not possible to be completely precise as to the exact number of its countries. Is Chechnya, for example, a breakaway region (as the Russia government insists) or an independent country as it claims? While Slovenia and Croatia have escaped from the wreckage of the former Yugoslavia, can post-Dayton Bosnia actually be described as an independently functioning state?

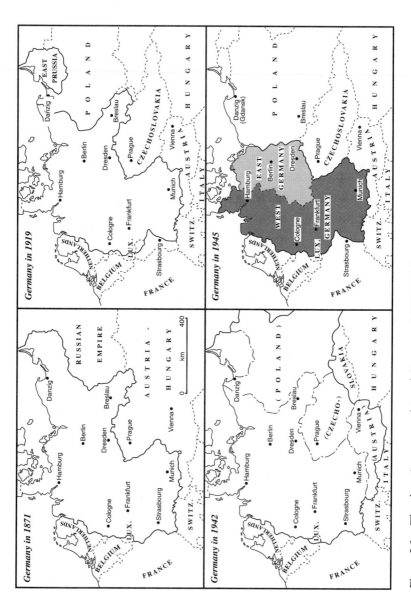

Figure I.3 The changing geographies of Germany, 1871–1990
Source: After Jess and Massey (1995, p. 166)

Figure I.4 The countries of Europe

Structure of the book

Part I: Europe and Europeanness

In addressing the contemporary implications of this heterogeneous continental geography and its diversity of peoples and identities, the book is divided into four parts, three of which are broadly defined by the overlapping and intermeshing themes detailed above: the interweaving of past and present; integration and diversity; and the renegotiation and contestation of space and place. The single chapter in Part I provides a conceptual context for the book and presents a detailed discussion of the themes themselves. It examines the layerings of identity that have derived from – and been defined by – the endless interplay between past and present that has created Europe's diagnostic diversity. The weakness of a European dimension to identity is discussed, given that the forces of continental integration are largely political and economic. It is argued that Europeanness should not subsume national, regional and local identity, but instead add yet another dimension that reflects the complex dismantling of the synonymity of territoriality, sovereignty, nationalism and the state in the new Europe. It would embrace and accentuate

· Table I.1 Population of European countries, 1997

Country	Population (estimated in millions, 1997)	Country	Population (estimated in millions, 1997)
Albania	3.4	Lithuania	3.7
Austria	8.2	Luxembourg	0.4
Azerbaijan	7.6	Macedonia (FYR of)	2.2
Belarus'	10.3	Malta	0.4
Belgium	10.2	Monaco	0.03
Bosnia-Herzegovina	3.8	Netherlands	15.7
Bulgaria	8.4	Norway	4.4
Croatia	4.5	Poland	38.6
Cyprus	0.8	Portugal	9.8
Czech Republic	10.2	Moldova	4.5
Denmark	5.2	Romania	22.6
Estonia	1.5	Russian Federation	147.7
Finland	5.1	San Marino	0.025
France	58.5	Slovakia	5.4
Georgia	5.4	Slovenia	1.9
Germany	82.2	Spain	39.7
Greece	10.5	Sweden	8.8
Hungary	10.0	Switzerland	7.3
Iceland	0.3	Turkey	62.8
Ireland	3.6	Ukraine	51.4
Italy	57.2	United Kingdom	58.2
Latvia	2.5	Yugoslavia	10.4
Liechtenstein	0.03		
		Total	805.4

Note: Populations for the Russian Federation and Turkey include Asian territories of those countries

Source: United Nations (1997): http: //www.un.org/depts/unsd/social/popltion.htm

notions of multiculturalism and of complex and overlapping – rather than intersecting – layers of identity, while striving to be inclusive of all the continent's peoples, constructs ultimately necessary to validate and legitimate political and economic integration.

Part II: The past in the present; the present and the past

The chapters in Part II explore three aspects of the ways in which Europe's contemporary diversity of peoples, places and identities have been produced by the interaction of numerous sets of past processes. The analysis begins in Chapter 2 with Michael Dunford's study of the long-term time changes – and their superimposition one upon the other – which have created the complex synthesis that is the economic geography of contemporary Europe. The outcome is a disparate mesh of trajectories of modernization, which are, and always have been, constrained by circumstances inherited from the past. Thus, to a significant extent, explanations for the patterning of contemporary economic contrasts in Europe are to be found in diverse pasts, defined by an array of processes with different temporalities and durations, divergent spatial extents and varying degrees of durability in their effects.

In Chapter 3, Michael Heffernan examines the impact of war on the political and cultural identity of modern Europe in a study which addresses the paradox that, although European unity has always been evoked in the name of peace and harmony, the unification of the continent – prior to the advent of the EU – has hitherto been achieved only by violence and then only momentarily. Nevertheless, the very idea of Europe emerged from the attempts to create an international system in which conflict and warfare between emerging states could be, if not eradicated, then at least controlled. The discussion demonstrates that the concept of European unity and integration, as the only real solution to the endemic violence produced by the system of nation-states, is not a recent phenomenon but one with a long history.

Continuing the theme of the endless interweaving of past and present, present and past, in Chapter 4 Denis Shaw analyses the relationship between ethnic identity and the formation of nationalism in the particularly complex circumstances of Central and Eastern Europe. He shows that the idea of Europe is itself a much contested enterprise, and that perceptions of what is 'West', 'Central' and 'East' have changed through time. The collapse of Communism at the end of the 1980s has 'shifted' the former Eastern European states, including Hungary, Poland and the Czech Republic, into Central Europe, the East now being defined by the successor states to the USSR. But wherever the countries are located, their present and past instability reflects the imposition of territorial boundaries on a mosaic of ethnic groups and cultures, over which was then superimposed the superficial veneer of centrally planned Communist economics. The chapter discusses the legacy of this era and the survival of ethnic nationalism, and considers the problems of contemporary nation-building in a post-Communist era that also demands the redefinition of Europe.

Part III: The nature of European integration and the consequences of diversity

The analysis in Part III begins in Chapter 5 with Richard Grant's account of the primarily political and economic nature of the structures designed to weld together the mutual solitudes of the individual states into an integrated Europe. It is this very narrow definition of integration which undermines the notion of a European dimension to identity. The chapter argues that integration is primarily concerned with the growing amount of European law governing the activities of the EU's citizens, whereas, conversely, the legitimacy of the project has received comparatively little attention. Although it is visualized as a nested hierarchy of local, regional and national identities, the EU currently lacks the tangibility and intelligibility that would enable it to capture the imagination, and therefore gain the voluntary support, of its citizens. The institutions and decision-making processes through which the EU operates have been created by bureaucratic élites, largely independent of the will of national populations. Decision-making processes – which depend on consensus within this élite – are veiled in secrecy, a lack of transparency that combines with its 'democratic deficit' to undermine the legitimacy of the EU. None the less, the Union remains the world's most successful example of institutionalized policy co-ordination and a unique supranational entity that ultimately depends on the thesis of multiple loyalties and identities at the core of this book.

If Chapter 5 illustrates the restricted, partial and often ineffectual basis of European integration, Chapters 6 and 7 richly demonstrate how the balance lies with centripetal powers of economic and social diversity. In the former, Mark Hart analyses the powerful forces that promote economic diversity, notwithstanding the centrifugal propensities of socio-economic convergence and cohesion policies which lie at the centre of the EU project. To a large extent, contemporary patterns of economic advantage and disadvantage reflect the varying trajectories of modernization discussed in Chapter 2. Such spatial variations in wealth have been instrumental in promoting the idea of a 'Europe of the regions', implicitly and explicitly reflecting patterns of uneven development. The discussion addresses the tension between greater economic and political integration and this apparent resurgence of regional identity and power. Even if the concept of 'Europe of the regions' is regarded as a post-nationalist exaggeration, and it is recognized that much decision-making remains concentrated at the national level, it is shown that decentralized and competent regional governance is a key element in cohesion processes, those aimed at improving the quality of European citizenship through measures that combine the promotion of solidarity, mutual support and social inclusion with sustainability of economic growth. Thus the forces of European integration may themselves be engendering further diversity in enhancing the regionalization of Europe at the expense of the nation-state.

In Chapter 7, Colin Williams turns to the implications of social and cultural diversity for the political structures of integration in a situation in which the territorial state no longer appears a sufficient means of organizing political space and commanding allegiance. In addressing the question as to how we might construct a European dimension to identity, he moves the focus of the discussion in the direction of the central concern of Part IV, the construction and renegotiation of the meanings attached to places in ever-mutating multicultural societies. Clearly this involves the recognition of cultural diversity as a key element of social and political life. In exploring these issues of cultural pluralism, the management of cultural diversity and conflict, and the nature of social integration, the chapter concentrates on the study of language. It concludes that while social diversity – and the concomitant propensity to conflict – might be managed through multicultural policies, these do not yet exist at the European scale.

Part IV: Identity and the renegotiation of the meanings of European place

Finally, Part IV is largely concerned with issues of belonging and the ways in which the ever more complex layering of identity demands the continued reinterpretation of the meanings attached to Europe and European places. To a very considerable extent, this involves the renegotiation of representations of nationalism essentially formulated in the eighteenth and nineteenth centuries to underpin the legitimacy of newly evolving states. In Chapter 8, John Agnew uses the example of Italy to explore the creation of representative national landscapes, demonstrating how diversity and integration are two sides of the same process because national identity does not sweep away all other identities. He addresses the dilemma that although the ideal of Europe is currently undergoing something of a revival in the guise of enlarging and deepening the EU, the reality of Europe has long been of making differences between Europeans 'on the basis of certain common characteristics that have been given distinctive casts in different places'. Although such constructions form one of the principal legacies of the past, the discussion demonstrates how landscape ideals are historically contingent and must therefore be subjected to continuous renegotiation, particularly as identity and economic organization become spatially more diffuse.

A similar theme is pursued in Chapter 9, in which John Tunbridge focuses on the heavily contested nature of European heritage. He argues that so much of the continent's heritage – that which we choose to select from the past and invest with meanings framed in the present – is compartmentalized into mutually unintelligible packages based on widely divergent interpretations of history. But political boundaries are not necessarily fixed through time and thus history as meaning is a process of continuous re-evaluation of events and places and of displaced minorities. Using the example of North-Central Europe, the chapter explores how the political and ideological fluidity of the region has led to a mosaic of identity that requires the continuous reconstruction of meanings of the past. The argument centres on the concept of dissonance, the idea that a discordance or lack of agreement and consistency in its definition ensures that heritage is central to the contested constructs of inclusion and exclusion that are characteristic of Europe's contemporary multicultural societies.

The final two chapters address the implications of multiculturalism and ever more fragmented layerings of identity for the renegotiation of urban and rural space in Europe. In Chapter 10, Gregory Ashworth concentrates on the idea that the European city as a physical structure can be regarded as the most prevalent, engaging and pervading cultural symbol of modern Europe. He sees the city as a text that can be read in different ways. Clearly the reading may be informed by a nationalist trope of identity, but the same urban landscape features can also be interpreted in other, more diverse ways. The difficulty lies not in accepting that urban landscape conveys meanings of place identity, but in determining which identities are being shaped by which communications. The chapter concludes that the messages of the European urban text can be plurally encoded by socially pluralist societies into a pluriform heritage, although this is not to underestimate the difficulties created through these meanings being inevitably distasteful to some degree to some people.

In Chapter 11, Hugh Clout discusses the European countryside as contested space. Continuing the theme of the renegotiation of place, he examines the ways in which our conceptualization of the rural world has changed through time. In particular, the countryside can no longer be seen as synonymous with agricultural production but rather must be viewed as an arena of harsh competition and markedly unequal societies and economic opportunities. Since the mid-1960s, the fundamentalist belief that farming was pivotal to the rural economy, and even to national identity, has lost ground. Moreover, it is not simply a contemporary renegotiation of rural space because the historical geography of the European countryside is structured around temporal pulsations in food production and spatial fluctuations in land use. Thus the present rural landscape is a result of past remakings of place in the light of constantly mutating political, economic, social, demographic, technological and political circumstances.

References

Braudel, F. 1972: *The Mediterranean and the Mediterranean world in the age of Philip II*. London: Collins.

Colley, L. 1992: *Britons: forging the nation*. London: Yale University Press.

Davies, N. 1996: *Europe: a history*. Oxford: Oxford University Press.

Douglas, N. 1997: Political structures, social interaction and identity changes in Northern Ireland. In Graham, B. (ed.), *In search of Ireland: a cultural geography*. London: Routledge, 151–73.

Graham, B. 1997: The Mediterranean in the medieval and renaissance world. In King, R., Proudfoot, L. J. and Smith, B. (eds), *The Mediterranean: economy and society*. London: Arnold, 75–93.

Guibernau, M. 1996: *Nationalisms: the nation state and nationalism in the twentieth century*. Oxford: Polity Press.

Jess, P. and Massey, D. 1995: The contestation of place. In Massey, D. and Jess, P. (eds), *A place in the world? Place, cultures and globalization*. Oxford: Open University/Oxford University Press, 133–74.

Parker, W. H. 1968: *An historical geography of Russia*. London: University of London Press.

Said, E. 1978: *Orientalism*. New York: Columbia University Press.

Said, E. 1993: *Culture and imperialism*. London: Chatto & Windus.

Taylor, R. 1996: The double headed eagle: Russia – East or West. In Bideleux, R. and Taylor, R. (eds), *European integration and disintegration: East and West*. London: Routledge, 252–80.

Times Atlas of World History 1986: Times Books: London.

Wilson K. and van den Dussen, J. (eds) 1993: *The history of the idea of Europe*. London: Routledge.

P A R T

I

Europe and Europeanness

1

The past in Europe's present: diversity, identity and the construction of place

BRIAN GRAHAM

Introduction: contested geographies

The function of this chapter is to elaborate upon the conceptual basis which underlies the book's consideration of Europe's contested geographies, and to explore the three themes that structure the subject matter of the book as a whole. To reiterate, these are the interweaving of past and present; integration and diversity; and the renegotiation and contestation of space and place. The chapter concludes with an examination of the ways in which national identities might be renegotiated to admit an element of Europeanness, which, it is argued, is necessary to validate and legitimate political and economic integration.

While the book's content reflects recent conceptualizations put forward within the broad fields of cultural, economic and historical geography, it also responds to the immediacy which these ideas have in any understanding of the axes of conflict and unagreed alignments of identity that fracture contemporary European society. Like identity, culture is a notoriously elastic concept but is defined here as a signifying system through which 'a social order is communicated, reproduced, experienced and explored' (Williams 1982, p. 13). It involves the conscious and unconscious processes through which people live in – and make – places, while giving meaning to their lives and communicating that meaning to themselves, each other and the world beyond (Cosgrove 1993). The cultural realm must not be visualized merely in social terms because it also reflects and interacts with political ideology and the economic domain. As this book constantly demonstrates, the cultural – or representational – landscape and its contestation are implicated in processes of empowerment, while one of the fundamental causes of cultural diversity is to be found in the differing trajectories of economic change and development that characterize Europe's complex human geography.

Inevitably, therefore, culture – like geography itself – invokes contestation. The key argument in this respect is to be found in Livingstone's ideas (1992) that geography, as is any other knowledge, is created within specific social, economic and political circumstances. Because it must be situated in this way, the nature of geography is always negotiable, subject to change through time and across space as social and intellectual circumstances alter. Geographical texts and contexts exist in a reciprocal relationship. Thus regions may have no existence outside the consciousness of geographers, 'who, by their elo-quence, are able to create place' (Tuan 1991, p. 693) but, in turn, geogra-phers and their geographies are products of particular social conditions and times. As is the nature of things, these circumstances are unlikely to be agreed, any one society being characterized by axes of ideological discord, which may – or may not – be contained by the structures of government and social control. Contemporary human geography is also much concerned with the manipulation of cultural landscape, which Cosgrove (1993) describes as a complex social construction contested along multiple and overlapping axes of differentiation. In this context, all places are imaginary in the sense that they cannot exist for us beyond the socially constructed images which we form of them in our minds (Shurmer-Smith and Hannam 1994).

We can isolate three dimensions to this general argument which are rele-vant to the discussions in this book. In the first instance, cultural landscapes can be regarded as allegories of meaning, as multivocal and multicultural texts, which are implicated in the construction of power within a society and are capable of being read in a variety of conflicting ways (Cosgrove 1984). These texts interact with social, economic and political institutions and can be regarded as signifying practices 'that are read, not passively, but, as it were, rewritten as they are read' (Barnes and Duncan 1992, p. 5). Because they can be interpreted in different ways, however, by competing social actors involved in the continuous transformation of societies, the meanings attached to these texts remain negotiable. As narratives, they are 'culturally and historically, and sometimes even individually and momentarily, variable' (Barnes and Duncan 1992, p. 6).

Second, the concept of socially constructed place is intrinsic to renditions of individual and group identity, which often embody particular readings or narratives of a people's interaction with their human landscape in all its man-ifestations. One of the most potent realizations of this process is provided by the formulation of nationalist ideologies, which depend on simplifying synec-doches of particularity, vested in place, to summarize and signify very much more complex social structures, and to erect criteria of social inclusion and exclusion. Baker (1992) identifies two essential ways in which cultural land-scape becomes a framework through which such ideologies and discourses can be constructed and contested. In the first instance, manipulated depictions of landscape offer an ordered, simplified vision of the world. Again, the sacred symbols of a landscape, rich in signs of identity and social codes, act as a system of signification supporting the authority of an ideology and emphasizing its holistic character.

Finally, however, nationalism is but one symbolic geography, other scales and dimensions of personal and group identity – reflecting the contestation of societies along axes that include regionalism, locality, class, gender, ethnicity and material well-being – being equally important. These, too, are concerned with additional criteria of inclusion and exclusion, which interact in complex and diverse ways with nationalistic tropes of identity. One of the most useful conceptualizations in this regard is provided by Duncan's idea (1990) of the representative landscape, best visualized as a collage encapsulating a people's image of itself. It symbolizes the particularity of territory and a shared past which helps define communal identity, and plays an active part in the reproduction and transformation of any society in time and space (Graham 1997). As such, a cultural or representative landscape can be visualized as a powerful medium in expressing feelings, ideas and values, while simultaneously being an arena of political discourse and action in which cultures are continuously reproduced and contested.

It follows that any landscape signifying the cultural and political values of a dominant group can be viewed as symbolic of oppression by those subservient to – or excluded from – these hegemonic values (Johnson 1993). Official renditions of cultural landscape may attempt to elide many of the social complexities emanating from class, gender and ethnicity. These authorized landscapes and places can be viewed as cultural capital – expressions of dominant ideologies (Ashworth 1993; see Chapter 10), which embody the values and aspirations of that ideology. Officially defined cultural landscapes are therefore implicated directly in the processes which validate and legitimate power structures. Official discourses of place often represent the values of a dominant ethnic group at the expense of minority interests and promote the interests of social élites, while concealing class, gender, ethnic and other inequities.

The past in the present; the present in the past

Socially constructed pasts

If place is socially constructed, so too is the past. History and heritage – that which we opt to select from the past – are used everywhere to shape emblematic place identities and support particular political ideologies (Ashworth and Larkham 1994; see Chapter 10). The past is often described as a chronological, modernist progression in which linear narratives link past events with a present that is generally depicted in terms of its superior development. As with any other construction of the past, such perspectives are actually highly selective filterings, constructed in the present to justify, validate and legitimate contemporary circumstances and, as such, subject to change. Thus the Renaissance (literally meaning rebirth), which originated in early fifteenth-century Florence under the patronage of the city's wealthy merchants, encapsulated a revival of interest in the ways in which the values and artistic forms of classical Rome and Greece could be used to establish new principles of social organization and thought. 'It was the spiritual force which cracked the mould of medieval civilization, setting in motion the long process of integration which gradually gave birth to "modern Europe"' (Davies 1996, p. 471). In the sixteenth century, it was followed by the Reformation, the destruction of the hegemony of Latin Christianity, which had been Western Europe's guiding ideology throughout the Middle Ages.

The perspective of the Renaissance – itself a highly selective rendition of the past – was responsible for the representation of a classical world ended by the Dark Ages, in which successive barbarian tribes irrupting out of the Asian steppelands – Europe's nemesis in the east – swept away the vestiges of Roman civilization. Many constructions of the European past reflect a deep-seated notion that modernity is defined by the presence of the universal state. Equally, there exists a persistent assumption that urbanization is a synonym for civilization. Consequently, much of the debate on the nature of the Dark Ages has centred on the somewhat arcane question of post-Romanic urban continuity. Henri Pirenne (1925, 1939), perhaps Europe's most influential urban historian, argued that although Mediterranean commerce and towns continued despite the invasions of Europe that succeeded the fall of the empire, urban life elsewhere was largely expunged until a revival occurred in the ninth and tenth centuries. More recent explanations, however, place far more emphasis on the conversion to Christianity of many of the Romans' successors, and the role of the Church as a sponsor of continuity of urban settlements, albeit perhaps in an attenuated form. Urban origins in Northern Europe are also now placed considerably earlier than envisaged by Pirenne (Hodges 1982; Verhulst 1989, 1994). The Dark Ages, therefore, have been revised, in the process becoming, as it were, shorter and less dark.

The secular culture conceived in the Renaissance ultimately evolved into the eighteenth-century Enlightenment, the Age of Reason with its belief in individuals' ability to think and act for themselves. It is to this period, also, that we owe our enduring allegiance to empirical, deductive theorizing. Like the Renaissance before it, we can see the Enlightenment being framed by its particular rendition of the past, one in which the values and traditions of European Christianity were depicted as the principal reactionary forces in European society. But in turn, the Age of Reason spawned its antithesis in a nineteenth-century Romanticism that emphasized the irrational, the reification of nature and, in terms of constructs of the past, a renewed fascination with the medieval world, which had been marginalized by both Renaissance and Enlightenment thinkers. The Gothic architectural style of Northern Europe's great medieval cathedrals was among the principal enthusiasms of the Romantics, as was the notion of conserving the past (see Chapter 10).

Thus, our contemporary constructs of the past are framed less by events than by the selective interpretations placed upon these in wider constructs of society. The past is – and always has been – constructed in present circumstances and its meanings are attached in that context. Inescapably, therefore, interpretations of the past are essentially transient, social constructions that are reflective largely of the particular time and place of their genesis and of the axes of social conflict characteristic of that epoch. In turn, these become part of the contested discourses of future generations, the past effectively being a resource that can be multi-interpreted (and multi-sold because it is an economic resource too). In order to illustrate something further of the articulation of past and present, I turn to the example of the medieval pilgrimage to the Galician city of Santiago de Compostela (Graham and Murray 1997; Murray and Graham 1997).

The pilgrimage to Santiago de Compostela

Ostensibly, pilgrimage is a religious phenomenon in which an individual – or group – sets forth on a journey to a particular cult location to seek the intercession of God and the saints of that place in an array of concerns (Nolan and Nolan 1989). But, as is characteristic of all manifestations of knowledge, pilgrimage is a social construction and, inevitably, a form of cultural production linked to the specific social, political and historical contexts in which it originally occurred. Consequently, any journey to a cult location – in whatever epoch – represents a resolution of conflicting ideals, both spiritual and profane. Thus Eade and Sallnow (1991) argue for a pluralist, heterogeneous model of pilgrimage in which competing religious and secular discourses vie for supremacy, a perspective that describes historical and contemporary representations of the pilgrimage to Santiago de Compostela.

If pilgrimage is defined as a journey to the sacred, then the even now relatively isolated location of Santiago de Compostela underscores the personal commitment of the individual pilgrim. Early medieval tradition held that St James the Apostle was the first Christian evangelizer of Spain. He was martyred on returning to the Holy Land, and his body was miraculously transported back to Galicia to remain undiscovered until the end of the eighth century. Virtually all Iberia had by then fallen to the Islamic Moors, who first crossed the Straits of Gibraltar from Tangier to Tarifa in 711 to defeat the crumbling Visigoth kingdoms of Iberia. In less than a decade, the Moors had crossed the Pyrenees, where Merovingian France almost suffered the fate of its Visigothic neighbour, before Charles Martel eventually defeated them near Poitiers in 732. This reverse did not prevent the consolidation of Muslim control over Iberia, and the establishment of the glorious medieval civilization of al-Andalus, centred on Córdoba and Seville, which were among Europe's largest cities during the Middle Ages. Only the mountain kingdoms of the far Atlantic north of the peninsula remained beyond the

Moorish orbit. It was here in Asturias that opposition to Islam first began and, almost certainly, the rediscovery of the tomb of *Sant'Iago* was related to these events.

By the beginning of the twelfth century, a complex network of pilgrimage roads, originating in Paris, Vézelay, Le Puy and Arles, led across northern Spain to Santiago de Compostela (Figure 1.1). The journey to *Sant'Iago* ranked with those to Rome and Jerusalem among the great pilgrimages of the medieval European Church, although it is difficult now to do more than glimpse the personal motivations of those who travelled to the shrine. In general terms in an uncertain medieval world, characterized by plague, war and disease, pilgrimage was about penitence. By undergoing the hardships and privations of the journey, the pilgrim could seek indulgences and the remission of sins. If, however, the medieval cult of St James is located within its own cultural and political milieu, a reading emerges that points to myriad non-official, personal motivations being shaped by and subsumed within official representations of the pilgrimage that emerged from the appropriation of the phenomenon by powerful institutions – not least the Latin Church – engaged in wider contestations of society.

One manifestation in particular stands out. This concerns the linkages between the pilgrimage to the shrine of St James and the *Reconquista*, the Christian reconquest of Iberia (Figure 1.2). The cult of St James was central to the effective mobilization of this process, the rhetoric and artistic iconography of the times presenting the *Reconquista* as a Holy War or crusade, a Christian equivalent to Islamic *jihad*. The medieval representations of the pilgrimage to Santiago de Compostela, primarily contested between official (Church, political) and unofficial (pilgrim) renditions of the process, is reflected in the extraordinary Janus-nature of the medieval St James (Figures 1.3 and 1.4). One version of his iconography served as the patron saint of the *Reconquista*. This was the remarkable representation of *Santiago Caballero* or *Santiago Matamoros* – St James the Moorslayer – riding a white charger through the skies above the *Reconquista* battlefields, wielding a bloody, dripping sword and surrounded by decapitated infidels. The final defeat of the Moors occurred only in 1492 (the year when Columbus set sail to [re]discover the New World) with the final capitulation of Granada to the Catholic monarchs, Ferdinand (of Castile) and Isabella (of Aragon). This symbolic event may not have created a Spanish polity in the fifteenth century, but the imagery of a state forged from the north by five centuries of holy war against the infidel under the blood-stained banner of *Santiago Matamoros* later became fundamental to orthodox constructions of Spanish nationalism. In the twentieth century, for example, such representations, accompanied by much imperial verbiage invoking the memory of Ferdinand and Isabella, were employed to exalt and reinforce the notion that the Falangist (Fascist) leader General Franco was the heroic leader of a crusade to liberate Spain from the godless hordes of Moscow (Preston 1993).

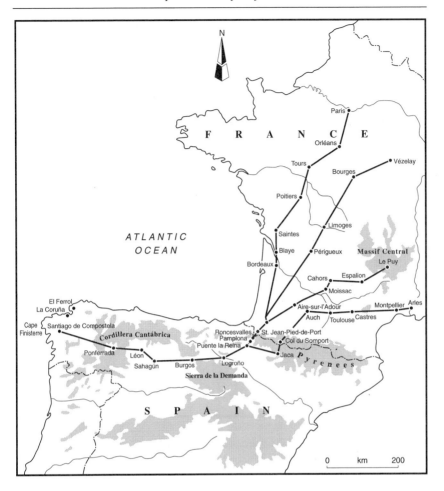

Figure 1.1 The medieval pilgrimage routes to Santiago de Compostela

Revisionist history, however, renders a less Catholic and conspicuously pluralist representation of medieval Spain, in which the barrier between Moor and Christian is depicted in very much more porous terms. Incorporating an imagery of Christian, Islamic and Jewish multiculturalism, this discourse integrates Spain much more coherently within the wider medieval European experiences of colonization of marches, urbanization and slowly maturing state formation. Far more meaningful to this context is the other face of James as *Sanctus Jacobus* or *Santiago Pelegrino*, often portrayed wearing the characteristic pilgrim garb of cloak and broad-brimmed hat, carrying a staff and adorned with the cockle-shell emblem of his pilgrimage. In this version, he becomes the antithesis of his warlike representation, an 'everyman', sharing the hardships of his humble devotees as they trudge across northern Spain from all the myriad corners of Europe.

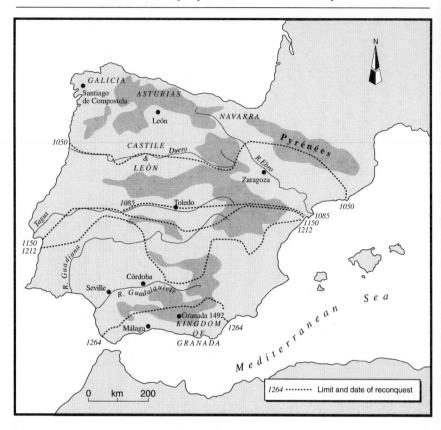

Figure 1.2   The Christian *Reconquista* of Iberia
Source: Adapted from Holmes (1988, p. 204)

A contemporary revival of interest in the pilgrimage is equally contested between rival appropriations. To the Catholic Church, it remains a symbol of the ascendancy of the spiritual over the secular and rationalist values of the modern world. For the Council of Europe, what is now a signposted routeway – the *Camino de Santiago* – becomes a European cultural itinerary, a symbol of the ideal of European integration. More prosaically, for the regional authorities of northern Spain, the *Camino* is a tourism resource and useful emblem for a region not well represented in conventional marketing images of the country. Unsurprisingly, the concept of the roads of Europe, symbolically focusing on Santiago de Compostela, has been adopted enthusiastically by the Galician government in its promotion of regional identity. Finally, the meaning of pilgrimage and the behaviour of pilgrims in a modern age remain bitterly contested between walkers, cyclists and other users of the *Camino*. For some – paralleling the medieval equation of faith with personal privation – the suffering of walking the physical Way to Santiago becomes a metaphor of the search for the

Figure 1.3 *Santiago Matamoros*

inner self (for example, Hoinacki 1996), something that cannot be revealed to the car-borne tourist. To the latter, the *Camino* is simply cultural tourism, while to others, for whom sport is a substitute for religion, it is a long hike or cycle ride. In its many simultaneous manifestations – icon of spiritual and/or religious renewal, recreation resource, economic

Figure 1.4   *Sanctus Jacobus*

commodity in urban and rural tourism promotion – the modern journey along the *Camino de Santiago* demonstrates again the heterogeneity and diversity of the articulation of past and present and the endless remaking of the past as it is contested between differing requirements and sets of values.

Integration and diversity

The essential conclusion to be drawn from exploring the interweaving of past and present – through the example of the various manifestations of the pilgrimage to Santiago de Compostela – is that the past is socially constructed and continuously remade. In turn, this points towards an understanding of the geographical and cultural diversity that is almost the diagnostic characteristic of contemporary Europe. In the past decade, for instance, the concept of a regional Europe has received much attention as many argue that the nation-state represents but one transient compromise, imposing no more than a spurious homogeneity on more fundamental spatial variations. But there has always been a Europe of the regions in the sense of what Massey (1995) refers to as the interdependence of a multiplicity of places and scales. By diversity of place, we allude effectively to the contested array of meanings of place. For example, the EU itself is dependent on the concept of multiple identities in which a nested hierarchy of local, regional and national loyalties is supposed to fit neatly together. At any one time, a place is implicated in a variety of meanings, which operate at a mix of scales – ranging from the global to the local – for an array of motives. Such diversity, however, is not unconstrained, or lacking in limits. An enduring theme in European historical and geographical thought is that of the cycle of regularities operating over relatively long-term time-spans, which interact with local particularities in the social construction of many different places. Therefore, somewhat paradoxically, one cause of diversity is to be found in structural regularity. Perhaps the best-known expression of such ideas occurs in Braudel's concept of *la longue durée*, the 'long duration' in which history is conceptualized in slow motion, revealing permanent values that can be set against a relatively unchanging physical environment (Braudel 1972; Butlin 1993). (These ideas are explained in more detail in Chapter 2.)

Medieval fragmentation

The concept of a 'Europe of the regions' – defined somewhat differently from a variety of perspectives – is dependent less on a postmodern fragmentation of the state than on the argument that the European past was structured in long cycles of change in which the ancient world was transformed *c*. AD 1000 by the 'crystallisation of another set of structures . . . which would be known as "feudal" ' (Bois 1992, p. 162). Endlessly mutating, five centuries later this system was giving way in another slow transformation that resulted in the eventual evolution of mercantile and – later – market capitalism. The nature of medieval society and its spatial organization remain a powerful influence on the contemporary diversity of Europe.

Although a much debated concept (feudalism is inherently a *post hoc* rationalization, an imposition of intellectual order upon the past), some agreement exists that the social structure which we designate as such had two fundamental features. First, it was a decentralized, hierarchical political order that evolved in the early medieval period because of the weakness of central authority and its inability to prevent the rise of local warrior aristocracies (Bloch 1961). Accordingly, the system was characterized by fragmented and often weak sovereignty and political power. Dodgshon (1987, p. 173) refers to 'the fissiparous tendencies of feudalism', reflecting the way in which power was devolved downwards rather than to the centre, as in the sovereign territorial state. Although this mechanism was the only way in which a king's will could reach all his subjects, feudalism as a mode of economic organization and social integration was inherently contradictory. The centre was forced to concede power in order to govern at all, but once secure in their geographical niches, feudal lords sought to maintain the effective independence of their territories from that locus of power (Elias 1982).

Second, feudalism was an economic order involving estate or peasant family production and the appropriation by the warrior class (and Church) of the agricultural surpluses produced by the unfree peasantry (serfs or vassals). To achieve this, the élite had to impose political and economic control over resources (such as land, forest and game) and monopolies (including mills and small town markets). These processes enabled nobles at each level, in a hierarchical chain that descended from monarchs and the Church to dukes, barons and lesser nobles to grant fiefs (which involved property rights and revenues) to their immediate dependants in return for homage and fealty (involving payments, advice and military service). Further, as Duby (1974, p. 187) writes, feudal – or seigneurial – lords were constantly thinking 'in terms of ensuring new sources of profit for themselves', one of the most ubiquitous means being the clearing of wastelands and the creation of new villages and small towns. Although northern France was the heartland of the feudal system, such processes ensured that the system was carried to the more remote corners of Europe, particularly as medieval population growth ensured an ever-increasing demand for land. Thus, in the twelfth and thirteenth centuries, for example, much of medieval Ireland was colonized by Anglo-Norman lords who constructed some hundreds of villages and small towns to attract colonists (Graham and Proudfoot 1993). Similar processes occurred in Eastern Europe as Germanic peoples settled beyond the Elbe (Bartlett 1993).

The significance of small town foundation, which occurred all over Europe, particularly in the twelfth and thirteenth centuries, lies in the intensely unequal nature of medieval society. While people owed obedience and loyalty to their immediate superiors in a hierarchy of authority at the head of which was a king, many members of society – in particular the inhabitants of chartered boroughs or small towns – were personally free (Hilton 1992), a characteristic often used as an incentive to attract colonists to the new lands being cleared throughout medieval Europe. In addition, the major cities maintained an ambiguous relationship with feudalism. Their mercantile élites serviced the feudal economy but, to some extent, stood apart from it. Consequently, both the territorial fragmentation of the feudal fief and the rise of the city-state acted as impediments to the concentration of power. The most complete breakdown of centralized power occurred in Italy, leading – particularly in the eleventh century – to the emergence of city-states such as Pisa, Lucca and Florence, which exercised complete control over their own affairs (Barber 1992). In visualizing this more complex medieval world, Reynolds (1994), in a notable critique of the feudal model, depicts a society with three essential strata. At the top were the nobles and gentry (including the higher clergy) and, at the bottom, the 'plough pushers', the unfree peasants or manual labourers who owed work and rent services to the top stratum. But in between were the 'not noble but free', primarily the inhabitants of the small town and mercantile city worlds. Reynolds therefore regards medieval society as one of infinite gradations or layers rather than one of wide social gulfs. The significance of this middle world lies not only in the freedom which it introduced to medieval society, but also in the much debated role of urban élites in the transition from feudalism to mercantile capitalism.

The conceptualization of diversity

Feudalism, therefore, was intensely territorial, resulting in highly complex and shifting alliances of fiefdoms in a singularly confused and hetero-geneous political geography. If we take the example of the feudal heartland of France, *c.* 1200, it was essentially no more than a loose federation of greater and lesser domains (Figure 1.5). The direct writ of the Capetian kings had yet to extend beyond the Île de France around Paris, although they had the allegiance of most of the fief-holders of eastern France. Conversely, virtually all western France from Normandy to the Pyrenees was directly or indirectly under the control of the Angevin kings of England, while parts of the Midi were controlled by the kings of Aragon. But these broad divisions conceal an even more finely differentiated het-erogeneity. While major feudal lords, such as the Counts of Toulouse, exer-cised some degree of hegemony over the Midi west of the Rhône, their fiefs also replicated the roles and functions of the royal centres to which they were nominally subject. Thus the County of Toulouse was itself subdivided into lesser fiefdoms (Figure 1.6) and characterized on a smaller scale by the same dubious bonds of loyalty that attended the relationships between kings and their major magnates.

Small wonder then that the conceptualization of diversity has largely been a French affair. The nineteenth-century historian Jules Michelet posed the central paradox: pointing to a diverse country in which unity was achieved only by force and civil war, he concluded: 'The material [of France] is essen-tially divisible and strains towards disunion and discord' (cited in Braudel

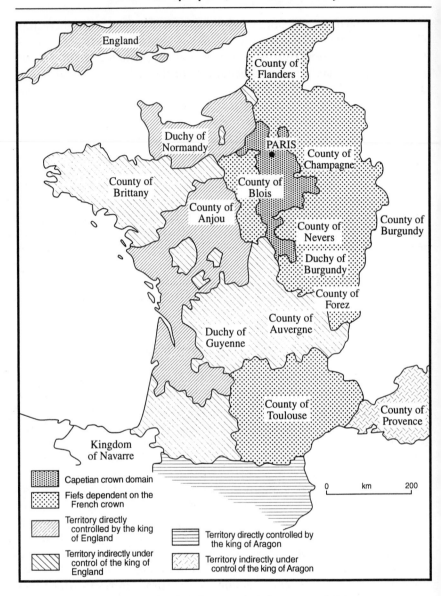

Figure 1.5 The fragmented political geography of France, *c.* 1200
Source: After Le Goff (1988, p. 99)

1988, p. 120). In *Tableau de la géographie de la France* (1903), Paul Vidal de la Blache strove to find a more positive model (Robic 1994). He argued that it is a diversity, rooted in the physical environment, which paradoxically provides France with its identity. The country's unity evolves from a beneficent force of commonality, derived from the ways in which social life combines with, and transcends, this diversity:

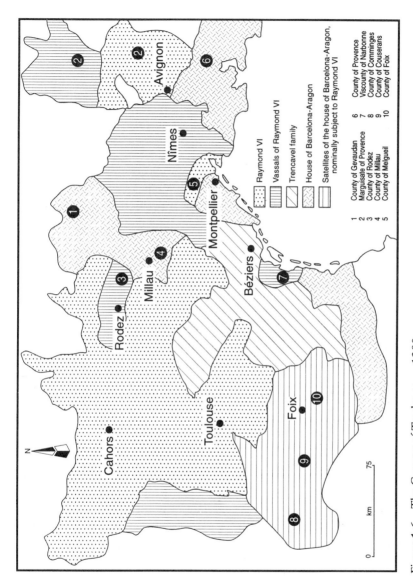

Figure 1.6 The County of Toulouse, *c.* 1200
Source: After Sumption (1978, pp. 26–7)

Legend:
- Raymond VI
- Vassals of Raymond VI
- Trencavel family
- House of Barcelona-Aragon
- Satellites of the house of Barcelona-Aragon, nominally subject to Raymond VI

1 County of Gevaudan
2 Marquisate of Provence
3 County of Rodez
4 County of Millau
5 County of Melgueil
6 County of Provence
7 Viscounty of Narbonne
8 County of Comminges
9 County of Couserans
10 County of Foix

Place names: Cahors, Rodez, Millau, Toulouse, Nîmes, Avignon, Montpellier, Béziers, Foix

N

km 0 75

> To be French was experienced at two levels: one was a member of a local or regional community and, because of that, was integrated into the national whole. Diversity appeared as a necessary component of . . . identity and was thus accepted and valued.
>
> (Claval 1994, p. 50)

Vidal emphasized the significance of ordinary people and their environment: to him, the region was not simply a convenient framework, but rather a social reality (Claval 1984). This relationship was encapsulated in the idea of the *pays* (an area with its own identity derived not only from divisions of physical geography, but also from ethnic and linguistic divisions imposed on a region by its history) as the geographical mediation of synthesis and continuity, the product of human interaction with the environment over many centuries.

In his monumental (if unfinished) investigation of the identity of France, Braudel (1988, p. 41) takes up the same theme. He sees a 'patchwork France', a jigsaw of regions and *pays* in which 'the vital thing for every community is to avoid being confused with the next tiny *"patrie"*, to remain *other*'. But in the course of his work, Braudel also becomes critical of the Vidalian idea that the uniqueness of France stems from its diversity. He accepts that fragmentation but recognizes that it is characteristic too of other European countries: 'Yes, France is certainly diverse . . . it is a diversity which breaks up, divides, and sets one region against another.' But it is not an 'unparalleled' diversity: 'Germany, Italy, Britain, Spain, Poland can all lay claim to diversity' (Braudel 1990, p. 669). So too can smaller countries. Pursuing the same symbiosis of physical environment and human society, Ireland's most influential geographer, Estyn Evans (1981), believed that even fractious Ulster could be reconciled as a single theme with many variations, its personality deriving from the fusion of many small *pays*. Through a unity defined by diversity, Evans sought a communality in Ireland's past to which all its present inhabitants might subscribe (Graham 1994).

Although Braudel and Evans saw the natural world as providing the backdrop to this diversity, both also recognized the decisive contribution of the economy and social intercourse. Moreover, diversity itself was more apparent because it often reflected what Braudel referred to as *l'histoire événementielle*, short-term episodic history or the history of events. Behind all that, however, lay *l'histoire structurelle*, the long-term changes of *la longue durée*, 'the slow speed of the secular trend', moving 'scarcely at all' (Braudel 1990, p. 678). The real essence of diversity lies in the ways in which these different histories, trajectories of time and social change and geographical particularities have fused in varying ways to produce distinctive places and peoples.

Regionalism and localism

The contemporary outcomes of diversity are rendered in several different ways. The brutal disintegration of Yugoslavia in the early 1990s is one tragic manifestation, as is the less publicized conflict in CIS republics such as

Georgia, Azerbaijan and Armenia. Again, a significant extent of devolved local government is found in some European countries (for example, the French *départements*), while Germany and Spain follow the federal state model. It has also been suggested that the concept of an economic 'Europe of the regions' provides a further schema of heterogeneity, although it is very difficult to find convincing examples of regional governance outside the federal states. French regions, for example, are subsidiary to the underlying structure of more local-ized *départements*. While the economic regions themselves are very diverse in terms of size and productivity, Harvie (1994), for example, points to the emer-gence of aggressive, urban-based and affluent regions, such as Lombardy, Baden-Württemberg, Catalonia and Rhône-Alpes, which contrast markedly with poorer, peripheral regions such as the Mezziogiorno, Andalusia and Galicia. There is nothing new in such urban regions, Europe's pre-nation-state economy having long been dominated by *de facto* or *de jure* city-states, includ-ing Venice, Bruges and Amsterdam.

Harvie concludes that the 'distinctive consciousness' of the contemporary city-regions lies more in their affluence than in any cultural identity, only Catalonia having a particular cultural identity, albeit largely that of Barcelona, its capital city. Together with Euskadi (the Basque Country), Catalonia might be referred to as a 'region-state', in which the regional gov-ernment portrays itself as the defender of the Catalan people against Madrid. The armed struggle waged by ETA in Euskadi is clearly aimed at secession from Spain. In neither region, however, do a majority support independence, even within the EU. It appears that both regions wish to retain their current economic prosperity within a highly decentralized Spain (Ross 1996). Again, the affluence of Lombardy was reflected in the emer-gence during the 1980s of a political movement, the Northern League, which portrays itself as a new clean start for Italy, based on northern pro-ductivity that contrasts markedly with the corruption of a southern Rome-based bureaucracy. Agnew (1995) argues, however, that while the League espouses a rhetoric of (often xenophobic) regionalism, it is still dragged into the axis of the state as a whole because decision-making remains concen-trated at that level. It is even possible that regional movements are more gen-erally in decline because regions need the nation-states in order to survive. In sum, therefore, the concept of regional – or local – identity appears to add additional layers to the diversity of meaning of place rather than providing substitutes for national unity.

The renegotiation and contestation of space and place

In addressing this third theme, we can adopt two separate perspectives. The first concerns the actual physical remaking of space through war and negoti-ation, the fluid and constantly altering political mosaic of past and present Europe. This is best illustrated by an examination of the nature of state formation and changing conceptualizations of sovereignty. In some contrast is the continuous representational remaking or renegotiation of place and identity, manifested particularly through nationalism and its contestation.

State formation and the renegotiation of space and sovereignty

Slowly, with many reverses and often only recently, the little cells of the Middle Ages have been forced together into territorial, centralized states. It is important, however, not to overstate the historical fragmentation of Europe. The Roman Empire achieved significant unity for half a millennium while, later, the Latin Christian Church provided some degree of commonality in the medieval West. Almost everyone understood something of its theology, even though illiteracy, the absence of standardized printed texts and difficulties in communication provided ample opportunities for heresy. Much of the liturgy and dogma of the Western church, including the sanctity of marriage, are eleventh-century additions, reflecting the interaction of large-scale changes in the structuring of feudal society and the Gregorian reforms of Latin Christendom (Duby 1984). These established the concept of the 'high-medieval papal monarchy', the ecclesiastical parallel to the system of absolute secular monarchy which was to be a principal characteristic of late-medieval Europe (Bartlett 1993). The Crusades against Islam in the Holy Land, which began in 1095, affirmed the Pope's claim to supreme territorial as well as spiritual power. Continuing intermittently until the end of the thirteenth century, the Crusades can be interpreted only within the context of the socio-economic and political changes occurring in Western Europe, in particular the emergence of assertive Latin Christianity under its centralizing papacy. They reflected also the Church's adoption of the warrior cult of the West and the tensions created by the growing importance of primogeniture in feudal society (inheritance of land by the eldest son only). The Holy Land provided a safety-valve, offering the lure of land to a surplus nobility.

Although the city-state can be seen as an ultimate expression of political fragmentation, urbanization can also be depicted as another form of unity. The medieval city began under the sponsorship of nobility and Church but, by the twelfth and thirteenth centuries, was increasingly declaring its independence of both. The merchant communities, investing in trade and nascent industry, were emerging as a third powerful élite constituency throughout Europe. Tilly and Blockmans (1994) argue that European states grew differently according to the character of the urban networks in a given region, the variable distribution of cities constraining the possibilities for state formation. The other factor was the rise of absolute monarchy in the sixteenth and seventeenth centuries. (Absolutist states are those in which absolute power unrestricted by any other governmental institution was vested in the hands of monarchs or other rulers.)

Thus the emergence of states did not impose unity on diversity: rather the state was another means of controlling diversity while adding a further layering to identity. Again, one must stress the long-term and volatile processes of state formation in Europe and the contingent nature of many contemporary political boundaries. The earliest states began to evolve through the medieval accretion of peripheries to urban-based core regions. The first centralized monarchy – based in London – emerged in twelfth-century England, although, being French, it probably lacked any sense of English identity. However, the later and fractious alliance between the Crown and the nobility meant that 'at the start of the fourteenth century, England was the most modern and the most stable Christian state' (Le Goff 1988, p. 98), one manifestation of this process being the invention of the professional army of longbowmen, which at the Battle of Crécy in 1347, and even more dramatically at Agincourt in 1415, allowed the victory of the 'common soldier over the mounted knight' (Keegan 1976, p. 78).

Conversely, the unification of France took centuries to achieve (de Planhol 1994). In the early thirteenth century, Capetian France, centred on the Île-de-France around Paris, allied with the papacy to launch the brutal Albigensian Crusade against the Cathar church in Languedoc. Ostensibly justified by the need to extirpate heresy, the underlying – and ultimately successful – political agenda sought to secure the submission of the Count of Toulouse to the French Crown and more fully integrate his territories within its domain and away from possible English influence. However, it took the divine intervention of Jeanne d'Arc and the Hundred Years War, which began in 1337 and lasted until the 1430s, to finally oust the English from their possessions in western France. That was not sufficient to unite the country, internecine dynastic struggles continuing well into the sixteenth century. The present territorial limits of the country – the 'hexagon' – were achieved only by a slow process of accretion and virtually endemic warfare (see Figure 3.1), and were not finalized until the accession of Nice in 1860. Even then, Alsace-Lorraine was annexed by Germany in 1871 and not returned until the Treaty of Versailles in 1919.

In Spain, as we have seen, the idea that the *Reconquista* created a unitary state is something of a fallacy, largely the perspective of a more recent nationalism. Ferdinand and Isabella held enormous power but it did not lead to political integration. The kingdoms of Spain preserved their particular institutional structures until the eighteenth century when a centralized monarchy, based in Madrid, was able to define the country as it now exists (Fernández Albaladejo 1994). Elsewhere in Europe, there was tremendous variation in the process of state formation:

> In [the] 'Age of Absolutism', absolutist states actually formed a minority. Between the completely decentralized, constitutional and republican confederation of Switzerland at one end of the scale and the extreme autocracies in Russia, the Ottoman Empire, and the Papal

States at the other, great variety flourished. Europe's republics were represented by Venice, Poland-Lithuania and the United Provinces; the constitutional monarchies at various times by England, Scotland, and Sweden; the absolute monarchies by France, Spain, and Austria. . . . Even greater variety can be found among . . . the hundreds of petty states. . . . There were miniature city-republics like Ragusa (Dubrovnik), Genoa, or Geneva; there were miniature principalities like Courland; ecclesiastical states like Avignon, and curious hybrids like Andorra.

(Davies 1996, p. 578)

The process of state formation has continued until the 1990s. The unification of Germany was not finally completed until 1871 and, as we have seen, its boundaries have altered dramatically since. Italy was largely unified by 1861, although Rome was acquired only in 1870 (see Chapter 8). A whole series of states was created in the aftermath of World War I (see Chapters 2 and 3); subsequently, however, some have experienced substantial changes, Poland providing one of the most dramatic examples (see Chapter 9). The present extent of the Federal Republic of Germany dates only to 1990, while Czechoslovakia split into Slovakia and the Czech Republic as recently as 1992.

Despite its centrality within European thought, the European state has often been a fragile and even transitory entity, and one that, historically, has been subjected to continuous transformation. The idea that sovereignty over everything is bundled into territorial state parcels reflects what Harvey (1989, p. 242) has called 'a radical reconstruction of views of space and time' that occurred only in the transition from pre-modern to modern societies. Agnew and Corbridge (1995) refer to the 'territorial trap' which equates state sovereignty and territorial space. As Ruggie (1993) argues, this emerged because of the collapse of a well-established system of hierarchical subordination such as that of feudal society and the Roman Church. Instead, social membership became exclusive and the 'identification of citizenship with residence in a particular geographical space became the central fact of political identity'. Consequently, the 'sovereign territorial state is not a sacred unit beyond historical time' (Agnew and Corbridge 1995, pp. 85 and 89). However, it has been so reified. Thus the renegotiation of place within the contemporary EU has to be located within the panoply of tensions created by a concept of sovereign territoriality, which cannot come to terms with the problem that many contemporary dilemmas of economy and society are incapable of being understood or addressed in terms of fixed territorial spaces.

A telling and powerful analysis of these processes is provided by Anderson (1996). He observes that the EU has been characterized as perhaps the world's first truly postmodern international political entity, distinct from the national and federal state forms of the modern era, but in some respects reminiscent of pre-modern territorialities (Ruggie 1993). This hypothesis, sometimes referred to as 'new medievalism', speculates that the growth of transnational corporations and networks, combined with substate nationalist and regionalist pressures, might produce overlapping forms of sovereignty analogous to the complex political arrangements of medieval Europe (Bull 1977). Sovereignty would cease to be a state monopoly. Moreover, at another scale, global 'time–space compression' is radically reconstructing our views of space, and leading to an accelerated unbundling of territorial sovereignty, with the growth of common markets and various transnational functional regimes and political communities which are not delimited primarily in territorial terms.

Anderson, however, is loath to overstate this argument. First, even in the EU, the unbundling is limited and partial, affecting different state activities very unevenly. The politics of economic development is the sphere where state power has been most affected by globalization, and – as we have seen with examples such as Lombardy and Catalonia – it is also generally a principal basis for regionalization. While territoriality is becoming less important in some fields (for example, financial markets), the state remains the principal spatial framework for many aspects of social, cultural and indeed political life, not least because of the enduring power of language differences. Second, Anderson argues, the EU, although a new political form, is itself territorial, and in many respects traditional conceptions of sovereignty remain dominant, whether exercised by Member States or by the EU as a whole. Thus it may be that although the political control of space is being renegotiated in contemporary Europe – as indeed it has always been – we are seeing *re*-territorialization rather than *de*-territorialization. But what also might be changing are the ways in which people identify with territory.

Contested representations of place and identity: nationalism and the nation-state

As Handler (1994) argues, the Western world has been accustomed for more than two centuries to think of identity as an object bounded in time and space, with clear beginnings and endings, and its own territory. The result is to be found in identity politics. If landscape can be depicted as a contested text or narrative, it is clearly implicated in a people's identity, itself embedded in particular intellectual and institutional and time contexts. Nowhere is this compression of time and space more apparent than in nationalist ideologies and movements, which politicize space by treating it as the distinctive and historic territory of the nation-state, 'the receptacle of the past in the present, a unique region in which the nation has its homeland' (Anderson 1988, p. 24). The rise of the nation-state in the eighteenth and nineteenth centuries was closely connected to Romantic notions of the mysticism of place and of notions of belonging and not belonging. Thus Colley (1992, pp. 5–6) argues that Great Britain can plausibly be regarded as an invented nation 'superimposed, if only for a while, onto much older alignments and loyalties':

It was an invention forged above all by war. Time and time again, war with France brought Britons, whether they hailed from Wales or Scotland or England, into confrontation with an obviously hostile Other and encouraged them to define themselves collectively against it. They defined themselves as Protestants struggling for survival against the world's foremost Catholic power. They defined themselves against the French as they imagined them to be, superstitious, militarist, decadent and unfree. And increasingly as the wars went on, they defined themselves in contrast to the colonial peoples they conquered, peoples who were manifestly alien in terms of culture, religion and colour.

(Colley 1992, p. 5)

National identity is thus created in particular social, historical and political contexts and, as such, cannot be interpreted as an immutable entity; rather, it is a situated and relational socially constructed narrative, capable of being read in conflicting ways at any one time and of being transformed through time. The power of a narrative rests on its ability to evoke the accustomed, a trope that works by appealing to 'our desire to reduce the unfamiliar to the familiar' (Barnes and Duncan 1992, pp. 11–12). The creation of landscape narratives facilitates this process by denoting particular places as centres of collective cultural consciousness. For example, the hegemonic image of the west of Ireland as the cultural heartland of the country was an essential component of the late-nineteenth-century construction of Irish nationalism (Nash 1993; Graham 1997). Strongly reinforced by the intellectual élite of early-twentieth-century Ireland, the 'west' became an idealized landscape, populated by an idealized people who invoked the representative, exclusive essence of the nation through their Otherness from Britain. The invented, manipulated geography of the west portrayed the unspoilt beauty of landscapes, where the influences of modernity were at their weakest and which evoked the mystic unity of Ireland prior to the chaos of conquest (Johnson 1993). As Agnew (1996) argues, such renditions of place are fundamental to a European tradition of over-simplifying space into idealized constructs of tradition and modernity.

According to Woolf (1996, pp. 25–6):

National identity is an abstract concept that sums up the collective expression of a subjective, individual sense of belonging to a sociopolitical unit: the nation state. Nationalist rhetoric assumes not only that individuals form part of a nation (through language, blood, choice, residence, or some other criterion), but that they identify with the territorial unit of the nation state.

In his influential book *Imagined communities: reflections on the origins and spread of nationalism* (1991), Benedict Anderson argues that any nationalist ideology is the work of the imagination, its communality in large measure self-delusory. It is imagined because the members of even the smallest nation will never know most of their fellow-members, it is imagined as a limited but sovereign entity; but, perhaps above all, it is

> imagined as a *community*, because, regardless of the actual inequality and exploitation that might prevail . . . the nation is always conceived as a deep horizontal comradeship. Ultimately it is this fraternity that makes it possible, over the past two centuries, for so many millions of people, not so much to kill, as willingly to die for such limited imaginings.
>
> (Anderson 1991, p. 7)

Thus nations are conceived as being internally homogeneous in terms of a shared cultural content. But, somewhat confusingly, an internal diversity of region, gender, class and ethnicity may be recognized – as in France or perhaps even Britain – or 'even celebrated as indicative of the nation's complexity and rich heritage'. None the less, 'in nationalist ideology, internal diversity is always encompassed by national homogeneity' (Handler 1994, p. 29).

But that subsuming of diversity, no matter how blatant, is not to deny the reality of the nationalist discourse, nor the fundamental contribution that significations of place contribute to this perception of such order. Nor are such relationships confined to nationalism alone. Both cultural nation and territorial state claim exclusive sovereign rights over, and access to, territory. Consequently, all states, whether nation-states or not, sponsor intensely territorial official state-ideas. This politicization of territory is achieved through its treatment as a distinct and historic land, nationalism and the state-idea always looking back in order to look ahead (Agnew 1987). As one mechanism in the processes that impose homogeneity upon diversity, cultural landscape is fundamental in validating the legitimacy of contemporary structures of authority which are derived not from the support of a numerical majority alone, but through renditions of plurality, largely fixed in the past, that transcend other social divisions and fix that imagined communality.

The renegotiation of the nation-state: towards a European identity and heritage

Nationalism has both ethnic and civic dimensions (see Chapter 4). Whereas the former are represented by cultural attributes, civic nationalism essentially equates with citizenship. The nation-state attempts to reconcile these nationalisms in the same territorial unit, which therefore claims sovereignty over all aspects of life. As we have seen, however, the unbundling of territoriality and the contemporary renegotiation of place occurring in Europe – on one hand, forces of globalization and even supranationalism, on the other the impact of regionalism and localism – have made the nation-state a very much more ambiguous concept. Paying due regard to Anderson's timely warning (1995) against exaggerating its death (and avoiding the rhetoric of 'post-nationalism'), we can still pose the question as to why civic and ethnic identity in Europe should be encompassed within the same territorial unit. Given the negotiated compromise of national sovereignty represented by EU membership and the various and spatially sporadic dimensions of devolution, and the uncontrolled loss of national authority invoked by the globalizing tendencies of capitalist time–space compression, the old bases of national identity are being rapidly undermined, leading Gillis (1994, p. 19), among others, to argue in favour of 'desacralizing the nation-state'.

It is important also to remember that identity cannot remain territorially fixed or exclusive through time. Throughout European history, mass migrations, legal and illegal, forced and voluntary, have created multicultural societies (and contrived the circumstances for genocide as in the Holocaust and the barbaric 'ethnic cleansing' that took place in the former Yugoslavia). Contemporary European cities, for example, are defined by increasingly heterogeneous populations, which in part reflect post-World War II labour migration from former colonies and countries like Turkey and Morocco, located on the fringes of affluent Europe (King 1993, 1995).

Thus it is apparent that nationalism is but one form of identity and that an individual can identify with imagined communities set in place at a variety of scales: local, regional, national, supranational, even global. Some of these scales may be combined, as in the imagery of the West as the synecdoche of Irishness. Nor is this simply some postmodern play of fragmented meaning. In the Middle Ages, for example, Latin Christendom provided a form of Western European identity that transcended the dominant locality of the age. The obvious question that arises from such reasoning is, why should a European identity be required? Again, if contemporary Europe is defined by diversity and multiculturalism – resulting in an increasingly complex fragmentation of identity and allegiance – can a European identity really be created?

At present a discordance exists between the political–economic integration and potential appeal of a unified Europe and the lack of legitimation accorded such a construct, perhaps because it lacks the validation of cultural consciousness vested in place. It is vital to remember, however, that the EU is far more than an economic union, a single market. Rather, it reflects an ideology, the central tenet of which lies in the creation of a social market economy which combines market forces with a commitment to the values of internal social solidarity and mutual support. This involves, for example, an explicit commitment to reducing regional disparities in income and opportunity and to promoting convergence and cohesion (Commission of the European Communities 1996) (see Chapter 6). It appears to be the case, however, that the wider ramifications of this agenda are not widely recognized or even appreciated, largely because of a lack of 'Europeanness' in the identity profiles of Europeans.

It can be argued, therefore, that the successful integration of Europe might demand an iconography of identity that would complement, but not necessarily replace, national, regional and local identities. Rose (1995) maintains that a sense of place does exist at the supranational level; the Renaissance, the Enlightenment and Judaeo-Christianity are seen as particularly European achievements. To go further, however, in reconciling diversity and Europeanness, we need to return to the issues raised above concerning the contested interpretations of the past. This latter – as we have seen with the example of the *Camino de Santiago* – is represented increasingly as heritage: that which we choose to conserve from the past. Samuel (1995) regards memory not as timeless tradition, but as a quality transformed from generation to generation through the contrived nature of heritage, which can be defined less as artefacts and traditions inherited from the past, than by the contested modern values and meanings that are attached to these objects. Tunbridge and Ashworth (1996, p. 20) conceptualize this process as the dissonance of heritage, 'a discordance or a lack of agreement and consistency' in its definition, and an intrinsic quality which ensures that heritage is implicated in the contested constructs of inclusion and exclusion that are characteristic of Europe's contemporary multicultural societies. The construction of a European heritage identity would involve the manipulation of heritage, demanding the addition of new layers of meaning to built environments and landscapes that are already fundamental symbols within national or regional iconographies and narratives (Ashworth and Graham 1997).

The difficulties are profound. If we accept that Europe does require a heritage of identity and legitimation to complement – and indeed validate – its role in economic and political decision-making, this is not to suggest that the function of such an iconography rests in the imposition of some contrived and spurious rendition of unity. Rather, a European heritage must accommodate the centrifugal heterogeneity of place and a multicultural diversity of peoples and cultures. It follows that multiculturalism provides the only viable basis for a meaningful axis of European identity.

The dilemma, however, is that much of the interaction between Europeans which could form the basis for multicultural themes is dissonant in some way or other to some Europeans. At its simplest the past is a contested resource, and thus the recognition of any one claim upon it may disinherit others, sometimes with grim results. This dichotomy is manifested in several serious ways in Europe. In the first instance, the nation-state is often perceived to be the outcome of a struggle for democratic freedom by a subordinated group demanding recognition. Such ideas are fundamental, for example, to Greek, Irish or Polish identity. Furthermore, the region-states or even less overt expressions of separatist identity within the nation-states are equally defined by histories of subjugation and demands for freedom – Euskadi, Languedoc, Brittany and Scotland are but a few among many examples. Legacies of war and atrocity feature large in all identities – the Cathar repression in Languedoc, Cromwell in Ireland, Turkish atrocity in the Balkans, the destruction of Warsaw in 1944 – and in a wider sense, it has to be remembered that virtually all the states of Europe and their nationalist narratives were forged by war or in its outcome.

Throughout Europe, evocatively dramatic memorializations of contested pasts are manifested in actual physical violence of battlefields, landscapes of death such as Verdun, Ieper (Ypres) and the Somme, and in their aftermath – the landscapes of memory that cluster so thickly across the low-lying plains that extend through Belgium and eastern France towards Paris; in Charles de Gaulle's words, 'the fatal avenue' in which 'we have just buried one-third of our youth'. The mortality endured by twentieth-century Europeans is no less than grotesque (see Figure 3.8). The absolute numbers remain contested but those cited here are unlikely to be overestimates. According to Davies (1996), military losses alone in the two world wars amounted to almost 23 million European deaths, while an estimated 27 million European citizens were killed during World War II, of whom 6 million died in the Nazi genocide of the Jews; *excluding* war losses, it has been calculated that 54 million people were killed in the Soviet Union between 1917 and 1953.

But these statistics, no matter how obscene, do not necessarily negate the notion of Europeanness. Although few places on Earth have been fought over so continuously, one of the primary ideological motivations behind the entire project of European unity is to ensure that such carnage does not occur again. Thus the memorials, graves and battlefields of Picardy, the Chemin des Dames and Flanders – clearly national heritage – might also be construed as a heritage of European reconciliation, testimony to the unacceptable that must never be repeated but cannot be forgotten. Heffernan (1995), for example, argues that the war cemeteries of the Western Front – muted, serene, peaceful and intensely moving – can be read as conveying no real sense of sacrifice to the nation-state. Instead, they are immortal, sacred landscapes – essentially apolitical. Above all, they symbolize the need to develop a new political culture in Europe that removes the possibility of war forever. Can other aspects of Europe's distasteful recent past be reconciled with the notion of European identity? Europeanness has to embrace the unacceptable: Srebrenica and Auschwitz as well as High Gothic cathedrals, romantic castles, utopian Renaissance town planning and symphonic music. The memorable history of Europeans embraces pogrom, persecution and prejudice, near-continuous internecine war, oppression and genocide. The twentieth century has seen mass death, carpet-bombing of cities and, above all, the Jewish Holocaust of 1933–45. This remains archetypically 'European' heritage, and arguably the most serious challenge facing contemporary European society in creating a sense of common identity. European Jews – ironically the principal European people not nationally defined – were deported and murdered by Europeans in Europe in pursuit of a European ideology. The physical relics of *shtetls* and camps remain throughout the continent, as do the memories of the victims, the perpetrators and the passive observers to haunt this and future generations of Europeans (Ashworth 1996; see Chapter 9).

Conclusion

It is apparent that unavoidable dissonance exists between the various but simultaneously present scales, meanings and interpretations of heritage throughout Europe, and that within this heterogeneity of meaning and motivation, the idea of a European heritage is poorly developed. But given the fragmentation of territoriality and sovereignty that is characteristic of the New Europe, it is important not to equate the creation of a European heritage with an all-encompassing, pervasive European sameness. Rather, the function of a European heritage is to create a European dimension that can be added to already complex formulations and layerings of identity. It might embrace common principles and places such as the struggle for democratic and civil rights and the attendant sites of war and atrocity, or even the European city (see Chapter 10), mutual themes that can then be related to the specific differences of nation and region. Above all, therefore, a European heritage must be manipulated as a mosaic of similarity – but also difference – that reflects the political reality of overlapping layers of empowerment. It cannot comprise that which is common to all; this would result merely in an anaemic and sanitized heritage in which the hard questions posed by the past are obscured by superficial narratives of progression and European cultural hegemony.

Agnew (1996) argues that European places and states are defined through a process in which blocks of space become labelled with essential attributes derived from a diagnostic time period: *backward* Italy, *Reconquista* Spain, the Dutch *Golden Age*, *Great* Britain. Europe needs to seize and define its own time–space block, not to suborn national or regional heritages, but to complement them in a postmodernistic diversity of plural identities and peoples which denies the political and commercial temptations that reduce interpretations of the past to insipid linear narratives. In other words, a European heritage has to portray dissonance and contestation as positive qualities, symbolic of the reality – to draw an analogy with Canada – that no one landscape or iconography can encompass Europe's diversity. An ideal European heritage would thus not be symbolic of Europeanness subsuming national and regional identity but instead reflective of the complex dismantling of the synonymity of territoriality, sovereignty, nationalism and the state in the new Europe. It would embrace and accentuate notions of multiculturalism and of complex and overlapping, rather than intersecting, layers of identity, while striving to be inclusive of all the continent's peoples, constructs ultimately necessary to validate and legitimize political and economic integration. It is unlikely that the dominant present unifying principles in Europe – those of a slightly constrained free-market capitalism – will prove a sufficient cement. In essence, this book explores the problems of creating a more enduring construct of Europeanness.

Acknowledgements

I am very grateful to Michael Dunford and Michael Heffernan for their comments on an earlier draft of this chapter. Greg Ashworth also provided very considerable assistance with the material on heritage.

References

Agnew, J. A. 1987: *Place and politics: the geographical mediation of state and society.* Boston: Allen & Unwin.

Agnew, J. A. 1995: The rhetoric of regionalism: the Northern League in Italian politics, 1983–1994. *Transactions of the Institute of British Geographers* 20, 156–72.

Agnew, J. A. 1996: Time into space: the myth of 'backward' Italy in modern Europe. *Time and Society* 5, 27–45.

Agnew, J. A. and Corbridge, S. 1995: *Mastering space: hegemony, territory and international political economy.* London: Routledge.

Anderson, B. 1991: *Imagined communities: reflections on the origins and spread of nationalism.* Revised ed., London: Verso.

Anderson, J. 1988: Nationalist ideology and territory. In Johnston, R. J., Knight, D. B. and Kofman, E. (eds), *Nationalism, self-determination and political geography.* London: Croom Helm, 18–39.

Anderson, J. 1995: The exaggerated death of the nation-state. In Anderson, J., Brook, C. and Cochrane, A. (eds), *A global world? Reordering political space.* Oxford: Open University/Oxford University Press, 65–112.

Anderson, J. 1996: The shifting stage of politics: new medieval and postmodern territorialities. *Environment and Planning A: Society and Space* 14, 133–53.

Ashworth, G. J. 1993: *On tragedy and Renaissance.* Groningen: Geopers.

Ashworth, G. J. 1996: Holocaust tourism and Jewish culture: the lessons of Krakow-Kazimierz. In Robinson, M., Evans, N. and Callaghan, P. (eds), *Tourism and culture towards the 21st century.* Newcastle upon Tyne: University of Northumbria, 1–13.

Ashworth, G. J. and Graham, B. 1997: Heritage, identity and Europe. *Tijdschrift voor Economische en Sociale Geografie* 88, 381–8.

Ashworth, G. J. and Larkham, P. J. (eds) 1994: *Building a new heritage: tourism, culture and identity in the new Europe.* London: Routledge.

Baker, A. R. H. 1992: Introduction: on ideology and landscape. In Baker, A. R. H. and Biger, G. (eds), *Ideology and landscape in historical perspective: essays on the meanings of some places in the past.* Cambridge: Cambridge University Press, 1–21.

Barber, M. 1992: *The two cities: medieval Europe, 1050–1320.* London: Routledge.

Barnes, T. J. and Duncan, J. S. 1992: Introduction: writing worlds. In Barnes, T. J. and Duncan, J. S., *Writing worlds: discourse, text and metaphor in the representation of landscape.* London: Routledge, 1–17.

Bartlett, R. 1993: *The making of Europe: conquest, colonization and cultural change, 950–1350.* Harmondsworth: Allen Lane.

Bloch, M. 1961: *Feudal society.* London: Methuen.

Bois, G. 1992: *The transformation of the year one thousand: the village of Lournand from antiquity to feudalism.* Manchester: Manchester University Press.

Braudel, F. 1972: History and the social sciences. In Burke, P. (ed.), *Economy and society in early modern Europe: essays from Annales.* London: Routledge & Kegan Paul, 11–42.

Braudel, F. 1988: *The identity of France,* vol. 1: *History and environment.* London: Collins.

Braudel, F. 1990: *The identity of France,* vol. 2: *People and production.* London: Collins.

Bull, H. 1977: *The anarchical society.* London: Macmillan.

Butlin, R. A. 1993: *Historical geography: through the gates of space and time.* London: Arnold.

Claval, P. 1984: The historical dimension of French geography. *Journal of Historical Geography* 10, 229–45.

Claval, P. 1994: From Michelet to Braudel: personality, identity and organization of France. In Hooson, D. (ed.), *Geography and national identity.* Oxford: Blackwell, 39–57.

Colley, L. 1992: *Britons: forging the nation.* London: Yale University Press.

Commission of the European Communities 1996: *First report on economic and social cohesion 1996.* Brussels/Luxembourg: CEC.

Cosgrove, D. E. 1984: *Social formation and symbolic landscape.* London: Croom Helm.

Cosgrove, D. E. 1993: *The Palladian landscape: geographical change and its cultural representations in sixteenth-century Italy.* Leicester: Leicester University Press.

Davies, N. 1996: *Europe: a history.* Oxford: Oxford University Press.

Dodgshon, R. A. 1987: *The European past: social evolution and spatial order.* Basingstoke: Macmillan.

Duby, G. 1974: *The early growth of the European economy.* London: Weidenfeld & Nicolson.

Duby, G. 1984: *The knight, the lady and the priest.* London: Allen Lane.

Duncan, J. S. 1990: *The city as text: the politics of landscape interpretation in the Kandyan kingdom.* Cambridge: Cambridge University Press.

Eade, J. and Sallnow, M. J. (eds) 1991: *Contesting the sacred: the anthropology of Christian pilgrimage.* London: Routledge.

Elias, N. 1982: *State formation and civilization,* vol. 2: *The history of manners.* Oxford: Blackwell.

Evans, E. E. 1981: *The personality of Ireland: habitat, heritage and history,* 2nd ed. Belfast: Blackstaff Press.

Fernández Albaladejo, P. 1994: Cities and the state in Spain. In Tilly, C. and Blockmans, W. P. (eds), *Cities and the rise of states in Europe, A.D. 1000 to 1800.* Boulder, CO: Westview Press, 168–83.

Gillis, J. R. 1994: Memory and identity: the history of a relationship. In Gillis, J. R. (ed.), *Commemorations: the politics of national identity.* Princeton, NJ: Princeton University Press, 3–24.

Graham, B. J. 1994: The search for the common ground: Estyn Evans's Ireland. *Transactions of the Institute of British Geographers* NS 19, 183–201.

Graham, B. J. (ed.) 1997: *In search of Ireland: a cultural geography.* London: Routledge.

Graham, B. J. and Murray, M. 1997: The spiritual and the profane: the pilgrimage to Santiago de Compostela. *Ecumene* 4, 389–409.

Graham, B. J. and Proudfoot, L. J. (eds) 1993: *An historical geography of Ireland.* London: Academic Press.

Handler, R. 1994: Is 'identity' a useful cross-cultural concept? In Gillis, J. R. (ed.), *Commemorations: the politics of national identity*. Princeton, NJ: Princeton University Press, 27–40.

Harvey, D. 1989: *The condition of postmodernity*. Oxford: Blackwell.

Harvie, C. 1994: *The rise of regional Europe*. London: Routledge.

Heffernan, M. 1995: For ever England: the Western Front and the politics of remembrance in Britain. *Ecumene* 2, 293–323.

Hilton, R. H. 1992: *English and French towns in feudal society: a comparative study*. Cambridge: Cambridge University Press.

Hodges, R. 1982: *Dark age economics: the origin of towns and trade, AD 600–1000*. London: Duckworth.

Hoinacki, L. 1996: *El Camino: walking to Santiago de Compostela*. University Park, PA: Pennsylvania State University Press.

Holmes, G. (ed.) 1988: *The Oxford illustrated history of medieval Europe*. Oxford: Oxford University Press.

Johnson, N. C. 1993: Building a nation: an examination of the Irish Gaeltacht Commission Report of 1926. *Journal of Historical Geography* 19, 157–68.

Keegan, J. 1976: *The face of battle*. London: Jonathan Cape.

King, R. (ed.) 1993: *Mass migration in Europe: the legacy and the future*. London: Belhaven.

King, R. 1995: Migrations, globalization and place. In Massey, D. and Jess, P. (eds), *A place in the world? Place, cultures and globalization*. Oxford: Open University/ Oxford University Press, 6–44.

Le Goff, J. 1988: *Medieval civilization*. Oxford: Blackwell.

Livingstone, D. N. 1992: *The geographical tradition*. Oxford: Blackwell.

Massey, D. 1995: The conceptualization of place. In Massey, D. and Jess, P. (eds), *A place in the world? Place, cultures and globalization*. Oxford: Open University/ Oxford University Press, 45–86.

Murray, M. and Graham, B. 1997: Exploring the dialectics of route-based tourism: the *Camino de Santiago*. *Tourism Management* 18, 513–24.

Nash, C. 1993: 'Embodying the nation': the west of Ireland landscape and Irish identity. In O'Connor, B. and Cronin, M. (eds), *Tourism in Ireland: a critical analysis*. Cork: Cork University Press, 86–112.

Nolan, M. L. and Nolan, S. 1989: *Christian pilgrimage in modern Western Europe*. London: Chapel Hill.

Planhol, X. de 1994: *An historical geography of France*. Cambridge: Cambridge University Press.

Pirenne, H. 1925: *Medieval cities: their origins and the revival of trade*. Princeton, NJ: Princeton University Press.

Pirenne, H. 1939: *Mohammed and Charlemagne*. London: Allen & Unwin.

Preston, P. 1993: *Franco: a biography*. London: HarperCollins.

Reynolds, S. 1994: *Fiefs and vassals: the medieval evidence reinterpreted*. Oxford: Oxford University Press.

Robic, M.-C. 1994: National identity in Vidal's *Tableau de la géographie de la France*: from political geography to human geography. In Hooson, D. (ed.), *Geography and national identity*. Oxford: Blackwell, 58–70.

Rose, G. 1995: Place and identity: a sense of place. In Massey, D. and Jess, P. (eds), *A place in the world? Place, cultures and globalization*. Oxford: Open University/ Oxford University Press, 87–132.

Ross, C. 1996: Nationalism and party competition in the Basque Country and Catalonia. *West European Politics* 19, 488–506.

Ruggie, J. 1993: Territoriality and beyond: problematicizing modernity in international relationships. *International Organization* 47, 139–74.

Samuel, R. 1995: *Theatres of memory,* vol. 1: *Past and present in contemporary culture.* London: Verso.

Shurmer-Smith, P. and Hannam, K. 1994: *Worlds of desire, realms of power: a cultural geography.* London: Arnold.

Sumption, J. 1978: *The Albigensian Crusade.* London: Faber.

Tilly, C. and Blockmans, W. P. 1994: *Cities and the rise of states in Europe, A.D. 1000 to 1800.* Boulder, CO: Westview Press.

Tuan, Yi-Fu 1991: Language and the making of place: a narrative-descriptive approach. *Annals of the Association of American Geographers* 81, 684–96.

Tunbridge, J. E. and Ashworth, G. J. 1996: *Dissonant heritage: the management of the past as a resource in conflict.* Chichester: John Wiley.

Verhulst, A. 1989: The origin of towns in the Low Countries and the Pirenne thesis. *Past and Present* no. 122, 3–35.

Verhulst, A. 1994: The origin and early development of medieval towns in Northern Europe. *Economic History Review* 47, 362–73.

Vidal de la Blache, P. 1903: *Le tableau de la géographie de la France.* Paris: Hachette.

Williams, R. 1982: *The sociology of culture.* New York: Schocken Books.

Woolf, S. 1996: Introduction. In Woolf, S. (ed.), *Nationalism in Europe, 1815 to the present: a reader.* London: Routledge, 1–39.

P A R T

II

*The past in the present;
the present and the past*

|2|

Economies in space and time: economic geographies of development and underdevelopment and historical geographies of modernization

MICHAEL DUNFORD

Introduction: economic disparities and pathways to modernization and development

Today the most developed parts of Europe are centred on a core of major international cities and advanced city regions largely located along a vital axis extending from Greater London through Benelux and the Rhinelands in the western half of Germany to northern Italy (Figure 2.1). Although there have been fundamental changes in the characteristics of the places that comprise this axis, the concentration of development in this part of Europe dates back at least to the medieval world, when Flanders and northern Italy were the major foci of European industry and commerce. Because of the growth of the historical capitals of Europe's major colonial powers (Amsterdam, London and Paris), this axis was reinforced and its centre of gravity moved northwards in the early modern era. In the nineteenth and twentieth centuries, wealth accumulated as a result of Europe's industrialization contributed further to the development of the axis and its north-western extension in Great Britain, Europe's first industrial nation. Not all areas that emerge as dominant poles of development retain their relative advantage. In the case of this axis of European development, however, there is clear evidence for the existence of long-term processes of circular and cumulative causation which have permitted the almost constant adaptation to changing circumstances of established cities with critical concentrations of people, economic infrastructures, enterprises, know-how and political power.

A series of orbital zones of relative underdevelopment are located around this axis and its adjacent regions. To the south, the Mediterranean's rich land and sea resources supported remarkable early developments in agriculture and trade and allowed it to emerge as the centre of a succession of hegemonic world economies. Five hundred years ago, however, when Ottoman power was at its height in the east, Spain and Portugal initiated the European conquest of the globe. As the Atlantic was opened up, the coastal cities of the Mediterranean (of which the most important were Venice, Genoa and Barcelona) slowly ceased to be the centres of global economic power and world decision-making. The centre of economic gravity shifted northwards and westwards, first to Amsterdam and later to Britain, while the subsequent industrial revolution, which created immense disparities between the economies that industrialized and those that did not, at first largely bypassed the Mediterranean, or locked its inhabitants into a subordinate role in wider divisions of labour (Dunford 1997). Of the countries and regions that surrounded the Mediterranean, northern Italy and Catalonia in north-east Spain

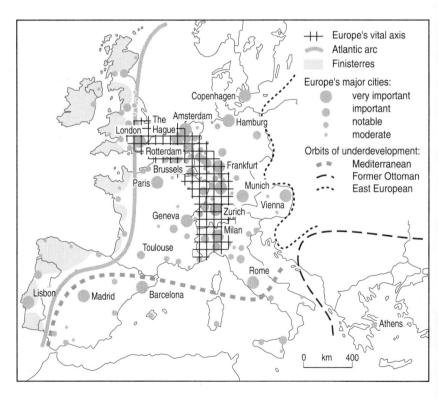

Figure 2.1 Europe's vital axis

were among the few examples of significant late-nineteenth-century industrialization. In the twentieth century, industrialization and modernization did occur more widely around the Mediterranean, but with differing degrees of delay and varying trajectories that are reflected in sharp contemporary inequalities around the 'inner sea'.

To the west, stretching from the Shetlands to Gibraltar, is a maritime Atlantic arc comprising regional and national economies which remain peripheral and have relatively low population densities and incomes per head. As with the Mediterranean, this situation reflects their particular trajectories of modernization and development. In particular, the rise of transatlantic sea commerce led to the rapid growth of a series of ports along Europe's western coasts. Growth, however, did not generally survive the subsequent decline in the relative importance of this commerce. In the nineteenth century, most of the areas in this arc failed to industrialize, specializing instead in the export of agricultural products and raw materials, sometimes owing to the loss of protection associated with imposition of the rules of free trade. In the absence of significant industrial growth, depopulation and emigration were common, although their scale and duration varied sharply. The most dramatic case is provided by Ireland: in the 26 counties that became the Republic of Ireland, population fell by almost 51 per cent between 1841 and 1951, although in the north-east corner (which became Northern Ireland), where there was a significant development of modern industries, it declined by just 17 per cent (Mjøset 1992; Munck 1993; Bradley 1996).

The fortunes of other small countries in North-Western Europe (such as Norway, Sweden and Denmark) were rather different in that emigration and population decline were relatively short-term processes, late industrialization leading to the twentieth-century development of successful export industries and related systems of innovation around their natural resource endowments. In Sweden, for example, industrial growth was centred on the export of forestry and wood processing products, iron and later steel, around which paper, cellulose and capital goods industries developed. Exports of agricultural products from Denmark led to increased incomes, which created market conditions for the growth of small-scale, import-substituting consumer goods industries (textiles, construction materials and processed food).

Greece and the Balkans to the south-east were part of a world centred on the Eastern or Byzantine Empire, founded in the early seventh century (Figure 2.2; see Chapter 4). Following the Great Schism of the Christian Church in 1054 (Figure 2.3), it was largely included in that eastern half of Europe that was Christian Orthodox. Following the collapse of Byzantium at the end of the twelfth century, the Balkans fell under Turkish Muslim Ottoman rule. At its peak, the Ottoman Empire extended as far as the gates of Vienna and into south-eastern Poland (Figure 2.4), and not until Napoleon Bonaparte sought to extend French influence throughout Europe did it start to crumble. Its legacy was one of relative underdevelopment, especially in the areas longest subject to Turkish occupation, the Ottoman Empire being

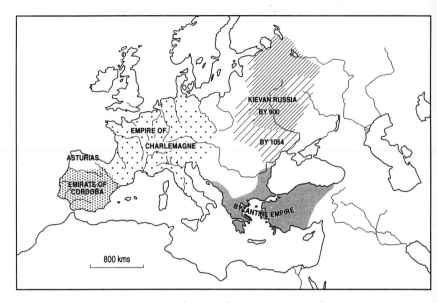

Figure 2.2 Europe, *c.* 814: the Carolingian Empire; the Byzantine Empire (610–1453); Kievan Rus (880–1054); the Emirate of Córdoba

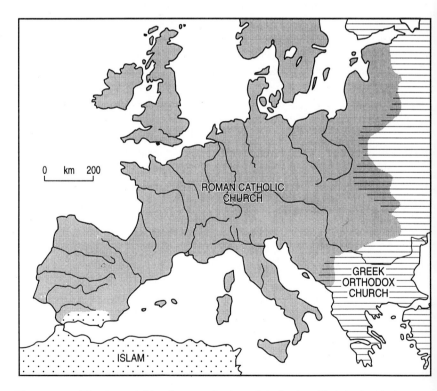

Figure 2.3 The Great Schism between Latin and Orthodox Christianity, 1054

characterized by a hierarchical, exploitative and economically – and techno-logically – unprogressive state system and mode of production (Berend and Ránki 1974). The dismantling of Ottoman rule took more than a century: not until the end of World War I was Turkey finally eliminated as a significant actor in the wider Mediterranean Europe.

Yugoslavia (the country of the southern Slavs) was one manifestation of this Ottoman legacy. Settled by Slavs (Slovenes, Croats, Serbs and Bulgars) in the sixth and seventh centuries, this part of Europe was dominated by three successive empires. It was first a part of the Christian Orthodox Byzantine Empire, albeit pressured from the north by Magyar migrations and the Kingdom of Hungary. After the Turkish defeat of a Serb coalition at the Battle of Kosovo Field in 1389, the region was incorporated into the Ottoman Empire. In the sixteenth century, the Catholic Austrian Habsburg Empire established control over its northern part, Slovenia and Croatia becoming the southern frontier of Christian Europe. Almost all the Turkish possessions in Europe – including Greece, Serbia and Romania – gained their independence during the nineteenth century, while the Austrian annexation of Bosnia-Herzegovina in 1908 precip-itated the events that led to the assassination in 1914 of the Archduke Ferdinand in Sarajevo and the start of World War I. Yugoslavia was one of the states created after 1918. It survived invasion by Nazi Germany in 1940, only to disintegrate in 1991 as a result of the secession of Slovenia and Croatia and the unsustainable tensions created by the interaction of structural adjustment, its political arrangements and complex ethnic geography.

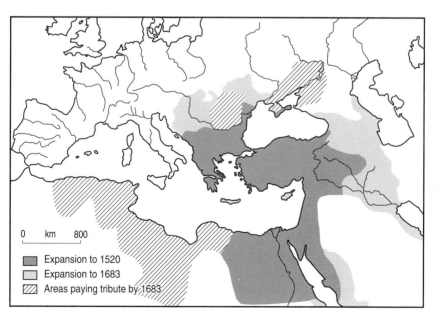

Expansion to 1520
Expansion to 1683
Areas paying tribute by 1683

0   km   800

Figure 2.4   Rise of the Ottoman Empire, 1300–1683

While South-Eastern Europe and the southern half of Central Europe were one of Europe's frontiers with the Islamic world and were subject to Turkish invasions and domination, Eastern Europe was the frontier with Central Asia. As such it was subject to successive waves of invasions by nomadic pastoral tribes from which it shielded the West, albeit at the expense of its own early agrarian development. The last of these invaders were the Mongol Tartars, who, after overrunning much of Eastern Europe and the Balkans, settled in the wedge-shaped western extension of the Eurasian steppelands in the Dnepr–Don–Volga area near the Caspian Sea (Figure 2.5), whence they exacted tribute from Russia from 1240 until 1480.

After 1500, when a new phase of expansion set in, the trajectory of the more densely peopled and developed Western Europe differed even more sharply from that of the East. The former began an acceleration towards a transition to capitalism, the creation of absolutist state systems and the evolution of an early modern world-economy. The East was very different (see also Chapter 4). First, after the elimination of feudal serfdom in the West, a 'second serfdom' was imposed in the East: in Brandenburg in 1494, in Poland in 1496, in Bohemia in 1497, in Hungary in 1492 and 1498, and in Russia in 1497 (Szücs 1988). Especially between 1550 and 1620, the German feudal *Gutwirtschaft*, established in areas beyond the Elbe, and large Polish estates cultivated by serf labour were increasingly integrated into Western Europe's economy through the export of grain: serfdom in the East accordingly served as a way of organizing production destined to serve Western markets. Second, the medieval processes of 'internal expansion' and settlement continued into the early modern period in the East

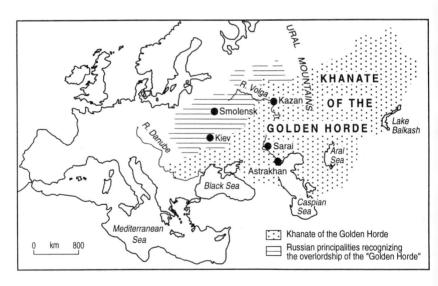

Figure 2.5   Western part of the Mongol Empire (1206–1696), *c.* 1300: the Khanate of the Golden Horde

(Szücs 1988). In 1480, the Russian Empire rejected the Mongol overlordship of the Khanate of the Golden Horde and began to push out southwards from Moscow, going on to conquer the khanates of Kazan in 1552 and Astrakhan in 1556, and subsequently annexing and conquering further territories to the west, south and east (see Chapter 4). Third, there were differences in the nature of Western and Eastern absolutism. In the West, absolutist states emerged after 1580 and survived in the more developed parts of Europe until the Enlightenment. As Anderson (1974a) (cited in Szücs 1988, p. 315) has argued, absolutism was 'a compensation for the disappearance of serfdom' in Western Europe, whereas in the East its roots lay earlier, in the Mongol invasion and conquest. In the West, absolutism outlived the eighteenth century only in Spain, Portugal and southern Italy, while in the East it was 'a device for the consolidation of serfdom', enduring in Russia until the twentieth century.

Despite some evidence of modernization, these factors contributed to the opening up of a large and persisting development divide between Eastern and Western Europe. Some areas of early industrial development were to be found in the eastern half of Germany, the Russian Empire and scattered through the extraordinary patchwork quilt of economies and nations that made up the Austrian Empire, especially after their governments embarked on state-led processes of late industrialization in the nineteenth century. More recently, the post-1945 victory of Communism was in part attributable to its being seen as a fresh way of attempting to close the gap between Eastern Europe and the advanced capitalist world. In spite of some early successes, however, it finally failed.

This sketch demonstrates that there were sharp contrasts in the timing, nature and speed of development and in the pathways to modernization within and between the peripheral and the core areas of European capitalism. The Europes that result can be divided up in a number of ways. Szücs (1988), for example, insists on the existence of three Europes and of two boundaries for Western Europe. The first is the Elbe–Leitha line, which defined the eastern extent of the Carolingian Empire (Figure 2.2) and, much later, also marked the Yalta division of Europe after 1945. The second line is further to the east, stretching from the Baltic to the Carpathians. This line separates *Europe Occidens*, influenced by Rome and Catholicism, from the Eastern Europe of Byzantium and Orthodox Christianity (Figure 2.3) (Bartlett 1993). Over the course of time, the relative importance of these two lines of cleavage has changed, leading Szücs to argue that Central Europe, which lies between them, is a distinct region of the continent, having more affinity with the West than with the Orthodox Christian East.

It is argued here that, to a significant extent, the causes of the geography of development disparities in contemporary Europe (see Chapter 6) lie in these contrasting pasts. More specifically, they are to be found in a number of processes with different temporalities or durations, with different spatial extents and with different degrees of durability of their effects. As

Marx claimed, people do make history but they do so in circumstances not of their choosing because of the constraints of their inheritance from the past.

Time and space: understanding historical systems

The diverse character of contemporary Europe is in part a result of recent events. As argued in Chapter 1, however, it is also rooted in the near and distant past and is in this sense a result of much longer-term economic and political developments. To analyse some of the longer-term processes, I shall draw further on the ideas of Braudel (1972) and Wallerstein (1988), which hold that there are different temporalities or different categories of space–time. Braudel distinguished several types of history. At one extreme was the short-term time of *l'histoire événementielle* (history of events or episodic history) and at the other, *l'histoire des savants* (history of the sages), which describes the use of the concept of time underlying the work of those universalistic and nomothetic social scientists who see invariant structures in the human social world. However, those histories which Braudel considered most important were the medium-term time of *l'histoire conjoncturelle* (the history of cycles) and the *longue durée*, the long-term time of *l'histoire structurelle* (the history of structures), both of which depend on the recognition of cyclical movements and evolutionary phenomena (Table 2.1). In developing Braudel's ideas, Wallerstein (1988, p. 291) argues that historical reality is

> the reality of enduring but not eternal sets of structures (what I would call historical systems), which have their patterned modes of operation (what I would call their cyclical rhythms), but also have a continuous slow process of transformation (what I would call their secular trends).

In Wallerstein's discourse the words 'crisis' and 'transition' are used to refer to changes that occur in structural time, or, as he prefers, structural 'TimeSpace'. At these points of crisis and transition, instabilities predominate, the determination imposed by the laws and functioning of a social order weaken, and a transformational TimeSpace exists in which all individuals and groups are able to exercise fundamental moral choice and choose a new order.

Braudel's and Wallerstein's notions of secular trends and, particularly, the idea that it is possible to identify historical systems that have undergone constant slow processes of transformation, but then experienced more radical transformation in occasional phases of crisis and transition, accords with the idea that European development has been associated with the development and crisis of a series of modes of production (classical, feudalism, merchant capitalism, industrial capitalism and Communism). The first ideas that I shall briefly develop concern the pertinence of these categories in explaining European modernization.

Table 2.1   Concepts of historical time and geographical space
(geohistorical time–space)

Fernand Braudel	Immanuel Wallerstein
L'histoire événementielle: (idiographic) episodic (short-term) history	Immediate (idiographic) geopolitical space
L'histoire conjoncturelle: cyclical (medium-term) history	Ideological space
L'histoire structurelle: history of economic and social structures that determine, over the *longue durée*, human collective action	Structural space
L'historie des savants: (nomothetic, 'too long-term') history of the sages	Eternal (nomothetic) space

Cyclical rhythms are shorter-term movements of a range of different durations, and their analysis is more concerned with transformations within a particular economic order (such as capitalism). The present discussion concentrates on longer-term cyclical movements and Kondratieff-type cycles in particular (Figure 2.6). Although their existence is queried by some economists, Kondratieff cycles play an important role in the accounts of historical change developed by Braudel and many other historians, and also in analyses of cyclical movements and structural change in industrial societies. The literature on industrialization contains several explanations of these cycles, the most influential of which are perhaps Schumpeterian and neo-Schumpeterian accounts of innovation and technical change. According to Schumpeter (1934), long-term cyclical movements in industrial societies occur because of a historical clustering of innovations in periods of slackening growth. As innovations are adopted and diffused throughout an economic system, the rate of growth accelerates until the scope for diffusion is exhausted, at which point growth slows down.

Neo-Schumpeterian accounts draw on these insights. Freeman (1988), for example, argues that there are four kinds of innovation. These include:

- incremental innovations, which produce a continuous flow of modifications to existing products;
- radical innovations, which involve qualitative shifts, as with the development of nuclear reactors for electricity generation, or the changeover from cotton to nylon;
- a change in 'technology system', which involves a constellation of technically and economically interrelated radical innovations that affect whole industrial sectors;
- the introduction of new technology systems, which have pervasive effects on the whole of economic life and involve major changes in the capital stock and skill profile of the population; such changes in 'techno-

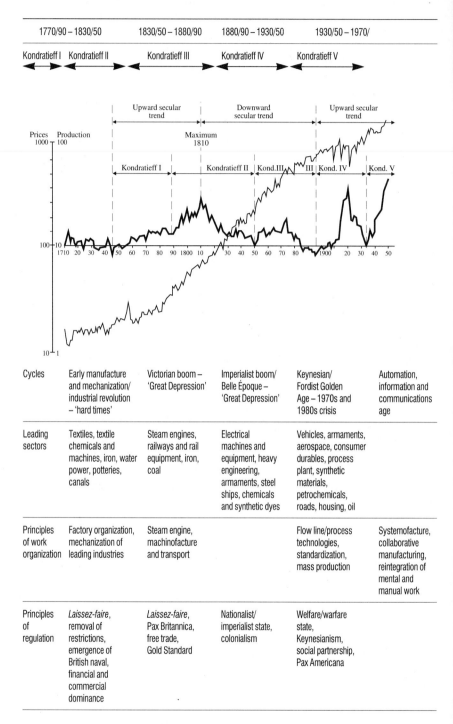

	1770/90 – 1830/50	1830/50 – 1880/90	1880/90 – 1930/50	1930/50 – 1970/
	Kondratieff I Kondratieff II	Kondratieff III	Kondratieff IV	Kondratieff V

Cycles	Early manufacture and mechanization/ industrial revolution – 'hard times'	Victorian boom – 'Great Depression'	Imperialist boom/ Belle Époque – 'Great Depression'	Keynesian/ Fordist Golden Age – 1970s and 1980s crisis	Automation, information and communications age
Leading sectors	Textiles, textile chemicals and machines, iron, water power, potteries, canals	Steam engines, railways and rail equipment, iron, coal	Electrical machines and equipment, heavy engineering, armaments, steel ships, chemicals and synthetic dyes	Vehicles, armaments, aerospace, consumer durables, process plant, synthetic materials, petrochemicals, roads, housing, oil	
Principles of work organization	Factory organization, mechanization of leading industries	Steam engine, machinofacture and transport		Flow line/process technologies, standardization, mass production	Systemofacture, collaborative manufacturing, reintegration of mental and manual work
Principles of regulation	*Laissez-faire*, removal of restrictions, emergence of British naval, financial and commercial dominance	*Laissez-faire*, Pax Britannica, free trade, Gold Standard	Nationalist/ imperialist state, colonialism	Welfare/warfare state, Keynesianism, social partnership, Pax Americana	

Figure 2.6 Kondratieff cycles and the secular trend

Leading economies	Britain, France, Belgium	Britain, France, Germany, USA, Belgium	Germany, USA, Britain, France, Belgium, Switzerland, Netherlands	United States, Germany, rest of EC, Japan, Sweden, Switzerland, rest of EFTA, Canada, AUSTRALIA, USSR	Japan, USA, Germany, Sweden, rest of EC and EFTA, East Asia

Figure 2.6   *continued*

economic paradigms' include the diffusion of steam and electric power in the past and the contemporary development of information and communications technologies.

Drawing on these distinctions, one can explain Kondratieff-type downturns in terms of the existence of contradictions or a mismatch between the development of the forces of production (in the shape of new techno-economic paradigms) and the social framework, social institutions and the social relations of production. (This explanation is close to Marxist accounts of historical change, which also dwell on the relationships between the forces and relations of production.) Kondratieff-type upturns, conversely, are explained by the diffusion of new technology paradigms permitted by the correspondence of forces and relations of production. This framework also offers a number of elements of an explanation of uneven development in its recognition that the extent of the mismatch differs from one nation to another, with the result that some economies (cities, regions and countries) are, in certain periods, far more successful than others.

Another, closely related interpretation of Kondratieff-type cycles is provided by theories of regulation. These theories rest on the view that capitalist development is characterized by phases of rapid and relatively successful growth and development (such as the Victorian boom, the *Belle Époque* and the Fordist golden age after World War II) punctuated by phases of crisis, such as the inter-war depression, marked by economic stagnation and mass unemployment (Figure 2.6). While each example has a particular explanation, crises are more generally explicable in terms of the contradictions and conflicts of interest inherent in capitalist economies. The most important are conflicts between capitalists and wage earners on the one hand, and co-ordination failures (such as simultaneous under- and over-investment, or inadequate demand resulting from attempts to squeeze wages to increase profits) on the other. As the existence of phases of successful growth shows, however, it is possible to 'regulate' these contradictions and conflicts, and to manage them in ways that are consistent with the dynamic expansion of the system (that is, with the establishment of a coherent model of development or regime of accumulation). As in neo-Schumpeterian approaches, each phase of growth is seen as embodying a particular technical paradigm/principles of work organization, and a particular set of institutions/modes of regulation. The establishment of a mode of regulation involves institutional reform and depends on a particular political compromise. An example is provided by the development of the welfare state and the managed economy after World War II. This involved a new social democratic or Christian Democratic compromise between industrial capital and wage earners which sought to resolve the contradictions that had led to mass unemployment in the inter-war years (Aglietta 1979, 1982; Dunford 1990). Thus similarities exist between theories of regulation and neo-Schumpeterian models. The main contrast, however, is that the former allocate social relations and economic mechanisms a much more central role (with the establishment of a new mode of regulation capable of regulating the contradictions of a capitalist order laying the foundations for a particular regime of accumulation), and technological factors a much more mediated one.

These theories indicate that societies characterized in structural terms by capitalist relations of production and exchange have assumed a number of historical forms. In each cyclical phase of development, the concrete expression of the social relations of a capitalist social order differ: capitalism has itself changed in order to survive. At the same time, changes in the technical foundations and social relations that underpin economic life have been accompanied by significant shifts in the map of economic development and the geographies of wealth and economic leadership.

Secular trends and the transition to capitalism

Cyclical movements and inequalities in development were a feature not just of the industrial era, but also of the medieval and early modern periods in Europe in which the foundations for later industrial development were laid (see Kriedte 1983; Braudel 1984; Wallerstein 1974–89). Moreover, short-term cycles and cycles of some 50 years' duration coexisted with much longer-term secular cycles lasting some two or three centuries. If we confine our attention to the more critical longer-term swings of the past millennium, a number of cycles can be identified. A long phase of internal and external agricultural and demographic expansion began *c.* 1050, accompanied by slow growth of commercial activity which lasted until *c.* 1250. This was reflected in Britain by considerable internal expansion and colonization, involving the clearing of forests, drainage of fens and reclamation of marshlands. In Germany, external expansion occurred with the extension of German settlement east of the Elbe, while the Spanish *Reconquista* saw the recovery of Spain from the Moors (see Chapter 1). At the same time much of Christian Europe was involved in more distant overseas expansion, especially as a result of the Crusades to the Holy Land. This long upswing was followed by a downswing from 1250 until 1450, which included the deep early-fourteenth-century crisis marked by famine and plague (transmitted in all probability as a result of the development of an extensive trading network with links to China, the Islamic world and the Mongols), population decline, settlement retreat and a fall in seigneurial incomes.

Coinciding with the discovery of America, the exploration of the sea route to India (via the Cape of Good Hope) and the start of European overseas expansion, the so-called 'long' sixteenth century (extending from 1450 to 1600 or 1640, depending on the parts of Europe considered) witnessed a phase of renewed expansion. This gave way to a new crisis during the seventeenth century, one marked by falling agricultural prices and slackening population growth. In the mid-eighteenth century, renewed upturn occurred, initiating a phase of expansion from which Britain emerged as the dominant global economic power, initially as the centre of an extensive world trading system and, in the latter part of the century, as the world's first modern industrial nation.

These movements were secular in the sense that they were long-term and also accompanied by a transition from one social order, feudalism, to another, merchant capitalism. There are several interpretations of this transition. One end of the spectrum is defined by theories of market emergence, which emphasize both the development of certain mentalities – in particular, human acquisitiveness – and also the emergence and protection of property rights and markets. Examples include Adam Smith's exegesis of the emergence of commercial society in terms of economic individualism and the pursuit of self-interest, together with North and Thomas's explanation of market emergence in terms of demographic movements, technological change and the creation by the state of a stable legal framework (for details see Holton 1985). At the other end of the spectrum are political economies of the transition to capitalism, conceived both as a market economy in which goods and services are produced for sale and as an economic order centred on a wage relation. The latter entails the creation of a wage-earning class, separated from ownership/control of land and the means of production and their concentration (via a process of primitive accumulation) in the hands of a relatively small propertied élite.

There is, however, considerable disagreement between the competing accounts of the transition to capitalism. In particular, this concerns the question as to whether the prime mover was the development of trade, commerce and the international division of labour (exchange relations theories), or the development of the wage relation (capitalist social relations and class conflict theories, which concentrate on the ways in which feudal social relations started to fetter economic and technological progress and were replaced by capitalist relations of production) (for a full account, see Hilton 1976). A more recent series of exchanges is included in the so-called Brenner debate between advocates of class relations explanations and the supporters of Malthusian-type models of economic change. The latter were developed specifically to account for the alternating phases of expansion and contraction of pre-industrial societies, but emphasize population growth as the principal dynamic factor leading to the cultivation of marginal land, diminishing productivity and subsistence, and demographic crises (Aston and Philpin 1985).

As capitalism is not merely a market system, the transition debate is more relevant than theories which confine their attention to the emergence of markets. It is argued here, however, that an explanation of the secular movements identified in this section, and of the transition from feudalism to capitalism, involves a synthesis of these competing perspectives. More specifically, I contend that the functioning of interconnected – but unequally developed – feudal societies led eventually to three developments:

- the establishment of the capitalist mode of agricultural production, most fully in England, and of more commercialized peasant capitalism in countries such as the Netherlands;
- an increase in the influence of merchant capitalists, financiers and mercantile companies;
- the establishment of the proto-industrial roots of Europe's industrialization.

One cumulative result was a secular transition from a feudal world dominated by knights and merchants to a merchant capitalist world of adventurers and companies, and the preparation of the ground for a further transition to an industrial capitalist world of generals and industrialists (Nolte 1992). Another was a series of shifts in European and global economic centres of gravity, reflected in part by the differential pattern of Europe's population growth (Table 2.2).

Geographies of modernization and of the transition to capitalism

Having concentrated on the dynamics of pre-industrial Europe and the main contours of the debate about the transition from feudalism to capitalism, I now turn to the social and geographical trajectories that resulted from the transformations of agriculture, commerce and industry. Early medieval Western Europe was diversified politically, economically and culturally, with multiple centres of population in fertile areas such as the Po Valley, the Rhinelands, the Paris Basin and southern England. In these areas of settlement, varying syntheses of elements inherited from late antiquity and Christianity, on the one hand, and the invading tribes on the other, resulted in the creation of demographically and economically varied societies. Attempts at unification failed, including Charlemagne's efforts to establish a unified (Holy Roman) Empire and centralized state in the early ninth century, and the tenth-century collapse of Kievan Russia (Figure 2.6). Thereafter, Western Europe was divided into a series of competing Christian polities, which were later consolidated – usually as dynastic states in the period from the 1540s to the 1690s, and as nation-states from the late eighteenth century onwards (Therborn 1995).

Table 2.2 Population changes in Europe, 1500–1800

	1500		1600		1700		1800	
	Population	Index	Population	Index	Population	Index	Population	Index
Northern Europe (Denmark, Norway, Sweden, Finland)	1.6	100	2.6	162	3.1	194	5	312
North-West Europe (British Isles, Netherlands, Belgium)	6.3	100	9.7	154	12.7	202	21.2	337
Western Europe (France)	17	100	17.9	105	20.8	122	27.9	164
Southern Europe (Portugal, Spain, Italy)	16.4	100	21.7	132	21.7	132	31.3	191
Central Europe (Germany, Switzerland, Austria, Poland, Czech parts of Czechoslovakia)	18.5	100	24	130	24.5	132	33.5	181
Total	59.8	100	75.9	127	82.8	138	118.9	199
Eastern Europe (European parts of Russia)	12	100	15	125	20	167	36	300
South-East Europe (Slovakia, Hungary, Romania, Balkan countries)	9.1	100	11.2	123	12.2	134	20.8	229
Total	21.1	100	26.2	124	32.2	153	56.8	269
European total	80.9	100	102.1	126	115	142	175.7	217

Source: Kriedte (1983, p. 3)

Agrarian change, the internal dynamics of European feudalism and the development of agrarian capitalism

Medieval Europe was defined by the emergence of a new form of society, feudalism, which, as discussed in Chapter 1, achieved its most developed form in North-Western Europe. At first, the internal dynamics of this system, together with a sequence of conjunctural events (including climatic conditions and the diffusion of diseases), resulted in a succession of short-term harvest cycles (caused by climatic conditions and the underdeveloped state of agricultural technologies) and, in the long term, secular phases of expansion and recession as identified in the last section. The internal dynamic of European feudalism was rooted, however, not in population growth but in the consumption and income requirements of the feudal nobility and the Church, and the needs of the state and nobility to finance wars (Bois 1978). As technological progress was limited, growth was accommodated by the extension of cultivation on to increasingly marginal land. As a result, productivity diminished, as did the rate of surplus extraction, leading to increasing seigneurial pressures on the peasantry and, eventually, Malthusian-type food and subsistence crises (Dunford and Perrons 1983).

At the same time, a secular transformation of these feudal systems was under way. As Bois (1984) has argued, the waves of expansion were associated with strong tendencies towards greater social differentiation. These involved the concentration of wealth in the hands of the more prosperous rural dwellers, and the proletarianization of small peasants. These conditions were conducive to the development of agrarian capitalism and, while the processes of change were held in check by feudal social relations, these limits were slowly but differentially pushed back. In northern France, which was the core of European feudalism, the fifteenth-century crisis led to a consolidation of the position of middling peasants as proprietors, enabling them to resist expropriation and helping ensure the survival of peasant farming until the middle of the twentieth century.

In England, the rights of the peasantry were less well established. Strong enough to resist any attempt to reimpose feudal obligations (in part because of the regeneration of England's urban economies and the strengthening of peasant rights), peasants had not established freehold control, thereby permitting lords to convert customary tenure to leaseholds, impose substantial rents and other monetary obligations on the peasant population, create large holdings and develop an improved and rationalized commercial agriculture. By the 1790s, great landlords and gentry controlled more than 80 per cent of agricultural land, and the share of agriculture in total employment had declined dramatically, sharply differentiating Britain from most other European countries.

In Eastern Europe, the outcome was profoundly different, as lords were able to reduce tenants to serfs working on large seigneurial demesnes. In the East, the 'second serfdom' endured until different points in the late eighteenth and nineteenth centuries (in Russia serfdom was not abolished until 1861), and was one of the reasons for the relative importance of agriculture in these societies, but also for the comparative underdevelopment of that agriculture (Anderson 1974b). Around 1800, 36 per cent of the working population were employed in agriculture, forestry and fishing in Britain, compared with some 65 per cent in Prussia. More comprehensive data for peripheral European countries between 1860 and 1910 are set out in Table 2.3. Even in the early 1930s agricultural employment accounted for 73 per cent of the workforce in Bulgaria, 35 per cent in Czechoslovakia, 52 per cent in Hungary, 61 per cent in Poland and 72 per cent in Romania.

Merchants, markets and the transformation of industry and commerce

Alongside the agricultural economy, an artisanal and guild-organized manufacturing sector developed in urban areas. Towns also developed as marketplaces in which merchants organized and co-ordinated local, regional and international trade. These developments in industry and commerce interacted with the development of agriculture: areas of concentrated manufacture depended on the supply of foodstuffs, while areas of more specialized agriculture generated a demand for manufactures, stimulating the growth of the domestic market. Greater specialization and an intensified division of labour also increased productivity.

Merchants played a decisive role in reshaping the map of economic development and in the transition to a capitalist order. Not simply did they encourage the commercialization of economic activities, but also they made profits by selling goods and services at a higher price than they bought them and, through the repetition of this cycle, accumulated commercial capital. In feudal societies, merchants were dependent on the feudal order: to maximize differences in prices they sought monopolies and privileges from feudal authorities, and there was always a tendency for merchants to invest in landed property, exploit ground rents and thus 'refeudalize' themselves. The accumulation of commercial capital was, nevertheless, a decisive factor in the development and transformation of industry and commerce.

Proto-industry and the transition to capitalism

At the start of the sixteenth century, uncontested leadership in terms of manufacturing lay with northern and central Italy, the southern Netherlands and the areas near Nuremberg and Augsburg in southern Germany. In subsequent decades, however, the northern parts of the Netherlands, together with England and France, moved to the forefront (Kriedte 1983). By then, a move-

Table 2.3   Agricultural employment and output in the European periphery, 1860 and 1910

	Agricultural labour force as a percentage of gainfully employed population		Percentage contribution of agriculture to national income	
	1860	1910	1860	1910
Denmark	55	36	48	30
Sweden	72	49	39	25
Norway	69	43	45	24
Finland	75	65	65	47
Portugal	73	57		
Spain	71	71		40
Italy	72	55	55	47
Greece	88	64	75	75
Hungary	75	64	70	62[a]
Russia	89	80	71	53
Romania	81	75		70
Serbia	89	82		79
Bulgaria	82	75		

Source: Berend and Ránki (1982, p. 159)

[a] Data for 1913

ment of industry into the countryside (often called proto-industrialization) had been set in motion. This transformation was attributable to the initiative of merchant capitalists. Commencing at the end of the Middle Ages, it accelerated into the sixteenth and seventeenth centuries and lasted until the nineteenth. Industrial cycles followed agricultural cycles, except that prices rose more slowly in upturns and fell less rapidly in downturns (as industrial production is not subject to diminishing returns and demand for industrial output is more elastic). One consequence was that industry was subject to irregular demand created by sharp short-term falls in demand during agricultural crises caused by poor harvests.

Faced with these cyclical conditions, merchants were anxious to increase the responsiveness of output to demand conditions without incurring significant fixed costs. They also desired to escape guild control, undercut urban monopolies, avoid urban taxes and levies, and reduce wage costs. To achieve such ends, merchants sought to employ a cheap rural labour force in stock-farming areas, which, resulting from the effects of agrarian change on the need for supplementary sources of income, they could draw into production for distant markets (often on a part-time basis). This process was associated with major changes in the map of industrial production and demographic growth in Europe and in the organization of industrial production. Two main types of system predominated: a *Kaufsystem,* in which independent artisans sold their produce to merchants and traders in public markets; and a *Verlagssystem,* in which urban merchant entrepreneurs put work out to dependent rural workers.

There were, however, numerous disadvantages associated with these rural cottage industries. These included the absence of supervision of productive work; the increase in circulation (including transport) costs associated with the extensive nature of growth; the irregularity of natural energy sources; the non-availability of some workers during harvest periods; and the need to increase wages as and where supplementary sources of income were eroded and wage dependence increased. In the eighteenth century, these disadvantages began to outweigh the advantages in some areas, with the result that rural proto-industrial activities started to give way to the factory system. Nevertheless, as the new technologies which were subsequently introduced had sectorally unequal effects on productivity growth, the growth of mechanized factory output at first led to an expansion of the cottage system in those sectors in which productivity did not at first increase.

Commerce and the development of a succession of world economies

Adventurers, and companies of merchant capitalists and financiers, played a major role in the development of a series of international trading systems and 'world economies', centred successively on Italy, the economic centres of the Spanish and Portuguese Empires, Amsterdam and Britain. In the fifteenth century, as the Ottoman world pressured Europe's eastern borders, Portuguese mariners sailed around Africa to India, and in 1492 Columbus set sail for America, initiating a second phase of European overseas expansion. (At the same time Russia was expanding to the east and south.) One result was that Europe's centre of economic gravity, which had been shifting northwards throughout the Middle Ages, was located first at Antwerp, which functioned as an outpost of the Spanish state (along with Genoa and merchant-banking centres in southern Germany). After the brief economic renaissance of Italy during the Eighty Years War between the Dutch United Provinces and the Spanish (1566–1648), the economic centre of the European world economy moved to Amsterdam, which remained the hegemonic centre of the European world economy for nearly two centuries. When it declined towards the start of the industrial era, leadership passed to Britain.

Each of these world economies was composed of interdependent but unequally developed geographical zones that made up a core–periphery structure within which there were relations of unequal exchange. In 1650, for example, the centre was Holland (now the western Netherlands) and Amsterdam, together with the region between Cologne, Paris and London.

In these regions agriculture was most developed, with enclosure, three-field rotations and specialized horticulture, and cities were important as centres of trade and homes to leading international companies. Much of the rest of Europe was made up of intermediate or secondary zones: the Baltic and North Sea states; the rest of England; the southern Rhine and the Elbe regions of Germany; the remainder of France, Portugal and Spain; and Italy north of Rome. As countries these areas were significantly militarized and politically independent. Economically, however, they were less developed than the core and occasionally dependent: agricultural progress was slower, and trade was sometimes in foreign hands. Scotland, Ireland and Scandinavia to the north, Europe east of a line running from Hamburg to Venice, Italy south of Rome and Europeanized America were all either semi- or wholly peripheral. These latter regions were often politically dependent, and frequently characterized by the existence of slavery and forced labour (much of the New World was based on slavery, and the outer reaches of Europe formed a second serfdom zone). The outer peripheries were involved also in the production of goods of high value but low weight, such as silver and gold, sugar and spices, furs and ivory (Braudel 1977; Nolte 1992).

Modern industrial development

The outcome of these developments that define the transition to capitalism was a progressively more rapid accumulation – especially in some of the core areas of Europe – of competences in science and in military, industrial and agrarian technologies. To these trends can be added the radical reorganization of production, trade and finance, and the accumulation of tangible material and financial wealth. Together these processes laid the foundations for a rapid acceleration of economic growth in the First Industrial Revolution, dramatic population growth as Malthusian constraints were lifted in the First Demographic Transition, and a new phase of European overseas expansion. As Nolte (1992, pp. 34–5) observes:

> Militarily it now became possible to conquer the interior of continents at relatively low cost; steamships, the telegraph and railroads made logistics possible across deserts and jungles, the rifle gave the firepower of a dozen musketeers to a single man, and whatever military problems might have remained were solved by the machine gun.

Military exploits and colonial expansion provided government-financed markets for European exports. Europe's cheap industrial products also displaced the former industrial exports from semi-peripheral and peripheral countries (such as Indian textiles or Russian iron), leading to deindustrialization and increased reliance on the export of raw materials (see Table 2.4 for indicators of the productivity advantage associated with the mechanization of

Table 2.4   Labour productivity in the cotton industry

Indian hand spinners (eighteenth century)	50 000
Crompton's mule (1780)	2 000
100-spindle mule (*c.* 1790)	1 000
Power-assisted mules (*c.* 1795)	300
Robert's automatic mule (*c.* 1825)	135
Most efficient machines in 1990	40

Source: Jenkins (1994, p. xix)
Note: The figures show the number of operator-hours required to process 110 lb of cotton

cotton spinning). Elsewhere, accumulated wealth and the products of new export-oriented primary goods production were exchanged for European industrial products, integrating large sections of the globe into a European-dominated international division of labour. To take but one example, Lancashire's cotton mills were supplied with cotton from the American Deep South, Egypt and Uganda, while England exported Indian cotton and later machine-made Indian textiles to China.

The geography of early industrialization

Early instances of modern industrialization were concentrated in a relatively small number of European regions, of which most were in Britain. Among the factors that explain the geography of early industrialization were the distribution of coal and mineral resources, and water as a source of power. Investment and growth also presupposed the availability of money wealth that could be transformed into capital to purchase factories, machines and materials on the one hand, and the labour power of a dependent wage-earning class on the other. As a result, industrial production ceased to be 'a mere accessory to commerce' (Marx 1981, pp. 440–55) and made commerce its servant. Also essential were certain general conditions of production (investments in transport, or a supportive legal framework) and access to potential markets. Once established, the development of external economies, through the creation of new infrastructures, supply industries, markets, knowledge of markets and skills, played an important role in an area's survival and further development.

In Britain between the 1760s and 1790s, there were some ten small islands of industrialization (Pollard 1981) (Figure 2.7). Of these, four soon declined: Cornwall, which specialized in tin and copper mining and smelting; Shropshire, with its early coal and iron industries; north Wales, with coal, slate, iron, lead and copper industries; and Derbyshire, which was an early centre for the cotton textile industry. Tyneside and Clydeside had emerged as centres of modern industry but survived as a result of later structural change. South Staffordshire was an important centre of coal, glass, chemical, metals, engineering and armaments industries. The two leading areas, however, were the West Riding of Yorkshire, with its specialization in the woollen and worsted industries, and south Lancashire, which was a centre for cotton textiles, together with metal-working and chemicals. In addition, London was a major industrial centre.

By the early nineteenth century, Britain besides, a number of significant industrial areas were located in inner Europe (Figure 2.7). These included the coal, iron, woollen, cotton and linen industries of the Sambre–Meuse and Scheldt valleys in Belgium, and the coal, woollen and cotton textile and iron industries of northern France. In addition, a series of modern industrial areas were to be found along the Rhine. These included the iron and textile industries of Rhineland and Westphalia; further north, the Ruhr, which developed later around the extraction and use of coking coal; to the south, Alsace had a traditional charcoal iron industry and a cotton textile sector; while in Switzerland the area from Basle to Glarus contained a concentration of textile industries, later replaced by chemicals and precision engineering. Further east, Saxony and Lusatia had a powerful textile industry, as well as metallurgical industries in the proto-industrial/manufacturing stages and, later, textiles, tobacco, railway engineering and printing. Silesia became a centre for coal and iron industries but its real expansion – as was the case in the Ruhr – occurred between the 1850s and 1870s. Normandy was an area of some potential with a significant cotton textile industry, but soon fell behind. Elsewhere in France, the Lyon region developed very significant textile industries, especially silk; Saint-Étienne concentrated on textiles and metallurgy; Le Creusot specialized in iron production; while the Upper Loire was another coal and iron region. In addition, smaller concentrations of modern industry were scattered throughout Europe, as in the Saarland with its coal and Württemberg with its textile industry. Finally, the cities of Paris and Berlin were, alongside London, important centres for industries needing access to large urban markets (Pollard 1981).

As this brief account of the early geography of industrialization indicates, the impact of the industrial revolution was – in several respects – very uneven. Geographically, it created wide variations in specialization and in rates of growth and/or decline. Sectorally, an unevenness in the development of industrial technologies and the means of transport was one reason for the initial very localized nature of modern industrial development. Finally, it was uneven in terms of methods of work organization, in that the spread of new methods of organization saw the expansion of factory production coincide with an expansion of employment in the domestic system.

Phases of industrial growth

This wave of early industrial growth was, however, just the first of a sequence of phases of industrialization, and its geography was only the first

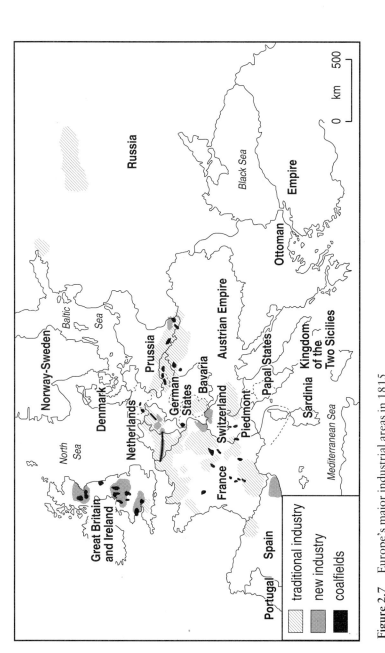

Figure 2.7 Europe's major industrial areas in 1815
Source: Based on Pollard (1981, p. xiv)

of a series of geographies of modern industrialization. In other words, growth itself was temporally uneven in that sustained phases of rapid growth alternated with enduring phases of crisis, marked by slackening growth, stagnation and even contraction. As Figure 2.6 shows, the first of the four periods of slower growth occurred in the period after the Napoleonic Wars, which saw, depending on the industrial or agrarian character of the country, the initial crisis of industrial capitalism or the last (Malthusian) crisis of the *ancien régime*. The second occurred in the Great Depression of the late nineteenth century at the end of the Great Victorian Boom, which followed the reforms of the 1830s and 1840s. The period between World Wars I and II, following the imperialist boom that had preceded 1914, marked the third crisis, while the fourth occurred in the years since *c.* 1970, which marked the end of the 'golden age' of fast growth and full employment that followed World War II.

As I suggested earlier, one important way of making sense of these temporal variations, and of the sectoral, organizational and geographical variations that accompanied them, is to acknowledge that – historically – industrialization involved the working out of a series of phases of accumulation. These have been characterized by changes in the nature of the dominant technologies; the leading sectors; the modes of organization of production and exchange; the ways of life and modes of consumption of wage earners; the character, role and functions of institutions and governance structures; the nature of the international order; the geography of economic activities; the relative standing of national and regional economies; and the location of global hegemony and leadership. According to this type of explanation, phases of stable growth are rooted in the emergence of a sequence of new development models, often centred on fundamental transformations of the preceding economic and social order. These take shape in phases of turmoil, crisis and rupture, when older socio-economic orders failed in economic terms and were rejected on political and social grounds. Competing social forces then agree on new compromises and new world-views (Dunford 1990, 1993). One pertinent example is provided by the post-World War II Fordist model (a system of production characterized by mass production, mass consumption, state intervention into the management of the economy and collective bargaining – indeed, the whole ideology of collectivism). Its roots lay in the inter-war struggle between social democratic and New Deal politics, Stalinism and Fascism. Each of these sought to resolve the contradictions of a failed liberal order, dependent on the centrality of a market rationality that was, as Polanyi argued in *The great transformation* (1944), one of the major causes of the savagery characteristic of the first half of the twentieth century.

Catching up and falling behind: uneven development

Industrialization and inequality

In the pre-industrial epoch, differences in gross marketed output per capita between the least and most developed countries in the world were of the order of only about 1 to 1.6 (Bairoch 1981). As I have argued, early modern industrialization was confined to a relatively small number of areas. After 1820, however, rates of growth accelerated more generally in Western Europe and especially in a number of new countries (the United States, Canada, Australia and New Zealand) settled by European emigrants (Maddison 1995) (Table 2.5). As growth accelerated, economic disparities increased sharply, with the result that increased wealth coexisted with great deprivation.

Table 2.6 records trends in inequality. The indicator used is the coefficient of variation of Gross Domestic Product (GDP) per capita. (The data are recorded in Table 2.5 as percentages of the US figure. As there are missing values in the sample, Table 2.6 also records the number of observations.) Column 2 of Table 2.6 indicates a sharp increase of inequality in Europe and the New World up to 1870, with a further increase up to 1913. The causes were twofold. First, early (Britain and Belgium) and late industrializers established a large lead over the rest of the continent. As the coefficients of variation for Western Europe show, an initial increase in inequality (from 21.6 to 31 per cent) was reversed as late industrializers started to catch up with the early leaders. As the data for Europe as a whole demonstrate, however, disparities in the wider Europe increased sharply as Eastern and Southern Europe were left behind (with the coefficient of variation increasing from 24.1 in 1820 to 40.8 in 1950). Second, inequality increased in the global group (Europe plus the new countries) because of the faster relative growth of the new countries: North America enjoyed a spectacular leap into a position of industrial superiority after the early 1890s, while Canada surged ahead in the years before World War I (from 67 to 79 per cent of the US figure) as a result of the prairie wheat boom (see Table 2.5).

Unequal development and late industrialization

At the root of these disparities lies the inequality inherent on the one hand in the dynamics of capitalist systems and, on the other, in the phasing of industrialization and in particular the process of late industrialization (Sylla and Toniolo 1991). Gerschenkron (1962) argues that industrialization is a process which exhibits certain uniformities, but whose characteristics vary with the degree of relative backwardness of a

Table 2.5 Comparative economic development of European countries, 1820–1992

	GDP per head											GDD multiplier			
	1820	1870	1900	1913	1929	1938	1950	1973	1979	1989	1992	1820–1913	1913–50	1950–73	1973–92
Twelve West European countries															
Austria	101	76	71	66	54	58	39	68	73	75	80	0.65	0.59	1.75	1.17
Belgium	100	107	89	78	72	77	56	72	73	75	80	0.78	0.72	1.28	1.11
Denmark	95	78	71	71	71	90	70	81	80	81	85	0.75	0.98	1.16	1.05
Finland	59	45	40	39	38	57	43	65	65	77	68	0.66	1.12	1.50	1.05
France	95	76	70	65	68	72	55	78	80	80	83	0.69	0.84	1.43	1.07
Germany	86	78	77	72	63	84	45	79	83	83	90	0.84	0.62	1.77	1.13
Italy	85	60	43	47	44	53	36	63	68	72	75	0.56	0.76	1.75	1.20
Netherlands	121	107	86	74	80	84	61	77	77	74	78	0.61	0.82	1.26	1.02
Norway	78	53	43	43	46	64	52	62	72	77	81	0.55	1.21	1.19	1.32
Sweden	93	68	63	58	56	77	70	81	80	81	79	0.63	1.21	1.15	0.97
Switzerland		88	86	79	90	103	93	108	96	98	98		1.18	1.16	0.90
United Kingdom	136	133	112	95	76	98	72	72	71	75	73	0.70	0.75	1.01	1.01
Arithmetic average	98	86	76	70	63	77	54	74	76	78	81	0.71	0.77	1.38	1.09
Four new countries															
Australia	118	155	105	104	74	92	75	75	74	76	75	0.88	0.73	1.00	1.00
Canada	69	66	67	79	69	70	74	82	88	91	84	1.14	0.93	1.12	1.03
New Zealand		127	105	98	77	106	89	76	67	65	65		0.91	0.85	0.85
United States	100	100	100	100	100	100	100	100	100	100	100	1.00	1.00	1.00	1.00
Arithmetic average	98	99	98	99	96	97	97	97	97	97	97		0.98	1.00	1.00

Five 'South' European
countries

Greece	74			31	35	44	20	47	49	47	48		0.67	2.30	1.02
Ireland[a]	72	61		51	42	51	37	42	44	47	54	0.70	0.71	1.15	1.28
Portugal		44	34	26	22	28	22	46	43	48	52		0.87	2.05	1.13
Spain	83	56	50	42	43	33	25	53	51	54	58	0.51	0.59	2.10	1.10
Turkey				18	14	22	14	16	18	18	21		0.74	1.21	1.24
Arithmetic average	67	56	30	33	31	31	21	36	36	36	38	0.49	0.64	1.72	1.06

Seven East European
countries

Bulgaria	66			28	17	26	17	32	34	29	19		0.61	1.84	0.59
Czechoslovakia	47	42		39	44	43	37	42	42	40	32	0.60	0.93	1.16	0.75
Hungary	52	41		40	36	36	26	34	34	31	26		0.66	1.30	0.78
Poland				40	31	20	36	26	32	32	26			1.26	0.68
Romania				17	17	20	12	21	22	18	12			1.70	0.57
Russia/USSR	58	42	30	28	20	35	30	36	35	32	22	0.48	1.06	1.23	0.59
Yugoslavia			19	19	20	22	16	26	31	27	18		0.83	1.58	0.71
Arithmetic average	55	39	23	24	23	32	27	35	34	31	21			1.27	0.62

Source: Elaborated from data in Maddison (1995, pp. 110–266)

Note: The figures for GDP per head are expressed in purchasing power standards relative to the United States, which in all cases is given a baseline of 10

a 'Southern' Europe is defined here in terms of its peripherality and relative underdevelopment, hence the inclusion of Ireland

Table 2.6   Trends in international inequality in Europe and the
New World, 1820–1992

Year	Europe and the New World		Western, Southern and Eastern Europe		Western and Southern Europe		Western Europe	
	CV[a]	Number	CV	Number	CV	Number	CV	Number
1820	23.7	18	24.1	15	21.5	13	21.6	11
1870	39.7	22	35.4	18	32.6	15	31.0	12
1900	37.4	22	37.1	18	33.1	15	30.6	12
1913	43.2	25	40.7	21	33.4	16	25.1	12
1929	45.7	27	46.2	23	33.7	16	25.2	12
1938	43.6	26	44.7	22	33.1	16	21.1	12
1950	52.8	27	51.1	23	40.8	16	28.9	12
1973	37.6	27	39.0	23	24.5	16	16.5	12
1979	35.8	27	36.8	23	21.8	16	10.7	12
1989	38.7	27	40.3	23	20.6	16	8.8	12
1992	44.2	27	47.5	23	19.3	16	9.6	12

Source: Elaborated from data in Maddison (1995, pp. 110–266)

[a] Coefficient of variation of real Gross Domestic Product per head at purchasing power standards

country at the point at which it starts to industrialize. Backwardness embraces factors including a country's wealth; its endowment with factors of production such as skilled labour, up-to-date technology and infrastructure; and the nature of its ruling class. This thesis diverges sharply from the ideas of economists such as Rostow (1961), who argue that there is a single path to industrialization, which involves a series of stages (and prerequisites) through which all societies must pass and along which each society can be located at any moment in time. In contrast, Gerschenkron contends that backward countries can establish substitutes for the preconditions for growth as these had been expressed in countries that had industrialized earlier. These substitute conditions would result in variations in the tempo and character of industrial growth.

On this basis, Gerschenkron analysed the situation in mid-nineteenth-century Europe in the following way. First, the degree of relative backwardness increased from the north-west to the east and south-east. Second, lags occurred in the timing of industrialization, with growth accelerating in Germany in the 1840s, in Hungary in the 1870s and in Italy and Russia in the 1880s (Figure 2.8). Third, the characteristics of these industrialization processes differed from the British model. Fourth, the determinants of industrialization also differed spatially, especially in relation to the role of joint-stock industrial credit banks (Germany) and the state (Russia) in the supply of industrial finance and the promotion of industrial growth. Ideologies of

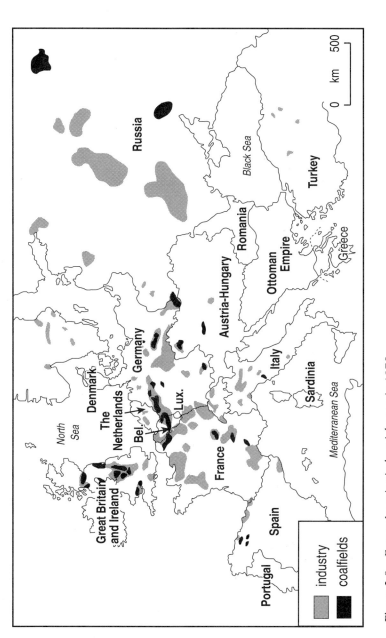

Figure 2.8 Europe's major industrial areas in 1875
Source: Based on Pollard (1981, p. xiv)

industrialization were also important. The general conclusion is that the pathway to industrialization depends not just on broad structures but also on a range of country-specific variables.

Modernization and emigration

Modernization

An explanation of these differences in the timing and nature of modernization lies in the trajectories of social and institutional relations, a factor which orthodox accounts such as Rostow's and, if to a lesser extent, Gerschenkron's fail to acknowledge. Attention has already been paid to the ways in which the interaction of the evolution of capitalist social relations and the forces of production underpinned the map of early industrialization. Once under way, the industrialization and modernization of the West (along with the political changes set in motion by the French Revolution) created a fundamentally new framework of opportunities, constraints and challenges for the more backward, peripheral countries of Europe (Berend and Ránki 1982). Two aspects of this challenge were particularly important. First, the industrialization and urbanization of the core countries led to an immense increase in the demand for food and raw materials. In the face of these trade opportunities, the more backward countries were encouraged to become exporters of raw materials, although the extent to which they could do so was seriously circumscribed by a range of internal obstacles to capitalist development (these included a labour force of serfs and sharecroppers; the absence of a modern credit and educational system; and the lack of a unified national market). Second, the modernization of the West posed a serious political and military challenge to the great-power status or independence of a number of countries of the periphery, while, for others, sovereignty and a weakening of ties of dependence were seen as central preconditions for socio-economic transformation. This combination of economic self-interest and power politics were powerful factors in prompting ruling élites to implement processes of political and institutional reform 'from above' (largely between 1820 and 1870), which, in different ways and to different extents, opened the way to partial and incomplete forms of capitalist development. Industrialization was an important economic consequence, although industrial growth involved particular difficulties owing to the weakness of earlier proto-industrial growth and the problems of infant industries faced with competition from already established industrial nations. Indeed, it was such obstacles, together with the more scientific foundation of later industrialization, that lay at the root of the more interventionist role of the state and the development of industrial credit banks noted by Gerschenkron.

As Berend and Ránki (1982) point out, the differences in the timing and nature of the development of a capitalist economic and political framework were striking. In the Scandinavian countries, which had long lost their roles as great powers, the feudal order disintegrated step by step through the seventeenth and eighteenth centuries, processes that transformed the peasantry of Sweden and Denmark into freehold farmers and led to the development of strong self-sufficient rural communities in Norway. Both sets of changes had significant implications for the later trajectories of these countries.

In the Mediterranean, the situation differed in several ways. Most important was the fact that even though earlier commercial successes had created space for bourgeois enterprise, the *ancien régime* remained particularly powerful. In Iberia and Italy, the development of a capitalist economic and political order involved a series of advances and reversals. These started with Napoleon's conquest (which led, for example, to the abolition of serfdom in the south of Italy) and the unfolding of a sequence of revolutionary insurrections and national struggles that led to reforms, which were subsequently reversed after defeats at the hands of reactionary forces. An important consequence was that the change which did occur involved only a modernization of the edifice of the *ancien régime*, rather than its complete overthrow (which itself was reflected in the handling of the land question and the preservation of great estates in the south of Spain, Italy and Portugal). Another repercussion was the limited degree of nineteenth-century industrialization, especially in Iberia and southern Italy. In Spain, the leading role was played by the textile industry, centred on Catalonia. The usefulness of Spain's enormous deposits of iron ore and non-ferrous metals was limited by its lack of coal. Domestic smelting did not start until the 1880s and, as late as the early twentieth century, virtually all Spain's iron ore was exported. In Greece the situation was less complicated in that wars of independence against the Turkish Empire went hand in hand with the struggle for transformation to market capitalism and representative politics. Industrial growth was, however, extremely limited. Growth depended largely on agricultural exports, although there were also (as was true of Portugal) a relatively large number of jobs in trade and shipping.

Central and Eastern Europe displayed a number of pathways to development. These included a Prussian, Russian and Polish path of reform from above; a Hungarian model involving revolution and the war of independence of 1848–9; and the struggle of Serbia and Bulgaria to shake off Ottoman rule. These trajectories had a number of features in common, however, in that they involved reform from above but failed to solve the land question and the problems of the peasantry. They were also associated with a slow development of representative democratic institutions, and strongly shaped by nationalism. The consequent processes of transformation lagged far behind those of the core countries of Europe, although by the end of the nineteenth century a free labour force, freedom of enterprise, security of private property and a credit system were largely in place. Nevertheless, countries such as Hungary and Russia assimilated only a subset of modern technologies in a few spheres of economic activity. As the majority of these remained very backward, and as industry accounted for a very small fraction of national wealth and income, 'even rapid gains in this sector did relatively little at first for total output or the standard of living' (Landes, cited in Berend and Ránki 1982, pp. 153–4). Still worse off were the countries of the Balkans, which proved largely incapable of moving beyond their pre-industrial state.

Consequently, the advantages realized by the core countries of Europe relative to the periphery, from their earlier capitalist modernization and industrialization, were largely reinforced during the second half of the nineteenth century. While Scandinavia was well on the way to joining the capitalist core, and northern Italy, Hungary and parts of Russia had started on the road to industrialization, the least developed countries of Eastern Europe, the Mediterranean (with the exception of northern Italy) and the Balkans 'remained in, or were pushed to, the periphery of the European division of labour' (Berend and Ránki 1982, p. 159).

Emigration

Economic change and the emergence of wide disparities in economic performance also fuelled a wave of mass migrations from Europe and in particular from some of its peripheries. Millions of those who possessed the minimum resources required to pay for their passage to the Americas chose emigration as an alternative to poverty and deprivation. At first the highest rates of intercontinental emigration were from Ireland (14.0 per thousand in 1851–60, 14.6 in 1861–70, 6.6 in 1871–80 and 14.2 in 1881–90) and Scotland (5.0, 4.6, 4.7 and 7.1 respectively), although England (2.6, 2.8, 4.0 and 5.6) and Norway (2.4, 5.8, 4.7 and 9.5) also had high rates (Baines 1991). After 1880, Germany (which included the Prussian parts of present-day Poland), Switzerland, Sweden, Denmark, Italy, Spain, Portugal, Austria-Hungary and Russia (including Russian Poland) joined the significant proportionate/absolute exporters of people. Between 1815 and 1930, more than 50 and perhaps as many as 60 million Europeans emigrated, mainly from (the peripheral parts of) Britain (11.4 million), Italy (9.9 million), Ireland (7.3 million), Spain and Portugal (6.2 million), Austria-Hungary (5 million), Germany (4.8 million) and Russia (3.1 million), Most went to the United States (32.6 million), but Argentina (6.4 million), Canada (4.7 million), Brazil (4.3 million) and Australia (3.5 million) were also significant recipients. Since the labour content of these migrations was very high, this outflux had a major impact on the distribution of the labour force, GDP per worker and wages, although it had a smaller impact on GDP per capita. This latter trend can be explained in that while emigration will raise GDP per capita by reducing the population, and may raise GDP per worker by offsetting diminishing returns in agricultural production, its selective character will work in the opposite direction, reducing output per capita, through taking away a disproportionate share of the labour force.

Twentieth-century economic growth

As Table 2.5 shows, the core economies of Western Europe continued to grow more slowly than the economies of the New World countries until 1929, and then, after an improvement in their relative position between 1929 and 1938, suffered a further setback as a result of World War II. After 1950, however, Western Europe closed this gap during the three decades of the 'Fordist' golden age and in the subsequent phase of crisis (marked by a halving of rates of productivity and output growth in developed capitalist economies).

The 'South' European countries, which started in second position in 1820, and Eastern Europe grew much more slowly than Western Europe until well into the twentieth century. The first group of countries to start to catch up were the Soviet Union in the 1930s and the Communist countries in the 1950s and 1960s. The 'South' European countries started to make up for their relatively slow growth only later, after 1960. As the 'South' did not suffer a relative slowdown of the same magnitude as that experienced in the Communist world in the late 1970s and 1980s, and did not experience the dramatic collapse that followed the fall of Communism, 'South' European economies, while less developed than those of the Western European core, have now restored their substantial lead over the East, creating the map of inequality with which this chapter started.

Conclusion

In the last section I argued that modern industrialization has witnessed a reproduction of inequalities whose roots go far into the European past. Simultaneously, there have been significant shifts in the map of European economic development. In the second half of the nineteenth century, the list of industrialized countries was enlarged and – within this expanded group – inequality diminished as newly industrializing countries caught up with, and sometimes overtook, the early leaders. But disparities between the economies that had been industrialized and the non-industrialized world increased very sharply at first, and continued to widen until the end of World War II. After 1945, Western and Southern Europe closed the gap on the United States (as did Japan and later Taiwan, South Korea, Thailand and other Asian countries). In the Communist world, convergence occurred after 1929 in the Soviet Union and, following World War II, in the countries that then fell under Soviet influence. In the early 1970s, however, slower growth and divergence set in.

More generally, I have argued that the geographies of modern Europe are a complex synthesis that emerges out of the articulation and superimposition of sequential structural and cyclical phases of development one upon another. At each stage, the trajectories of Europe's urban and regional economies depended on choices that were always constrained by circumstances inherited from the past. In turn, these very circumstances were themselves consequences of past choices and the structural constraints by which these were shaped. The ensuing geographies indicate that there are places whose inhabitants have managed to act in ways that have enabled them to survive over quite long periods as major concentrations of activity, and to remain relatively prosperous. At the same time, however, other communities that were once active and thriving, owing perhaps to an extraordinarily rapid development of a narrow range of activities, have declined, sometimes quite quickly, into devastated, derelict and depressed areas. Yet other areas have been locked into states of persistent relative poverty and underdevelopment over long periods of time. What underlies these differences is not simply processes of unequal development with their roots in mechanisms of differentiation (divergence) and equalization (convergence) but, in many cases, simultaneous and sequential over- and underdevelopment. While these processes reflect the interrelationships between the resource endowments of different places (which are themselves a result of previous development and not therefore 'natural') and the opportunities offered by wider economic and political contexts, the choices made are also strongly dependent on technological, productive, institutional and political structures and arrangements. The implication is that an appreciation of the changing geographies of Europe depends closely upon an analysis of the relation between spatial change and the changing articulation of productive forces and social and institutional relations of production.

References

Aglietta, M. 1979: *A theory of capitalist regulation*. London: New Left Books.
Aglietta, M. 1982: *Régulation et crises du capitalisme: l'éxpérience des Etats-Unis*, 2nd ed. Paris: Calmann-Levy.
Anderson, P. 1974a: *Passages from antiquity to feudalism*. London: New Left Books.
Anderson, P. 1974b: *Lineages of the Absolutist State*. London: New Left Books.
Aston, T. H. and Philpin, C. H. E. (eds) 1985: *The Brenner debate: agrarian class structure and economic development in pre-industrial Europe*. Cambridge: Cambridge University Press.
Baines, D. 1991: *Emigration from Europe 1815–1930*. London: Macmillan.
Bairoch, P. 1981: The main trends in national economic disparities since the Industrial Revolution. In Bairoch, P. and Lévy-Leboyer, P. (eds), *Disparities in economic development since the Industrial Revolution*. London: Macmillan, 3–17.
Bartlett, R. 1993: *The making of Europe: conquest, colonization and cultural change, 950–1350*. Harmondsworth: Allen Lane.
Berend, I. T. and Ránki, G. 1974: *Economic development in east-Central Europe in the nineteenth and twentieth centuries*. New York and London: Columbia University Press.
Berend, I. T. and Ránki, G. 1982: *The European periphery and industrialisation, 1780–1914*. Cambridge: Cambridge University Press.

Bois, G. 1978: Against the neo-Malthusian orthodoxy. *Past and Present* 79, 60–9.

Bois, G. 1984: *The crisis of feudalism: economy and society in eastern Normandy, c. 1300–1550.* Cambridge: Cambridge University Press.

Bradley, J. 1996: *An island economy: exploring long-term economic and social consequences of peace and reconciliation in the island of Ireland.* Dublin: Forum for Peace and Reconciliation.

Braudel, F. 1972: History and the social sciences: the *longue durée.* In Burke, P. (ed.), *Economy and society in early modern Europe: essays from Annales.* London: Routledge & Kegan Paul, 11–42.

Braudel, F. 1977: *Afterthoughts on material life and capitalism.* Baltimore and London: Johns Hopkins University Press.

Braudel, F. 1984: *Civilization and capitalism, 15th–18th century,* vol. 3: *The perspective of the world.* London: Collins.

Dunford, M. 1990: Theories of regulation. *Environment and Planning D: Society and Space* 8, 297–321.

Dunford, M. 1993: Regional disparities in the European Community: evidence from the REGIO databank. *Regional Studies* 27, 727–43.

Dunford, M. 1997: Mediterranean economies: the dynamics of uneven development. In King, R., Proudfoot, L. J. and Smith, B. (eds), *The Mediterranean: economy and society.* London: Arnold, 126–54.

Dunford, M. and Perrons, D. 1983: *The arena of capital.* Basingstoke: Macmillan.

Freeman, C. 1988: *The factory of the future: the productivity paradox, Japanese just-in-time and information technology.* PICT Policy Research Papers, 3. London: Economic and Social Research Council.

Gerschenkron, A. 1962: *Economic backwardness in historical perspective.* Cambridge, MA: Harvard University Press.

Hilton, R. H. (ed.) 1976: *The transition from feudalism to capitalism.* London: New Left Books.

Holton, R. J. 1985: *The transition to capitalism.* London: Macmillan.

Jenkins, D. T. 1994: *The textile industries.* Oxford: Blackwell.

Kriedte, P. 1983: *Peasants, landlords and merchant capitalists: Europe and the world economy, 1500–1800.* Leamington Spa: Berg.

Maddison, A. 1995: *Monitoring the world economy, 1820–1992.* Paris: OECD.

Marx, K. 1981: *Capital,* vol. 3. Harmondsworth: Penguin.

Mjøset, L. 1992: *The Irish economy in a comparative institutional perspective.* NESC Report no. 93. Dublin: The Stationery Office.

Munck, R. 1993: *The Irish economy: results and prospects.* London: Pluto Press.

Nolte, H.-H. 1992: Europe in global society to the twentieth century. *International Social Science Journal* 54, 23–40.

Polanyi, K. 1944: *The great transformation: the political and economic origins of our time.* Boston: Beacon Press.

Pollard, S. 1981: *Peaceful conquest: the industrialization of Europe, 1760–1970.* Oxford: Oxford University Press.

Rostow, W. 1961: *The stages of economic growth: a non-Communist manifesto.* Cambridge: Cambridge University Press.

Schumpeter, J. A. 1934: *The theory of economic development: an inquiry into profits, capital, credit, interest and the business cycle.* Cambridge, MA: Harvard University Press.

Sylla, R. and Toniolo, G. 1991: Introduction: patterns of European industrialization during the nineteenth century. In Sylla, R. and Toniolo, G. (eds), *Patterns of European industrialization: the nineteenth century*. London: Routledge, 1–26.

Szücs, J. 1988: Three historical regions of Europe: an outline. In Keane, J. (ed.), *Civil society and the state: new European perspectives*. London and New York: Verso, 291–332.

Therborn, G. 1995: *European modernity and beyond: the trajectory of European societies, 1945–2000*. London: Sage.

Wallerstein, I. 1974–89: *The modern world-system*. 3 vols. San Diego and London: Academic Press.

Wallerstein, I. 1988: The inventions of TimeSpace realities: towards an understanding of our historical systems. *Geography* 73, 289–97.

3

War and the shaping of Europe

MICHAEL HEFFERNAN

Introduction: geography and war

The study of war, which has a long and often disturbing history (Gaillie 1978, 1991), has generated a vast literature informed, essentially, by two opposing projects. For some, studying war is a moral imperative, the necessary first step towards the eradication of war in the future. This pacifist belief rests on the conviction that harmonious coexistence is the natural human condition. War is an irrational, exceptional state; an aberration caused by political failings, which can be identified and corrected through careful investigation and analysis. Once the conditions which produce war are fully understood, the possibility exists to avoid war completely.

For others, knowledge of war is a strategic imperative. Their objective is to study war in order to wage it more efficiently. The key document here is Carl von Clausewitz's *On war*, originally published in 1832, which sought to 'normalize' war by establishing the predictable rules by which it should be governed. War, Clausewitz famously claimed, is 'nothing but a continuation of political intercourse, with a mixture of other means' (Clausewitz 1968, p. 402; Howard 1983). Strategic analyses of war are often straightforwardly partisan, the aim being to strengthen the military capability of a particular state or armed force. But these studies are also justified on higher moral grounds. If one accepts war as a 'natural' human condition, a biologically inherent impulse and the inevitable consequence of life on a crowded planet, then belligerence can never be expunged from human existence. The best we can hope to achieve is to limit the incidence of war by evolving military capabilities which ensure that the costs of war will outweigh the gains. Through the development of 'intelligent' forms of warfare, the length and destructiveness of wars can be reduced and controlled. This argument reached its apogee at the high point of super-power nuclear proliferation during the Cold War. The doctrine of Mutually Assured Destruction held that world peace rested on the ceaseless development of more and better weapons of mass destruction, coupled with a full understanding of the terrible implications of such arsenals.

The literature on war is mainly associated with the disciplines of history, political science and international relations but there is a long tradition of geographical research on war and peace (Lacoste 1976; O'Sullivan and Miller 1983; Pepper and Jenkins 1985; O'Loughlin and van der Wusten 1986, 1993; Kliot and Waterman 1991). 'War', claimed Sir George Taubman Goldie (1907, p. 8), 'has been one of the greatest geographers'. Geography, asserted G. B. Mackie (1917, p. 498), has 'more military significance than all the other branches of the standard curricula taken together'.

There are three distinct traditions of geographical writings on war, each reflecting either the moral or the strategic imperatives described above. The first, and most overtly strategic, tradition is military geography, a particular form of 'applied' geography widely taught in military academies around the world since the eighteenth century (Maguire 1899; Peltier and Pearcy 1966; Bateman and Riley 1987). Here the technical, cartographic and other interpretative skills of the geographer are explicitly addressed to military needs and objectives. Military geography has rarely found a place in the curricula of civilian educational institutions in time of peace. During wartime, however, all forms of geographical expertise and research tend to be 'militarized' (Balchin 1987; Stoddart 1992; Godlewska 1994; Kirby 1994; Heffernan 1995a, b; Harris 1997). Some claim that the advent of geographical information systems (GISs), remote sensing and other 'geographical' technologies have continued this 'militant' tradition (Pickles 1995). A principal concern of this approach has been the assessment of the physical environment as a factor influencing the conduct of warfare (Mahan 1980; Doyle and Bennett 1997).

Research in the second tradition has considered the reverse relationship: war as an agent of landscape change. Work has included studies of the impact of warfare on rural settlement patterns and landscape design (Sutton 1981; Ashworth 1991), of the destructive power of war in particular places (Lacoste 1977; Hewitt 1983), and of the rebuilding of lives and landscapes in the aftermath of war (Clout 1996). Several historiographers have considered the dialectical relationship between warfare and the environment: war shaped by landscape and landscape shaped by war. An excellent example is provided by Fernand Braudel's sparkling account of Mediterranean warfare in the sixteenth century (Braudel 1972; Pryor 1988). Some, but by no means all, of the work carried out in this tradition is informed by an anti-war perspective.

A third, 'geopolitical', tradition, emerging from the relationship between geography and international relations, deals with larger theoretical questions. The principal concerns here include the causes of war, the role of war within the international system, and the influence of war on the wider political landscape. This tradition has been influenced by both moral and strategic imperatives in almost equal measure. Throughout the first half of this century, the European 'geopolitical' movement, particularly (but not exclusively) in Germany and Italy, developed geographical arguments about war which were openly motivated by 'strategic' national projects of imperial expansion or revanchism in the aftermath of 1918 (see below). Countering this was a rival geopolitical agenda that sought to develop general theories of war in order to facilitate a moral programme which could lead to the total and lasting cessation of all human conflict (for example, Taylor 1946). The recent development of 'critical geopolitics' (a radical critique of all earlier forms of geopolitics and an emerging empirical project on the social, cultural and ideological factors which determine the relationships between societies and states) can be seen as a continuation, in a very different guise, of earlier forms of 'geopacifics' (Ó Tuathail 1996).

This chapter operates within the third tradition. Informed by the moral imperative discussed above, its theme concerns the impact of war on the political and cultural identity of modern Europe. This is an enormous and overwhelmingly important topic. Europe has relinquished, hopefully for ever, its status as the most violent and war-torn of the earth's regions. But this must not blind us to the dangers of war in the future. Paradoxically, the ending of the Cold War has increased the threat of warfare in Europe, as the tragic events in the former Yugoslavia make depressingly clear. Nor should we forget the ongoing importance of previous wars: Europe's external 'limits', its internal geopolitical order and the identities of its different peoples probably owe more to the periodic outbursts of violence than to the intervening periods of stability and peace. Even if we reject the comforting belief that by studying war we may one day eliminate it, we must surely accept the need to confront the role which war and conflict have played in shaping our own identities, both individual and collective (Fussell 1975; Eksteins 1989; Winter 1995).

There is a final, compelling reason to consider the impact of war on the human geography of Europe. Twentieth-century demographic statistics recount a sombre story of Europe's 'missing' human dimension. To emphasize the contested nature of the statistics already quoted in Chapter 1 should not detract from their awful magnitude. One eminent historian has calculated somewhat conservatively that the 31 years between 1914 and 1945 witnessed the premature death through warfare of nearly 60 million Europeans (Bullock 1991). Another has estimated that of the 110 million lives claimed by violence around the world from 1900 to 1972, 80 million were lost in Europe (including Russia) (Elliot 1972). These 80 million people would have changed and potentially enriched their surroundings had their existence not been cut short. There are, then, 80 million twentieth-century reasons to consider the impact of war on Europe's human geography; 80 million lives whose imprint on the continent's landscapes must never be forgotten.

War and the idea of Europe

'The origins of Europe', it has been claimed, 'were hammered out on the anvil of war' (Brown 1972, p. 93). This statement refers to the outburst of violence after the collapse of the Roman Empire but might equally apply to the post-Renaissance geopolitical debate which produced the first modern definitions of 'Europe'. The idea of 'Europe' as a secular geopolitical concept (distinct from the earlier Christian notions of Christendom or *Respublica christiana*) emerged from the early modern attempt to create an international system in which conflict and warfare between emerging nation-states would be, if not eradicated, then at least controlled. The very word 'Europe' entered the political lexicon as part of a wider discussion about war. The 'critical years' were those between the Treaty of Westphalia in 1648 and the Treaty of Utrecht in 1713 (Hazard 1990). By the mid-seventeenth century, the origins of the modern system of nation-states were firmly in place, the product of 150 years of intensifying capitalist production and the dramatic expansion of 'Old World' power into the 'New World' (Shennan 1974; Tilly 1975, 1989). This 'system' was associated with a new intensity of warfare, qualitatively different from earlier medieval conflicts, in terms of both the numbers of people involved and the technology deployed (gunpowder, musket, cannon) (Bean 1973; Midlarsky 1975; Levy 1983; Howard 1976; McNeill 1982; Keegan 1993). The Reformation and Counter-Reformation had shattered the fragile unity of the Christian Church (which had given a measure of cohesion to medieval Christendom), thereby adding a combustible religious factor to the secular tensions between Europe's rival dynasties and emerging states. Religious wars decimated France in the sixteenth century and erupted with a particular savagery across Central Europe during the Thirty Years War (1618–48).

The concept of the 'balance of power', first discussed in regard to the Renaissance city-states by Niccolò Machiavelli in *The prince*, published in 1513, was advanced as an ideal arrangement by which rival states might control their apparently in-built tendency towards violence and blood-letting (Sullivan 1973; Pocock 1975). Once balance could be established between several states of broadly comparable power, a relatively stable system seemed likely to endure. Any pre-emptive action by one state would be counteracted by alliances between other states to restore the overall balance (Wright 1975; Sheehan 1996).

This concept grew in popularity through the seventeenth century and was a central objective of the Treaty of Westphalia, the 'foundation treaty of the inter-state system' (Taylor 1996, p. 21). This was the first major accord to be framed by reference to 'natural' (what would later be termed 'international') law. It was signed by all the major European powers except Poland and England to bring an end to the Thirty Years War. By c. 1700, the idea of the balance of power was regularly invoked as the defining characteristic of 'Europe', a secular term which had largely replaced the older notion of Christendom. According to Denys Hay (1968, p. 116), it was 'in the course of the seventeenth and early eighteenth centuries . . . [that] Christendom slowly entered the limbo of archaic words and Europe emerged as the unchallenged symbol of the largest human loyalty'. Europe was thus the political arena in which a balance of power between rival states might conceivably operate and where peace, religious freedom and commercial liberty might be assured. The equation of Europe with the idea of the balance of power was invoked with particular insistence in the Protestant states of Northern Europe in their struggle against an expansive, Catholic France. From 1667 to 1713, France was in a state of virtually permanent war around its eastern 'borders', leading to a dramatic expansion of French territory (Parker and Smith 1978) (Figure 3.1). These wars assumed new and more alarming forms. Enormous armies were raised, the maintenance of which required a wholesale revolution in government and a massive extension of the state's fiscal and taxation powers. The deliberate use of murder and famine following the French occupation of the Palatinate in 1680 sent shock waves around Europe's capitals; this was sheer terrorism directed at civilians rather than

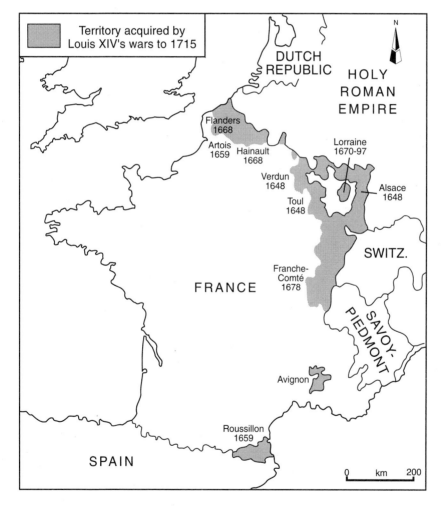

Figure 3.1   The expansion of France, 1648–1715
Source: After Merriman (1996, p. 318)

opposing armies. French aggression was also unleashed against its own Protestant population in the wake of the 1685 revocation of the Edict of Nantes, which had previously guaranteed religious freedom. Louis XIV cloaked his territorial ambitions in an older Catholic rhetoric, presenting himself as the defender of *Respublica christiana*, divinely ordained by God. His Protestant enemies, on the other hand, invoked the new, secular rallying cry of 'Europe' in their anti-French propaganda. In England and the Netherlands, constitutionally united following the Glorious Revolution of 1688, France was depicted as having seized the dangerous mantle relinquished by Habsburg Spain after the collapse of the Spanish Armada in 1588. The 'Sun King' was the principal threat to the balance of power and, therefore, the enemy of the new concept of 'Europe'.

From its inception, then, the idea of Europe was equated with a stated desire to avoid war through concerted international action based on the balance of power. 'Europe' emerged as a liberal slogan with which to attack absolutism and defend the cause of international peace and religious and commercial freedom. Schmidt (1966, p. 178) states that 'the term Europe established itself as [an] expression of supreme loyalty in the fight against Louis XIV. It was associated with the concept of a balanced system of sovereign states, religious tolerance, and expanding commerce'. The word 'Europe', though regularly deployed in the seventeenth century, only found real currency and influence after *c.* 1700 (Burke 1980). Significantly, the Treaty of Utrecht, signed in 1713 at the end of the War of the Spanish Succession, was the last accord to make reference to *Respublica christiana* (Hay 1968).

Ever since the mid-seventeenth century, Europe has been organized, with rare interruptions, according to the doctrine of the balance of power or its later variations (the Concert of Europe in the nineteenth century or the 'balance of terror' after 1945). Despite the comforting rhetoric, this system has manifestly failed to prevent bilateral wars, these occurring with monotonous regularity. More disastrously still, it was incapable of preventing the two eras of total devastation which scarred the first half of the twentieth century. The absence of war since 1945, when the levers of military power have been controlled from outside Europe, is sometimes cited as evidence of the balance of power operating effectively. Yet this was only possible because the costs of war were so high, involving likely global annihilation. Moreover, the idea that the permanent, emotionally debilitating threat of nuclear apocalypse warrants the term 'peace' is itself highly dubious. In general, the balance of power has failed to prevent war and may well have had the opposite effect.

This is scarcely surprising, for it is difficult to equate the balance of power with a full-blooded desire to eradicate war. The real objective has been much more limited: to control the cost and length of warfare. It could even be argued that periodic episodes of war were necessary if the system was to operate successfully. The assumption that all states naturally seek comparative advantage or hegemony over their rivals implies that, eventually, a frustrated power will resort to violence to achieve its objectives. It is only through counter-violence that balance can be restored to an inter-state system under threat from such an ambitious power. Thus, it can be argued that, without some other mechanism to settle international grievances, the balance of power has tended to encourage war. The desire to improve or overthrow the balance of power in Europe has a history as long as the idea itself. Two distinct alternatives have been offered, one based on an appeal to absolute peace and the total eradication of war, the other based on an appeal to a new European order achieved through force of arms and violence.

Europe built on peace?

A number of schemes have been advanced, usually during or immediately after periods of intense warfare, which have sought not merely to limit war, but to avoid it altogether through the creation of Europe-wide governmental institutions (Hinsley 1967; Heater 1992). Some of these schemes, the intellectual precursors of the post-1945 project of European unity, pre-date the Treaty of Westphalia. The most famous early proposal was conceived by the Duc de Sully, Maximilien de Béthune (1558–1641). Sully was a far-sighted Protestant who rose to become French finance minister under Henri IV (Buisseret 1968). Following the assassination of Henri in 1610, Sully spent his retirement devising a 'Grand Design' for European peace and unity, which was issued in various forms during the middle decades of the seventeenth century, albeit usually attributed to his murdered patron (Ogg 1921).

Sully's scheme was not entirely without a narrow, nationalist motivation. He was clearly interested in enhancing the power and reputation of the French Bourbons and ensuring the eclipse of Habsburg authority. In this sense, his design was compatible with the balance of power. Yet his commitment to peace on both moral and practical commercial grounds was genuine. Sully's objective was to reorder the political landscape 'in such a manner that none of . . . [the European powers] . . . might have cause either to envy or fear from the possession or power of others' (Ogg 1921, p. 41). He envisioned 15 of the existing states providing proportional quantities of men and resources to establish a common European army. Each state would also send representatives to a general council to be located at a suitable city on the tense borders of the Frankish and Germanic territories. Sully's vision of Europe was also secular: Russia, though part of the Christian world, was excluded from Europe largely on cultural grounds. The Russians, Sully claimed, were really an Asiatic people.

A generation later, in the midst of the late-seventeenth-century crisis, the English Quakers William Penn and John Bellers offered their own intriguing views on ways of eradicating war and establishing European unity. Penn (1644–1718), scion of a wealthy aristocratic family who embraced non-conformism as a student at Oxford, fled to America in 1682 to avoid persecution. Here he established a Quaker colony in the state – Pennsylvania – which still bears his name (Wildes 1977). In 1693, Penn wrote a passionate 8000-word essay on the 'groaning state of Europe', which expressed the urgent need for a European Diet or Parliament to remove the possibility of war. Penn's abhorrence of war is manifest throughout this extraordinary document: 'He must not be a man but a statue of brass or stone', he begins, 'whose bowels do not melt when he beholds the blood tragedies of war' (Penn 1936, pp. 5–6). His proposed Diet would meet at least once a year in a symbolic circular room, located somewhere in Europe, to settle grievances and disputes by negotiation and majority voting. Its composition would be based on the scientific calculation of each state's population and productivity (an early estimate of GDP). Penn believed this would produce a council comprising 10 representatives from France and Spain, eight from Italy, six from England, four each from 'Sweedland', Poland and the Netherlands, three each from Portugal, Denmark and Venice, two from Switzerland, and one from the Dukedoms of Holstein and Courland. The 'Turks and Muscovites', firmly excluded by Sully, would also be welcomed 'as seems but fit and just' and would be allocated 10 places each (Penn 1936, p. 16). Each state would be responsible for nominating its own representatives and the language of debate would be either French ('very well for civilians') or Latin ('most easy for men of quality') (Penn 1936, p. 19).

The rest of the document dealt with the fears Penn imagined such an arrangement would provoke. The end of war in Europe would not 'endanger an effeminacy by such a disuse of soldiery,' he insisted, 'because each sovereignty may introduce as temperate or severe a discipline in the education of youth as they please'. A common European army would still be able to resist external aggression. In words which echo down the centuries, he also claimed that the Diet would not compromise the domestic sovereignty of nation-states, except in so far as military power was concerned: 'And if this be called a lessening of their power, it must be only because the great fish can no longer eat up the little ones, and that each sovereignty is equally defended from injuries' (Penn 1936, p. 22). Thus would war come to an end:

> by the same rules of justice and prudence by which parents and masters govern their families, and magistrates their cities, and estates their republics, and princes and kings their principalities and kingdoms, Europe may obtain and preserve peace among her sovereignties.
>
> (Penn 1936, p. 30)

John Bellers (1654–1725) developed his own far-reaching version of Penn's scheme 17 years later (Bellers 1710; Clark 1987), arguing that the 1707 Act of Union between England and Scotland was a model for a future Europe. He advocated accepting whatever geopolitical order emerged from the ongoing War of the Spanish Succession, but that, once a European Parliament was established, its main task should be to redraw the political map of Europe. The objective was '100 Equal Cantons or Provinces, or so many, that every Sovereign Prince and State may send one Member to the

Senate at least' (Clark 1987, p. 140). This arrangement could only evolve gradually but, once in place, each canton would be required to contribute men and resources to a common European army. Only through this rational reorganization of Europe's political geography could war be avoided in the future: 'For nothing makes Nations, and People more Barbarous than War; so Peace must be the first step, to fit Mankind for Religion' (Clark 1987, p. 152). Like Penn, Bellers insisted that Russia and the Ottoman Empire should be welcomed into a reordered European state.

Two years later, during the midst of the negotiations for the Treaty of Utrecht, another Frenchmen, the Abbé de Saint-Pierre, produced a hugely expanded version of Sully's Grand Design (Saint-Pierre 1927, originally 1713). Saint-Pierre's well-meaning but pedantic style attracted cruel comments from his contemporaries, and the verdict of posterity has been no less charitable: one authority calls him 'the great bore of eighteenth-century France' (Heater 1992, p. 66). Like that of Sully, Saint-Pierre's vision was compatible with the balance of power but, in common with earlier formulators, he argued that balance should be maintained by parliamentary dialogue rather than periodic warfare. Saint-Pierre's scheme, with its emphasis on 'natural' law, was based on the ideas of the Dutch jurist Hugo Grotius (1583–1645) whose *De jura belli et pacis* had proposed legally binding codes by which warfare might be conducted (Cutler 1991). Like his English Quaker contemporaries, Saint-Pierre was happy to admit Russia into his scheme and was also markedly more sympathetic towards the Muslim world than Sully. Such a utopian view was widely mocked: Frederick the Great wrote sarcastically to Voltaire that: 'The Abbé de Saint-Pierre has sent me a fine work to re-establish peace in Europe. The thing is very practicable: all it lacks to be successful is the consent of all Europe and a few other small details' (Heater 1992, pp. 84–5). Jean-Jacques Rousseau was far less dismissive, however, and he revived the Abbé's scheme in an elegant abridgement and re-examination of 1761 (Roosevelt 1990). As far as Rousseau was concerned, European unity was already well established: 'there are not today any more Frenchmen, Germans, Spaniards, even English. There are only Europeans. They all have the same tastes, the same feelings, the same customs because none has received a national shape by any exclusive institutions' (Joll 1980, pp. 11–12).

Further schemes for perpetual peace were advanced towards the end of the eighteenth century, despite the shadow of the French Revolution and the violence it sparked off. The two most famous are probably Jeremy Bentham's *Plan for an universal and perpetual peace* (1789), the fourth and final essay in his *Principles of international law* (1780–9) (the work in which Bentham first coined the word 'international'), and Immanuel Kant's *Zum ewigen Frieden* (1795) (Heater 1992).

After the Revolutionary and Napoleonic interlude, a fresh and characteristically nineteenth-century vision of European peace was proposed by the utopian crypto-socialist philosopher, Claude-Henri de Saint-Simon. His was one of the first philosophies of industrialism as the engine of human progress. Unlike the predominantly anti-industrial European Romantics, Saint-Simon welcomed the emerging industrial order and regarded Britain, the first industrial state, as an exemplary image of the future. Like Marx a generation later, Saint-Simon believed that the new industrial order would transform the role of government into a purely administrative function. The old nation-states were destined to 'wither away', ultimately coalescing on a European scale to form larger and larger confederations.

In October 1814, as Europe's statesmen gathered in Vienna to redraw the map of the continent in the wake of Napoleon's defeat, Saint-Simon (together with his secretary, Augustin Thierry) produced a remarkable pamphlet on the 'reorganization of European society' (Saint-Simon and Thierry 1814). This argued for an Anglo-French political union (both nations now sharing a common form of 'liberal' constitutional monarchy), the core alliance in a general European government based on the British two-chamber model. The European 'House of Commons' would be elected by members of new professional associations representing science, industry and commerce, law, and the civil service. Economic integration was already gathering pace and this would be facilitated by developments in transport, notably canals. Saint-Simon's scheme was based on a complete rejection of the balance of power, which he saw as stimulating rather than preventing war. Conflict could be fully eradicated only through European economic and political integration. If Europe operated as a single economic unit, war would become counter-productive, a form of self-mutilation. Saint-Simon's belief that commerce, industry, communications and technology could bind together a political space has been enormously influential, especially on the agenda of European union since 1945. According to Hinsley (1967, p. 102), Saint-Simon's vision was 'a more far-reaching proposal for the federal organization of Europe than anyone had ever proposed'.

Sadly, the concept of the balance of power triumphed once again at the Congress of Vienna and re-emerged to determine the political landscape of Europe through the dangerous years of the nineteenth century, when confrontational European nationalism and imperialism rose steadily to reach their pernicious high-points on the eve of World War I. The outbreak of war in 1914, and the unprecedented carnage it unleashed, demonstrated the redundancy of balance of power reasoning. To some extent, the new geopolitical order, negotiated at the Paris Peace Conferences in 1919 and incorporated into the Treaty of Versailles, recognized this fact, most notably by the creation of the League of Nations as the first forum through which international disputes could be resolved. The League was, however, far too weak to carry the burden placed upon it in the absence of several important states (notably the United States) and any convincing legal powers. The idea of the balance of power was to retain its predominant position in the inter-war years.

Like earlier European crises, the terrible events of 1914–18 provoked a further flurry of debate about war, peace and the European condition. Many believed that Europe was doomed to a fundamental collapse of civilization, akin to the eclipse of the Roman Empire, unless drastic remedial action were taken to secure unity and remove the prospect of war (Demangeon 1920; Spengler 1980; Stirk 1989, 1996). Among the more intriguing prophets of peace and European unity was another remarkable Quaker, Lewis Fry Richardson. He produced dozens of extraordinary articles and books on the nature and causes of war which, though touchingly naive to the cynical modern mind, reveal the faith of an early-twentieth-century scientist (Richardson was a mathematician, physicist and meteorologist) in the power of reason and logic to solve the most intractable political problems. His project was startlingly ambitious: to develop a general mathematical and predictive model of war. By analysing the frequency and nature of previous conflicts, Richardson hoped to predict when and where warfare was likely to erupt in the future, thus allowing preventive action to be taken (see Richardson 1919, 1960a, b). Richardson also wrote a remarkable unpublished essay in 1915 (while serving in France as an ambulance driver) entitled 'The conditions of a lasting peace in Europe'. Education was, he believed, the key to removing the war impulse: 'A great step will have been taken when schoolboys are taught to think Livingstone a greater hero than Wellington' (Richardson 1915, p. 10).

Far more influential was Richard Coudenhove-Kalergi, the driving force behind the Pan-European Union (PEU), established in the mid-1920s and easily the most successful inter-war organization devoted to European unity (Wiedemer 1993). The PEU adopted a technological and economic strategy, similar to that proposed by Saint-Simon (O'Loughlin and van der Wusten 1990). It represented, according to one authority, 'an astonishing mixture of large-scale Utopianism, potent political analysis and clear-sighted pragmatism' (Bugge 1993, p. 101). Coudenhove-Kalergi saw no place in Europe for Russia as long as it remained under the Bolshevik yoke, nor for Britain while it remained locked into its extra-European imperial realm. Divested of these unfortunate distractions, both countries would be welcomed into the new European order, particularly Britain, which could provide a cultural bridge between Europe and the United States (Coudenhove-Kalergi 1926). The PEU's proposals were supported by leading politicians, notably Aristide Briand in France and Gustav Streseman in Germany. The former (as French Foreign Secretary) went so far as to issue a formal memorandum on 17 May 1930 inviting all the European powers (particularly Britain) to begin the process of European economic and political unification (Pegg 1983).

These various schemes share many common characteristics: a clear commitment to end war, a dissatisfaction with the balance of power, and a belief that European unity was the only solution to the endemic violence produced by the existing system of nation-states. The most striking common feature, however, was their almost total failure to influence the course of history. Despite the fleeting 'success' of the PEU, all these blueprints foundered on the rocks of old-style national antipathies and under the pernicious influence of powerful military–industrial complexes in each state, whose very survival depended on a permanent state of preparedness for war, coupled with occasional episodes of real conflict.

Europe built on war?

This raises a central paradox at the heart of the European project. Although European unity has always been invoked in the name of peace and harmony, the unification of the continent has hitherto been achieved only as a result of violence and warfare, and even then momentarily. The three modern episodes of general European warfare – 1792–1814, 1914–18 and 1939–45 – all demonstrate this ugly truth. (Some would also argue that the post-1945 process of European unification, both East and West, was also the outcome of 'war', albeit 'cold' rather than 'hot'.) These three periods of increasingly bloody warfare were associated with audacious attempts to destroy the existing balance of power by force, and to create a new European order to the advantage of one state. Viewed optimistically, the failure of all three attempts underscores the effectiveness of the balance of power. Interpreted pessimistically, that these attempts took place at all, despite increasingly high levels of death and destruction on each occasion, demonstrates the extraordinary resilience of faith in war as a solution to perceived injustices within the international system.

The political landscape of nineteenth-century Europe was largely determined by the French Revolution and the subsequent Napoleonic Empire. The reconceptualization and reorganization of space was a key feature of the Revolution, a theme only now receiving due attention (Vovelle 1993). According to Jules Michelet, the nineteenth century's most sympathetic historian of the Revolution, 1789 liberated not only France but the whole of Europe: 'geography itself was annihilated. There were no longer any mountains, rivers, or barriers between men . . . time and space, those natural conditions to which life is subject were no more' (White 1973, p. 151). The most obvious manifestation of this reinvention of space was the administrative reorganization of France and its newly won imperial territories into a large number of small *départements*, each administered by a prefect nominated by the central authority in Paris (Konvitz 1987; Vic-Ozouf Marignier 1989). The objective was to maximize central power over the hostile provinces of France. Once these were 'pacified' (itself a bloody business), the same spatial strategy was 'exported' to the rest of the Empire which was carved out by Napoleon and the other imperial offspring of the Revolution.

By 1812, the high-point of Napoleonic imperial power, most of Europe had been reordered on French rationalist lines (see Figure 3.2). Beyond the formal Empire, which stretched to the Rhine in the east and deep into the Italian peninsula to the south, lay a new European political geography conceived in the image of its French conquerors (Woolf 1991; Broers 1996). Sustained by a new form of 'scientific' bureaucracy and statecraft based on a

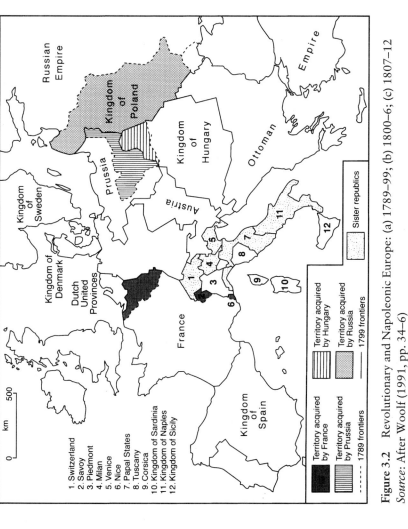

Figure 3.2   Revolutionary and Napoleonic Europe: (a) 1789–99; (b) 1800–6; (c) 1807–12
Source: After Woolf (1991, pp. 34–6)

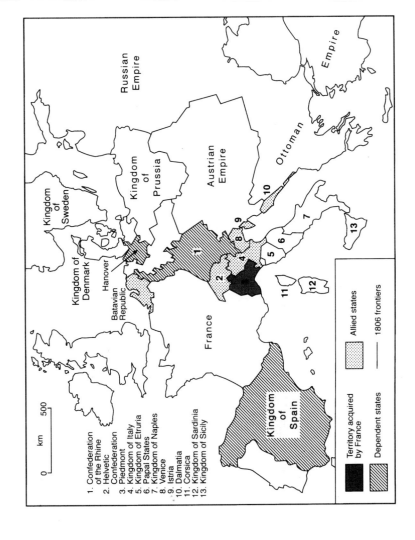

Figure 3.2b

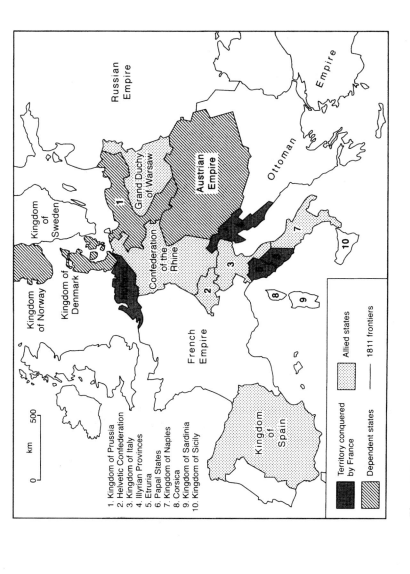

Figure 3.2c

revolution in the collection and use of statistics, censuses and surveys, the Napoleonic Empire was an attempt to create a united Europe as an extension of revolutionary France (Perrot and Woolf 1984). France's national culture, born of the political and economic conflicts of the French Enlightenment, was elevated to a European, even a universal, scale. The idea of the balance of power obviously had no place here; this was a Europe built on a single, revolutionary model in the name of a national civilization which presumed to speak for all. The new idea of France was equated with a new idea of Europe, while acts of imperial conquest were translated into acts of liberation (Schama 1977).

All this was achieved – or at least accompanied – by violence and warfare on an unprecedented scale. The Revolution had created the first 'people's army' of conscript soldiers and *sans-culottes* to protect the Republic from both internal 'subversion' and external monarchist threat (Cobb 1987). Once established, this military machine grew ever more powerful and was ultimately to seize control of the political process under Napoleon. This was the first modern military dictatorship in which all state institutions were directed towards a single objective: the greater martial glory of France. In 1789, the French army was probably no larger than 165 000 men; by the summer of 1793, following the *levée en masse*, in which all males aged 18 to 25 were made liable to serve for a limited period, there were around 750 000 men *sous le drapeau*. By May 1813, in the midst of Napoleon's disastrous Russian campaign, which was to destroy his Empire, the imperial armies across Europe probably numbered 1.1 million men (Woolf 1991).

The human cost of Napoleonic adventurism was enormous. Each year of French expansion saw battles on a larger scale. A 'mere' 40 000 French troops conquered Egypt in 1798; by the winter of 1805, 73 400 French troops faced 85 400 Austrians and Russians at Austerlitz, the northern gateway to Vienna. For his ill-fated invasion of Russia, the Emperor amassed a combined army of over 600 000 men, only a small fraction of whom returned alive (see Figure 3.3). By 1814, when conscription ceased, at least 400 000 of the 2 million men who served in Napoleon's armies from 1806 had been killed and a further 600 000 were registered as either prisoners of war or *disparu*. Only a proportion of the latter groups survived. Some 84 000 died in Spain and Portugal, 171 000 in Russia and 181 000 in Germany. Twenty per cent of all Frenchmen born between 1790 and 1795 had been killed in battle by the end of 1814 (Parker 1995).

Even then, horrifically, this is a less gruesome total than the 25 per cent of the 1891–5 French cohort who were claimed in World War I. The 51 months between August 1914 and November 1918 witnessed the premature death of nearly 8 million Europeans, twice the number of casualties produced in all previous conflicts between 1789 and 1914 (Mosse 1990; Gilbert 1994). Nearly 2 million Germans died, 1.7 million Russians, almost 1.4 million French, 1.2 million from Austria-Hungary and 750 000 from Britain. In some European regions, 15 per cent of the labour force was killed.

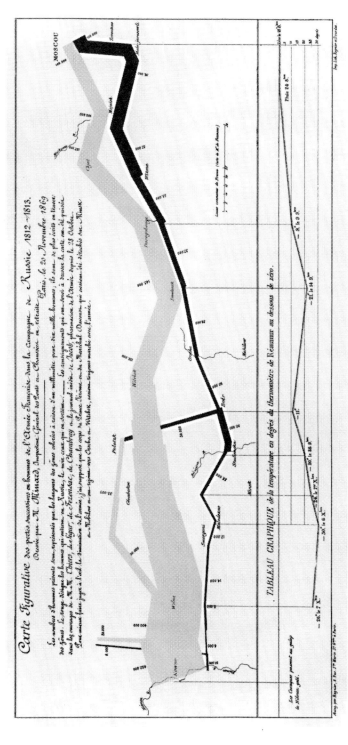

Figure 3.3 Napoleon's invasion of Russia, 1812–13. This famous cartographic representation, produced in 1869, of the French advance to, and retreat from, Russia by Charles Joseph Minard (1781–1870) is a pioneering image in the history of thematic mapping. The width of the 'flow-line' indicates the rapidly diminishing number of French troops on the outward and return journey. Beneath the map, Minard included a temperature scale indicating the freezing conditions during the French retreat which claimed most lives.
Source: Reproduced in Tufte (1983)

Although its causes are still hotly debated, World War I represented the greatest and most costly failure of the European balance of power (Winter 1988). The traditional focus on the play of national rivalries, the calculations of the key personalities, and the influence of war-hungry military machines (Fischer 1967; Massie 1991) has been supplemented by accounts which emphasize how an international system that developed in relation to one form of industrial technology was simply overwhelmed by the speed of events and the flow of information under new, twentieth-century technologies in transport and telecommunications (Taylor 1969; Kern 1983; Adas 1989).

In discussing the 'deep' ideological and intellectual causes of 1914, several commentators have also referred to the climate of bourgeois opinion in which war was seen not only as an inevitable consequence of human relations, but as a motor of human progress and a desirable, morally uplifting force (Nef 1950; Eksteins 1989; Carlton 1990; Pick 1993). The major mid-nineteenth-century wars (the American Civil War, the Crimean War and the Franco-Prussian War) had already fuelled an earnest debate about the wider implications of war. Under the influence of an all-pervasive social Darwinism, many agreed with the German political geographer Friedrich Ratzel that nation-states were living organisms which, like their counterparts in the natural world, needed constantly to struggle in order to survive and expand (Smith 1980; Bassin 1987). Periodic bouts of warfare between nations were inevitable and also necessary to the 'natural' struggle for life. War allowed the 'fittest' races to survive and preserve their vitality, energy and 'virility'. Although this view reached its disturbing peak in the blood-curdling writings of Friedrich von Bernhardi (1914) and Heinrich von Treitschkte (1916) in Germany, it was common throughout Europe. In Britain, for example, the physician Sir Arthur Keith wrote of war as 'nature's pruning hook', regularly removing 'dead wood', and ensuring the vigorous regrowth and development of a given race (Crook 1994; Hynes 1990; Stepan 1987).

Considering the wider cultural and intellectual roots of the 1914–18 war takes us beyond the narrow concern to establish cause and effect, guilt and responsibility, and into the morally ambiguous realm of ideas. Some have argued, most recently Modris Eksteins (1989), that World War I should be viewed as a clash between rival bourgeois visions of the world, represented on the one hand by German *Kultur* and on the other by Anglo-French *civilisation*; the one dedicated to changing the world, the other to maintaining the status quo. The idea of Europe which developed in Germany after the establishment of the new state in 1870–1 reflects this ambitious yearning for change. Prior to the middle decades of the nineteenth century, Europe was usually divided by a simple but geographically variable binary line, sometimes north–south, more usually east–west. From *c.* 1850, a quintessentially Germanic idea of Europe emerged in which the continent was divided into at least three zones: western, central and eastern (Schultz 1989; Okey 1992; see

Chapter 2). The intervening zone, *Mitteleuropa* – initially understood in economic terms, as a trading zone bound together by German capital and railways – soon acquired a deeper cultural and political significance in the German geopolitical imagination. After 1870, this variously defined region was advanced increasingly as Germany's legitimate arena of expansion and hegemony, a 'manifest destiny' implicit in the new state's unrivalled economic and demographic potency. Here lay the new Germany's *Lebensraum* or 'living space'. Perhaps the most widely read version of this thesis was that of Friedrich Naumann (1915), which argued that the contradiction between German and Anglo-French–Russian perspectives on the world had become so stark that the only solution was a permanent division of the continent into three hermetically sealed zones, demarcated by two great 'Chinese walls' (Naumann 1916). Such extreme views, albeit seriously debated, were never official policy in Germany. The image of Europe developed in war-time Berlin, however, was clearly informed by the idea of a German-dominated Central Europe (Blouet 1996).

The failure of the German bid for enhanced power in Europe in 1918 left the entire continent shattered. Russia was now under the control of the Bolsheviks and sealed off from the rest of the world, while the Western powers had been massively diminished by a war which had cost the equivalent of 6.5 times the global national debt from 1700 to 1914 and which was to set back real levels of European industrial output by at least a decade (Kitchen 1988). Inter-war Europe was also more complex geopolitically and economically: an extra 20 000 km of international frontier had been established at the Paris Peace Conferences and twice the number of national currencies were in circulation (Figure 3.4).

The rise of European Fascism, first in Italy in the early 1920s and then in Germany after 1933, destroyed the liberal-democratic vision of Europe and the fragile balance of power which had been delicately put in place after 1918. The tragic collapse of Republican Spain at the hands of the Nazi-backed 'Nationalist' (Fascist) forces of Franco in the mid-1930s, a civil war which claimed 600 000 lives, was a harbinger of the greater violence to come (Figure 3.5; Thomas 1977). Much has been written on the Fascist vision of Europe, particularly in respect to German geopolitics (Heske 1986, 1987; Smith 1986; Burleigh 1988; Fahlbusch *et al.* 1989; Kost 1989; Rössler 1989; Korinman 1990; Herb 1997). While academic geopolitical writings in Germany and Italy (under the dominating influence of Karl Haushofer and Ernesto Massi respectively) became more voluminous, the principal pre-1914 concerns of *Mitteleuropa* and *Lebensraum* retained their talismanic significance (Sinnhuber 1954; Meyer 1955; Brechtefeld 1996). Once German control of Central Europe was assured, as it must be according to Nazi logic, then the new European order would be decided by a life-or-death struggle for living space between Germany and Communist Russia. Hitler was thus able to present himself to the German people in two distinct ways in respect to Western and Eastern Europe. In the West, he was the

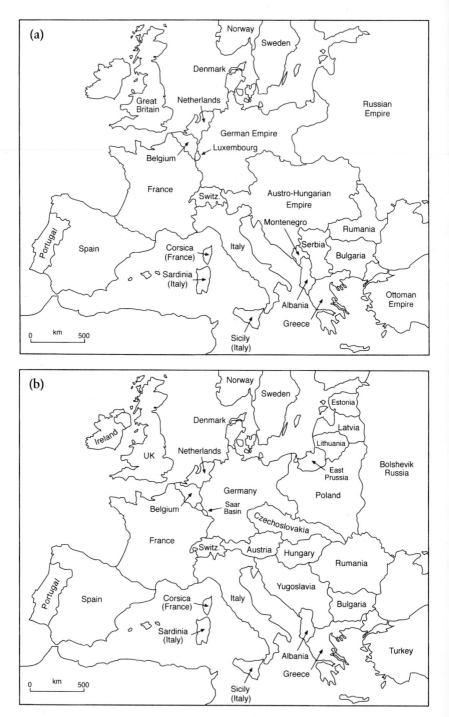

Figure 3.4 Impact of World War I on Europe's political geography: (a) 1914; (b) 1925

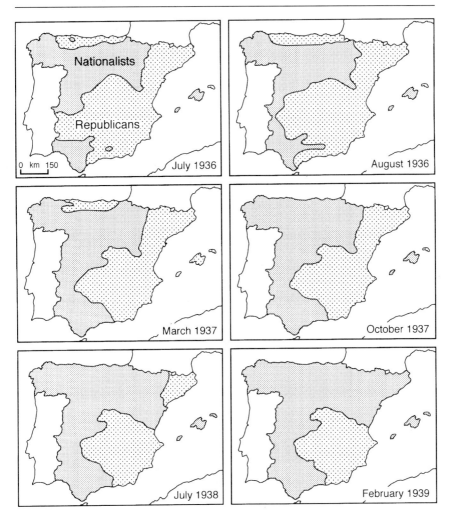

Figure 3.5 The Spanish Civil War: the collapse of Republican Spain
Source: After Thomas (1977, pp. 256, 402, 610, 732, 833 and 885)

defender of a reborn German *Kultur*, which had been betrayed in November 1918 by a weak-willed political establishment in Berlin, and then further humiliated by the conservative forces of Anglo-French *civilisation* at the peace conferences. In the East (once the cynical Hitler–Stalin pact had come to an end in the summer of 1941), he could pose as the standard-bearer of all Europe against the evils of Communism: 'This is not a war we are fighting just on behalf of the German people', he once said; '[it is] a struggle for the whole of Europe and the whole of civilized humanity' (Bugge 1993, p. 110).

The full force of the *Wehrmacht* was unleashed against the Soviet Union in Operation Barbarossa in June 1941. This was the greatest military campaign ever waged on European soil: 200 German divisions attacked along a 1500 km front from the Baltic in the north to the German–Slovakian border in the south (Gilbert 1989). Sixteen months later, in November 1942, German territorial dominance of Europe reached its furthest extent, destined, nevertheless, to collapse in an even more brutal rerun of the Napoleonic nemesis 130 years earlier (Figure 3.6).

In the midst of this desperate gamble, the Nazi vision of a reordered Europe was set tragically in motion in those areas which fell under German control. References to abstract geopolitical concepts such as *Lebensraum* and *Mitteleuropa* must not conceal the fact that the Nazi agenda for Europe was ultimately racial. Race had been an increasingly dominant theme in debates about the internal order of Europe and about the continent's position within the wider world for at least two hundred years. The twin ideologies of nationalism and imperialism were, of course, largely built upon racial theories. After the Great War, the racial question loomed large in discussion of the future geopolitical order in Europe, many liberals advocating the reorganization of the European political map into a larger number of ethnically cohesive states as the only solution to the chronic instability of larger geopolitical units (Bryce 1921; Simon 1939; Harvie 1994). But it was under the Nazis that the racial interpretation of Europe reached its evil pinnacle with the adoption, after 1941, of an 'eliminationist antisemitism' (Mosse 1978; Müller-Hill 1988; Weindling 1989; Goldhagen 1996). The Nazis systematically sought to obliterate from the face of the continent those groups – Jews, Gypsies and other minorities – who were deemed to have no place in the new order (Burleigh 1994). Six million Jews were murdered, either at the hands of wandering *Einsatzgruppen* and their civilian accomplices, or in the death camps (Figure 3.7; Gilbert 1982, 1986).

The number of those who died between 1939 and 1945 will never be precisely known; millions expired leaving no record of their passing. Most authorities agree, however, that at least 40 million men, women and children perished, more than half of whom were civilians. In absolute terms, the worst-affected country was Russia, which lost at least 21 million people, 11 per cent of the total population. Proportionately, however, the 6 million Poles who died represented a staggering 17 per cent of that country's 1939 population (Figure 3.8; Gilbert 1989; Bullock 1991). The Nazi plan for Europe was vastly more ambitious, and infinitely more terrifying, than the comparatively 'modest' 1914–18 war aims of a German-dominated *Mitteleuropa* and a larger share in world affairs. It was, in many ways, the twentieth-century equivalent of the Napoleonic vision, though structured by a corrosive racial theory. All three episodes of general European warfare shared a common impulse to overturn the existing balance of power and ensure a new European order in which an expansive state would predominate.

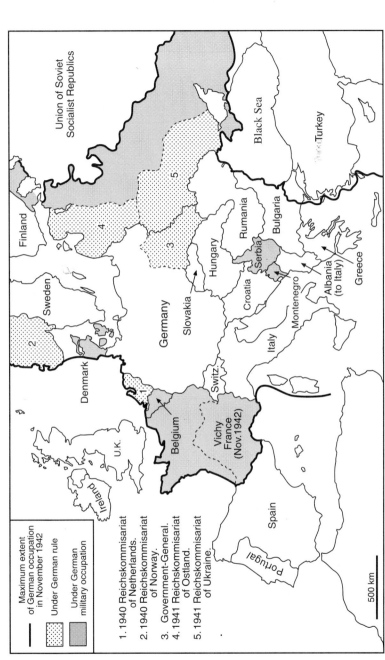

Figure 3.6 Maximum extent of German occupation, November 1942
Source: After Bullock (1991), pp. 816–17)

Legend:

Maximum extent
of German occupation
in November 1942

Under German rule

Under German
military occupation

1. 1940 Reichskommisariat
   of Netherlands.
2. 1940 Reichskommisariat
   of Norway.
3. Government-General.
4. 1941 Reichskommisariat
   of Ostland.
5. 1941 Reichskommisariat
   of Ukraine.

500 km

Union of Soviet
Socialist Republics

Finland
Sweden
Denmark
U.K.
Ireland
Belgium
Vichy
France
(Nov.1942)
Spain
Portugal
Germany
Switz.
Slovakia
Hungary
Croatia
Serbia
Montenegro
Albania
(to Italy)
Italy
Rumania
Bulgaria
Greece
Black Sea
Turkey

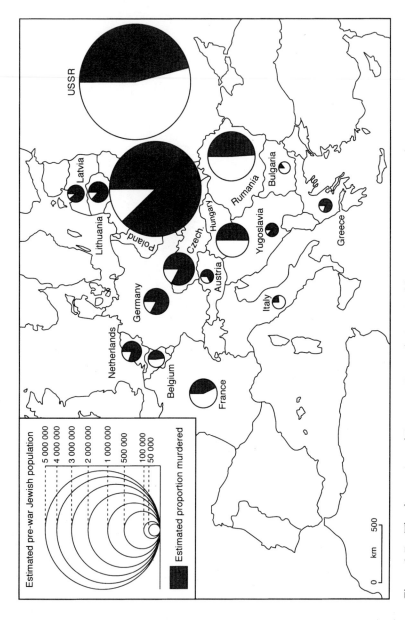

Figure 3.7   The destruction of European Jewry during World War II
Source: After Bullock (1991, pp. 1088–9)

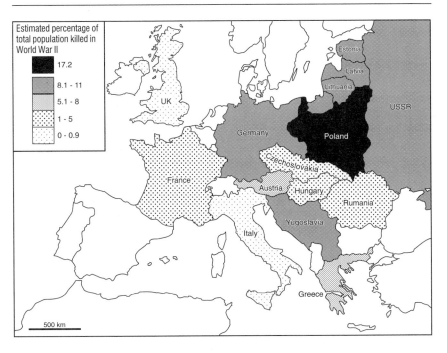

Figure 3.8 Estimated percentage population loss during World War II
Source: After Bullock (1991, pp. 1086–7)

Conclusion: war, memory and identity in Europe

The landscape of Europe bears the imprint of its war-torn history in many different ways, from the elegant, star-shaped beauty of the Renaissance forti- fications around Palma Nova in the Veneto to the sombre greyness of a World War II 'pill-box' bunker looming above the swaying wheat in an Essex field. Certainly the most poignant and moving landscapes of war are those created especially to commemorate the fallen. Scattered along the Western Front, for example, straddling the Franco-Belgian border, are the 1000 or so sublimely beautiful cemeteries and memorials to the British and Com- monwealth war dead of 1914–18, interspersed with the smaller number of vast cemeteries and memorials for the French, German and Belgian casualties (Figure 3.9; Winter 1995; Heffernan 1996). It is impossible to visit such places and remain untouched by the experience. As one stands in the midst of these endless rows of white headstones or iron crosses, or beneath the acres of chiselled names on the walls of the great memorials, the terrible legacy of European conflict is brought home with a force which a textual narrative can never achieve.

Yet, tragically, these reminders of the horrors of warfare have done little to prevent war, or the even more chilling genocidal urge which reared its

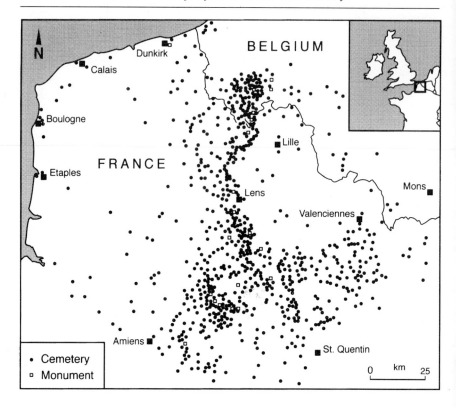

Figure 3.9   British and Commonwealth World War I cemeteries, Western Front
Source: 1/200 000 *Carte routière et touristique*, Michelin/Commonwealth War
Graves Commission, nos. 51–3

repugnant head during the recent outburst of 'ethnic cleansing' in the former Yugoslavia (Campbell 1997). It is clearly not enough merely to remember war, or to create special places in which those who perished in combat
are recalled and commemorated. An urgent need exists to develop a new
political culture within a changing Europe which removes the possibility of
war for ever. The assumption that European integration will achieve this is,
perhaps, mere wishful thinking for the EU which, like the old nation-states
that it seeks to replace, is still based upon a conventional form of territorial-
ized political identity. Behind the rhetoric of unity and peace, the idea of
Europe has always been (and remains) wedded to the geopolitics of exclusion and division; the drawing of lines between those within and those without. The time has surely come, in an increasingly globalized world, to look
beyond all forms of territorial identity and forge a new kind of politics based
on cosmopolitanism and internationalism; a politics in which the violence
and warfare associated with traditional territorial geopolitics will finally
become a thing of the past.

References

Adas, M. 1989: *Machines as the measure of man: science, technology and ideologies of western dominance*. Ithaca, NY: Cornell University Press.

Ashworth, G. J. 1991: *War and the city*. London: Routledge.

Balchin, W. G. V. 1987: United Kingdom geographers in the Second World War. *Geographical Journal* 153, 159–80.

Bassin, M. 1987: Imperialism and the nation state in Friedrich Ratzel's political geography. *Progress in Human Geography* 11, 473–95.

Bateman, M. and Riley, R. C. (eds) 1987: *Geography of defence*. London: Croom Helm.

Bean, R. 1973: War and the birth of the nation state. *Journal of Economic History* 33, 203–21.

Bellers, J. 1710: *Some reasons for an European state &c*. London: no publisher given.

Bernhardi, F. von 1914: *Germany and the next war*. London: Edward Arnold.

Blouet, B. W. 1996: The political geography of Europe: 1900–2000 AD. *Journal of Geography* 95, 5–14.

Braudel, F. 1972: *The Mediterranean and the Mediterranean world in the age of Philip II*. London: Collins.

Brechtefeld, J. 1996: *Mitteleuropa in German politics, 1848 to the present*. London: Macmillan.

Broers, M. 1996: *Europe under Napoleon, 1799–1815*. London: Arnold.

Brown, R. A. 1972: *The origins of Europe*. London: Constable.

Bryce, J. 1921: *Modern democracies*. London: Macmillan.

Bugge, P. 1993: The nation supreme: the idea of Europe 1914–1945. In Wilson, K. and van den Dussen, J. (eds), *The history of the idea of Europe*. London: Routledge, 83–149.

Buisseret, D. 1968: *Sully and the growth of centralized government in France, 1598–1610*. London: Eyre & Spottiswoode.

Bullock, A. 1991: *Hitler and Stalin: parallel lives*. London: HarperCollins.

Burke, P. 1980: Did Europe exist before 1700? *History of European Ideas* 1, 21–9.

Burleigh, M. 1988: *Germany turns westward: a study of Ostforschung in the Third Reich*. Cambridge: Cambridge University Press.

Burleigh, M. 1994: *Death and deliverance: 'euthanasia' in Germany, 1900–1945*. Cambridge: Cambridge University Press.

Campbell, D. 1997: *National deconstruction: violence, identity and justice in Bosnia*. Minneapolis: University of Minnesota Press.

Carlton, E. 1990: *War and ideology*. London: Routledge.

Clark, G. (ed.) 1987: *John Bellers, 1654 to 1725: Quaker visionary – his life, times and writings*. York: Sessions Book Trust.

Clausewitz, C. von 1968: *On war*. Harmondsworth: Penguin.

Clout, H. D. 1996: *After the ruins: restoring the countryside of northern France after the Great War*. Exeter: Exeter University Press.

Cobb, R. 1987: *The people's armies: the armées révolutionnaires: instrument of the Terror in the départements, April 1793 to Floréal Year II*. New Haven, CT: Yale University Press.

Coudenhove-Kalergi, R. N. 1926: *Pan-Europa*. New York: A. A. Knopf.

Crook, P. 1994: *Darwinism, war and history: the debate over the biology of war from the 'Origin of Species' to the First World War*. Cambridge: Cambridge University Press.

Cutler, A. C. 1991: The 'Groatian tradition' in international relations. *Review of International Studies* 17, 41–65.

Demangeon, A. 1920. *Le declin de l'Europe*. Paris: Payot.

Doyle, P. and Bennett, M. R. 1997: Military geography: terrain evaluation and the British Western Front 1914–1918. *Geographical Journal* 163, 1–24.

Eksteins, M. 1989: *The rites of spring: the Great War and the birth of the modern age*. London: Black Swan.

Elliot, G. 1972: *Twentieth-century book of the dead*. Harmondsworth: Penguin.

Fahlbusch, M., Rössler, M. and Siegrist, D. 1989: Conservatism, ideology and geography in Germany, 1920–1950. *Political Geography Quarterly* 8, 353–67.

Fischer, F. 1967: *Germany's aims in the First World War*. London: Chatto & Windus.

Fussell, P. 1975: *The Great War and modern memory*. Oxford: Oxford University Press.

Gaillie, W. B. 1978: *Philosophers of war and peace*. Cambridge: Cambridge University Press.

Gaillie, W. B. 1991: *Understanding war*. London: Routledge.

Gilbert, M. 1982: *Atlas of the Holocaust*. London: Raibird.

Gilbert, M. 1986: *The Holocaust: the Jewish tragedy*. London: Fontana/Collins.

Gilbert, M. 1989: *Second World War*. London: Fontana/Collins.

Gilbert, M. 1994: *The First World War*. London: Weidenfeld & Nicolson.

Godlewska, A. 1994: Napoleon's geographers (1797–1815): imperialists and soldiers of modernity. In Godlewska, A. and Smith, N. (eds), *Geography and empire*. Oxford: Blackwell, 31–53.

Goldhagen, D. J. 1996: *Hitler's willing executioners: ordinary Germans and the Holocaust*. London: Little, Brown.

Goldie, G. T. 1907: Geographical ideals. *Geographical Journal*, 29, 1–14.

Harris, C. D. 1997: Geographers in the U.S. Government in Washington, DC, during World War II. *Professional Geographer* 49, 245–56.

Harvie, C. 1994: *The rise of regional Europe*. London: Routledge.

Hay, D. 1968: *Europe: the emergence of an idea*, 2nd ed. Edinburgh: Edinburgh University Press.

Hazard, P. 1990: *The European mind: the critical years 1680–1715*. Harmondsworth: Penguin.

Heater, D. 1992: *The idea of European unity*. Leicester and London: Leicester University Press.

Heffernan, M. 1995a: The spoils of war: the Société de Géographie de Paris and the French empire, 1914–1919. In Bell, M., Butlin, R. and Heffernan, M. (eds), *Geography and imperialism, 1820–1940*. Manchester: Manchester University Press, 221–64.

Heffernan, M. 1995b: For ever England: the Western Front and the politics of remembrance in Britain. *Ecumene* 2, 293–323.

Heffernan, M. 1996: Geography, cartography and military intelligence: the Royal Geographical Society and the First World War. *Transactions of the Institute of British Geographers* NS 21, 504–33.

Herb, G. 1997: *Under the map of Germany: nationalism and propaganda, 1918–1945*. London: Routledge.

Heske, H. 1986: German geographic research in the Nazi period. *Political Geography Quarterly* 5, 267–82.

Heske, H. 1987: Karl Haushofer: his role in German geopolitics and Nazi politics. *Political Geography Quarterly* 6, 135–44.

Hewitt, K. 1983: Place annihilation: area bombing and the fate of urban places. *Annals of the Association of American Geographers* 73, 257–84.

Hinsley, F. H. 1967: *Power and the pursuit of peace.* Cambridge: Cambridge University Press.

Howard, M. 1976: *War in European history.* Oxford: Oxford University Press.

Howard, M. 1983: *Clausewitz.* Oxford: Oxford University Press.

Hynes, S. 1990: *A war imagined: the First World War and English culture.* London: Pimlico.

Joll, J. 1980: Europe – an historian's view. *History of European Ideas* 1, 7–19.

Keegan, J. 1993: *A history of warfare.* London: Hutchinson.

Kern, S. 1983: *The culture of time and space 1880–1918.* London: Weidenfeld & Nicolson.

Kirby, A. 1994: What did you do in the war, Daddy? In Godlewska, A. and Smith, N. (eds), *Geography and empire.* Oxford: Blackwell, 300–15.

Kitchen, M. 1988: *Europe between the wars: a political history.* Harlow: Longman.

Kliot, N. and Waterman, S. (eds) 1991: *The geography of peace and war.* London: Belhaven.

Konvitz, J. W. 1987: *Cartography in France, 1660–1848: science, engineering and statecraft.* Chicago: University of Chicago Press.

Korinman, M. 1990: *Quand l'Allemagne pensait le monde: grandeur et décadence d'une géopolitique.* Paris: Fayard.

Kost, K. 1989: The conceptions of politics in political geography and geopolitics in Germany until 1945. *Political Geography Quarterly* 8, 369–85.

Lacoste, Y. 1976: *La géographie, ça sert, d'abord, à faire la guerre.* Paris: Maspéro.

Lacoste, Y. 1977: An illustration of geographical warfare: bombing the dikes on the Red River, North Vietnam. In Peet, R. (ed.), *Radical geography: alternative viewpoints on contemporary social issues.* London: Methuen, 244–61.

Levy, J. S. 1983: *War in the modern Great Power system, 1495–1975.* Lexington: University Press of Kentucky.

Mackie, G. B. 1917: Geography in relation to war. *Scottish Geographical Magazine* 33, 498–507.

McNeill, W. H. 1982: *The pursuit of power: technology, armed force, and society since AD 1000.* Chicago: University of Chicago Press.

Maguire, T. M. 1899: *Outlines of military geography.* Cambridge: Cambridge University Press.

Mahan, A. T. 1980 edition: *The influence of sea power upon history.* London: Bison.

Massie, R. K. 1991: *Dreadnought: Britain, Germany, and the coming of the Great War.* New York: Ballantine.

Merriman, J. 1996: *A history of modern Europe,* vol. 1: *From the Renaissance to the age of Napoleon.* New York: W. W. Norton.

Meyer, H. C. 1955: *Mitteleuropa in German thought and practice 1815–1945.* The Hague: Martinus Nijhoff.

Midlarsky, M. 1975: *On war: political violence in the international system.* New York: Free Press.

Mosse, G. L. 1978: *Toward the Final Solution: a history of European racism.* London: Dent.

Mosse, G. L. 1990: *Fallen soldiers: reshaping the memory of the world wars.* Oxford: Oxford University Press.

Müller-Hill, B. 1988: *Murderous science: the elimination by scientific selection of Jews, Gypsies and others in Germany, 1933–1945.* Oxford: Oxford University Press.

Naumann, F. 1916: *Central Europe.* London: P. S. King.

Nef, J. U. 1950: *War and human progress: an essay on the rise of industrial civilization.* London: Routledge & Kegan Paul.

Ogg, D. (ed.) 1921: *Sully's Grand Design of Henry IV: from the memoirs of Maximilien de Béthune, Duc de Sully (1559–1641).* London: Sweet & Maxwell.

Okey, R. 1992: Central Europe/Eastern Europe: behind the definitions. *Past and Present* 137, 102–33.

O'Loughlin, J. and van der Wusten, H. 1986: Geography, war and peace: notes for a contribution to a revised political geography. *Progress in Human Geography* 10, 484–510.

O'Loughlin, J. and van der Wusten, H. 1990: The political geography of pan-regions. *Geographical Review* 80, 1–20.

O'Loughlin, J. and van der Wusten, H. 1993: Political geography of war and peace. In Taylor, P. J. (ed.), *Political geography of the twentieth century.* London: Pinter, 63–113.

O'Sullivan, P. and Miller, J. W. 1983: *Geography of warfare.* London: Croom Helm.

Ó Tuathail, G. 1996: *Critical geopolitics.* Minneapolis: University of Minnesota Press.

Parker, G. (ed.) 1995: *Cambridge illustrated history of warfare.* Cambridge: Cambridge University Press.

Parker, G. and Smith, L. M. (eds) 1978. *The general crisis of the seventeenth century.* London: Routledge & Kegan Paul.

Pegg, C. H. 1983: *Evolution of the European idea, 1914–1932.* Chapel Hill: University of North Carolina Press.

Peltier, L. C. and Pearcy, G. E. 1966: *Military geography.* Princeton, NJ: Van Nostrand.

Penn, W. 1936: *An essay towards the present and future peace of Europe by the establishment of an European Diet, Parliament, or Estates.* London: Peace Committee of the Society of Friends.

Pepper, D. and Jenkins, A. (eds) 1985: *The geography of war and peace.* Oxford: Blackwell.

Perrot, J.-C. and Woolf, S. 1984: *State and statistics in France, 1789–1815.* London: Harwood Academic.

Pick, D. 1993: *War machine: the rationalization of slaughter in the modern age.* New Haven, CT: Yale University Press.

Pickles, J. 1995: *Ground truth: the social implications of geographic information systems.* New York: Guilford Press.

Pocock, J. G. A. 1975: *The Machiavelli moment: Florentine political thought and the Atlantic Republican tradition.* Princeton, NJ: Princeton University Press.

Pryor, J. H. 1988: *Geography, technology and war: studies in the maritime history of the Mediterranean, 649–1571.* Cambridge: Cambridge University Press.

Richardson, L. F. 1915: The conditions for a lasting peace in Europe. Unpublished manuscript, Cambridge University Library, Manuscripts Room, Papers and Correspondence of Lewis Fry Richardson.

Richardson, L. F. 1919: *The mathematical psychology of war.* Oxford: W. Hunt.

Richardson, L. F. 1960a: *Arms and insecurity: a mathematical study of the causes of war.* London: Stevens.

Richardson, L. F. 1960b: *Statistics of deadly quarrels.* London: Stevens.

Roosevelt, G. G. 1990: *Reading Rousseau in the nuclear age.* Philadelphia: Temple University Press.

Rössler, M. 1989: Applied geography and area research in Nazi society: central place theory and planning, 1933 to 1945. *Environment and Planning D: Society and Space* 7, 419–31.

Saint-Pierre, C. I. C. 1927: *Projet pour rendre la paix perpétuelle en Europe.* Paris: Garnier.

Saint-Simon, C.-H. and Thierry, A. 1814: *De la réorganisation de la Société Européenne, ou de la nécessité et des moyens de rassembler les peuples de l'Europe en un seul corps politiques, en conservant à chacun son indépendance nationale.* Paris: Adrien Egron.

Schama, S. 1977: *Patriots and liberators: revolution in the Netherlands, 1780–1813.* London: Fontana.

Schmidt, H. D. 1966: The establishment of 'Europe' as a political expression. *Historical Journal* 9, 172–8.

Schultz, W. D. 1989: Fantasies of *Mitte: Mittelage* and *Mitteleuropa* in German geographical discussion in the 19th and 20th centuries. *Political Geography Quarterly* 8, 315–40.

Sheehan, M. 1996: *The balance of power: history and theory.* London: Routledge.

Shennan, J. H. 1974: *The origins of the modern European nation state, 1450–1725.* London: Hutchinson.

Simon, E. D. 1939: *The smaller democracies of Europe.* London: Victor Gollancz.

Sinnhuber, K. A. 1954: Central Europe – Mitteleuropa – Europe Centrale. *Transactions and Papers of the Institute of British Geographers* 20, 15–39.

Smith, W. 1980: Friedrich Ratzel and the origins of *Lebensraum. German Studies Review* 3, 51–68.

Smith, W. 1986: *The ideological origins of Nazi imperialism.* Oxford: Oxford University Press.

Spengler, O. 1980: *The decline of the West.* London: Allen & Unwin.

Stepan, N. L. 1987: 'Nature's pruning hook': war, race and evolution, 1914–18. In Bean, J. M. W. (ed.), *The political culture of modern Britain: studies in memory of Stephen Koss.* London: Hamish Hamilton, 129–48.

Stirk, P. M. R. (ed.) 1989: *European unity in context: the inter-war period.* London: Pinter.

Stirk, P. M. R. 1996: *A history of European integration since 1914.* London: Pinter.

Stoddart, D. R. 1992: Geography and war: the 'new geography' and the 'new army' in England, 1899–1914. *Political Geography* 11, 87–99.

Sullivan, R. 1973: Machiavelli's balance of power theory, *Social Science Quarterly* 54, 258–70.

Sutton, K. 1981: The influence of military policy on Algerian rural settlement. *Geographical Review* 71, 379–94.

Taylor, A. J. P. 1969: *War by timetable: how the First World War began.* London: MacDonald.

Taylor, G. 1946: *Our evolving civilization: an introduction to geopacifics – geographical aspects of the path towards world peace.* Oxford: Oxford University Press.

Taylor, P. J. 1996: *The way the modern world works: world hegemony and world impasse*. Chichester: John Wiley.

Thomas, H. 1977: *The Spanish Civil War*. Harmondsworth: Penguin.

Tilly, C. 1975: *The formation of the modern nation-states of Western Europe*. Princeton, NJ: Princeton University Press.

Tilly, C. 1989: The geography of European state-making and capitalism since 1500. In Genovese, E. D. and Hochberg, L. (eds), *Geographic perspectives in history*. Oxford: Basil Blackwell, 158–81.

Treitschkte, H. von 1916: *Politics*. 2 vols. London: Macmillan.

Tufte, E. 1983: *The visual display of quantitative information*. Cheshire, CT: Graphics Press.

Vic-Ozouf Marignier, M. 1989: *La formation des départements et la représentation du territoire français à la fin du XVIIIe. siècle*. Paris: Éditions de l'École des Hautes Études en Sciences Sociales.

Vovelle, M. 1993: *La découverte de la politique: géopolitique de la Révolution française*. Paris: La Découverte.

Weindling, P. 1989: *Health, race and German politics between national unification and Nazism, 1870–1945*. Cambridge: Cambridge University Press.

White, H. 1973: *Metahistory: the historical imagination in nineteenth-century Europe*. Baltimore: Johns Hopkins University Press.

Wiedemer, P. 1993: The idea behind Coudenhove-Kalergi's Pan-European Union. *History of European Ideas* 16, 827–33.

Wildes, A. 1977: *William Penn*. London: Macmillan.

Winter, J. M. 1988: *The experience of World War I*. London: Guild.

Winter, J. M. 1995: *Sites of memory, sites of mourning: the Great War in European cultural history*. Cambridge: Cambridge University Press.

Woolf, S. 1991: *Napoleon's integration of Europe*. London: Routledge.

Wright, M. 1975: *The theory and practice of the balance of power, 1486–1914*. London: Dent.

4

'The chickens of Versailles': the new Central and Eastern Europe

DENIS J. B. SHAW

Introduction

> Is the fact that the frontiers between Eastern and Western Europe were considered to run along virtually identical lines in Charlemagne's time and Winston Churchill's merely a coincidence? Or not?
>
> (Longworth 1992, p. 261)

The fall of the Berlin Wall in November 1989 will long be remembered as a defining moment in European history. From Winston Churchill's famous 'Iron Curtain' speech at Fulton, Missouri, in 1946, to the demolition of the Wall, the continent had been divided between a Communist-controlled East and a capitalist West, their relations tersely summed up by the term 'Cold War'. For the peoples of the East, the removal of the Wall heralded the possibility of what came to be known as a 'return to Europe', salvation from the increasingly moribund Communist system, and the chance to enjoy the freedoms and the wealth which were assumed to be typical of the West. For the West, too, the events of 1989 proved momentous. Many right-wing observers regarded them as final proof of the superiority of capitalism over socialism. But any easy assumption that the West's security would now be assured was soon to be disappointed. Not only did new economic difficulties begin to flow from the attempt by Germany, for example, to assimilate its restored eastern territories, but parts of Eastern Europe were soon engulfed in ethnic and political struggles of various kinds. The question arose as to whether such problems would spill over into the West. So long as it existed, the Iron Curtain had at least helped provide the West with a definite boundary – in a sense, with a shield. What happened beyond it was of relatively little concern to the West and often hardly known there. But 1989 changed all that. The West was now faced with new challenges and opportunities, and the problems of the East threatened to become those of the West also. A period of self-doubt ensued. In the circumstances it became legitimate to ask whether it made sense any longer to talk of 'West' and 'East', what boundaries were now important, and what the future relations between the two halves of the continent were to be. Looming behind such questions, of course, was the most significant issue of all: what actually is Europe?

One purpose of this chapter is to argue that such questions were not new in 1989 but have a long history behind them. Another is to suggest that the 'West' (however defined) has been far more successful in imposing its ideas of 'Europe' on the East than the other way round. Indeed, the very notion of the 'East' or 'Eastern Europe' is a construct of the West, finding little echo in the eastern half of the continent. Whatever the divisions between the countries of Western Europe, they are insignificant compared to those of the East, where there has been little sense of historic unity. For the peoples of the East, the main task in geopolitical thinking on Europe has been not so much to ponder the nature and extent of their part of the continent but rather to define their own relationship to something called 'Europe', seen as lying to their west. And, given the diversity of the East, there is a diversity of definitions. This then raises yet another issue: why do we feel the need to define Europe at all, and whose interests are such definitions meant to serve?

Definitions of Europe: historic dimensions

> As Ivan the Terrible correctly commented to Possevino: 'Our faith is not Greek, but Christian.'
>
> (Likhachev 1991, p. 116)

A major shortcoming of a once popular kind of regional geography was its assumption that regions are an essentially unchanging objective reality. But as Massey (1995) has reminded us, places and regions, far from being static, are perhaps best regarded as 'processes'. They need not be defined by boundaries, nor do they possess single, unique identities. These points must be borne in mind as we consider the nature of what has been described in the title of this chapter as 'Central and Eastern Europe' and its ever-changing relationships with the rest of the continent.

The imposition of Soviet-type Communism on the countries of Central and Eastern Europe in the years after 1945 led to their incorporation into what the West regarded as the 'Soviet bloc' (Figure 4.1). Their economies were centrally planned on the Soviet model and prolonged attempts were made to integrate those economies with that of the USSR. Under the aegis of such organizations as the Council for Mutual Economic Assistance (CMEA

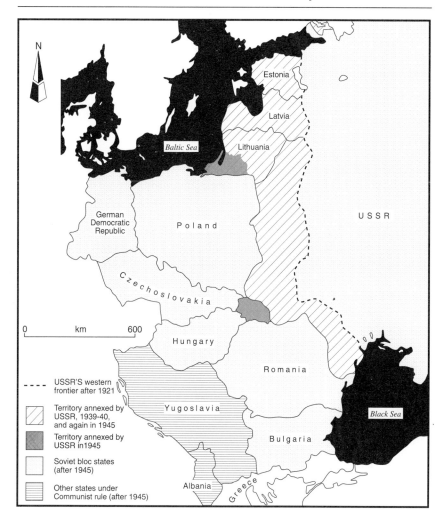

Figure 4.1   Central and Eastern Europe, 1921–89

or Comecon) and the Warsaw Military Pact, the seven European members of the Soviet bloc (German Democratic Republic – GDR, Poland, Czechoslovakia, Hungary, Romania and Bulgaria, together with the USSR itself) were meant to adopt a uniform mode of economic, social and political development. Only Yugoslavia and Albania, which had adopted Communist systems largely through their own efforts, escaped the Soviet net. Stalin, the Soviet leader, was in fact attempting to impose uniformity on a group of countries which differed among themselves on ethnic, linguistic, religious, cultural and political grounds, and which were at different stages of economic development. Not surprisingly, the attempt failed. Nevertheless, in Western eyes, at least, the fact that it was made was enough to obscure the

differences between these countries and to consign all of them to a single geographical region, which became known as 'Eastern Europe'. Interestingly
enough, Greece, which did not become Communist but is located in the same
part of the continent, was usually excluded from the generally accepted definition of 'Eastern Europe'.[1]

Since the fall of Communism, however, the former members of the Soviet
bloc have tended to deny their membership of a region called 'Eastern
Europe' and have argued instead for some other mode of regionalization.
This reflects their rejection of the old Soviet system, with its overtones of
domination by Moscow, in favour of a 'return to Europe'. It also represents a
revival of historical memory. Thus Czechs and Slovaks recall their pre-1918
status as part of the Austro-Hungarian Empire, decidedly a Central European
state, while in contemporary Poland, the historic antagonism towards the
Russians is seen as having geographical as well as historical significance.
Western scholars, too, are beginning to recognize the diversity of the region
in its post-Communist guise. One common device, for example, is to sub-
divide the former 'Eastern Europe' with the USSR into three: East-Central
Europe (Poland, the Czech Republic, Slovakia, Hungary); the Balkans; and
the states of the former Soviet Union (see Miall 1994). But this form of
regionalization also has its difficulties. The three Baltic states of Estonia,
Latvia and Lithuania, for example, have resolutely refused to join the CIS, the
inter-state organization which effectively replaced the USSR in 1991, and are
successfully restructuring their economies towards the West. These three
states thus stress their linkages with Northern and Central Europe, rather
than with the former USSR. Similarly, now that the former Czechoslovakia
has split into the Czech Republic and Slovakia, it is no longer clear that the
latter fits readily with the other three members into 'East-Central Europe'.
Other attempts to rethink the regionalization of the eastern part of the continent also create problems. As discussed in Chapter 3, the old concept of
Mitteleuropa, which is still favoured by some, raises in the minds of others the
unwelcome spectre of German domination (Meyer 1955; Droz 1960;
Naumann 1966; Brechtefeld 1996).

The concept of a 'return to Europe' implies a reunion with European civilization from which the countries of Central and Eastern Europe were unnat-
urally wrenched by the years of Communist domination. It is a notion which
fits neatly with the ideas which Western Europeans have held about themselves. Wolff (1994) argues that the secular concept of 'Europe' derives from
the eighteenth-century Enlightenment when 'European' was essentially synonymous with 'civilized'. The boundaries of Europe therefore corresponded
with the boundaries of 'civilized behaviour'. One result, Wolff argues, was a
definition of 'Eastern Europe' as a region where 'manners' and the way of life
generally were seen as less 'civilized' than in the West, but not so uncouth as
to be regarded as characteristic of 'Asia'. Consequently, Eastern Europe
emerges as a kind of halfway house on the ascent to civilization, with Western
Europe basking in sunlight at the peak (though leaving room for arguments

about which part of the West was located at the actual summit), and Asia and other parts of the world in the gloom of the valley.

Western European feelings of superiority towards the East long pre-date the eighteenth century. They are reflected, for example, in the sixteenth-century travel accounts of von Herberstein and Fletcher, and in the seventeenth-century work of Olearius (Baron 1991).[2] Wolff (1994) argues that the west–east conceptual division of Europe was preceded, during the Renaissance and perhaps earlier, by a south–north division in which the southern, Mediterranean world was seen as being culturally superior. But this is a partial view, very much that from the self-congratulatory core. As Longworth's quotation at the beginning of this chapter suggests, the west–east division of Europe has an even older lineage.

An examination of this pedigree entails going back as far as classical times, when commentators like the Greek historian Herodotus and the Roman poet Ovid drew attention to the differences between the world to which they belonged and that of the 'barbarians' to the north and the east. A significant moment occurred in AD 330 when the Roman emperor Constantine the Great, the first emperor to embrace Christianity, inaugurated the city of Constantinople (formerly Byzantium, now the Turkish city of Istanbul) as his imperial capital. Although Constantine managed to reunite the decaying Roman Empire, his establishment of a 'New Rome' to the east ultimately strengthened the already existing tendency for the empire to split into eastern and western halves. The barbarian invasions of the western part of the empire, which began in the fifth century AD, left the eastern half to develop as a separate entity, ultimately becoming known as the Byzantine Empire. After the reign of the Emperor Justinian (AD 527–65), this empire became more Greek than Roman, albeit continuing to claim its Roman heritage.

Even more significant for the future of Europe was the religious division of the erstwhile empire. By calling the Church Council of Nicaea in AD 325 to debate the Arian controversy, Constantine proclaimed the duty of the state to be concerned in matters of the Church. In time this gave rise to what in Eastern Christianity became known as the 'symphony' between Church and State, the two theoretically being in perfect harmony and supporting each other in their respective realms. Church and (Roman) Empire were respectively seen as being spiritual and temporal arms of authority, exercising oversight over a single Christian world. Unfortunately for that vision, the cleft in the Empire tended to split the Church. Thus over time, its western part developed into what became known as the Catholic Church, subject to the rule of the Pope in Rome, while the eastern part became the Orthodox Church under its patriarchs, of whom the Patriarch of Constantinople was the most important. The theory that there was a single Christian world, the *ecumene* as it was known in Greek, nevertheless continued. The Byzantine emperor, who in reality ruled over a diminishing part of Eastern Europe and neighbouring parts of Asia, claimed to exercise temporal power over Christendom as a whole. This situation changed dramatically at Aachen on Christmas Day 800, when the Pope crowned Charlemagne Holy Roman Emperor. This was regarded in Constantinople as a usurpation of the imperial power, which belonged to the Byzantine emperor alone. Exacerbated by doctrinal and other differences, the final split between Eastern and Western Churches came in 1054. Although there were several subsequent attempts to reunite the two churches, not least when the Byzantine Empire was being menaced by the Turks, the schism has lasted to this day (Obolensky 1950; Barraclough 1957; Meyendorff 1981; Kumar 1992).

Europe's Christian heritage has certainly been one factor underlying the historic tendency of Europeans to be intolerant of – and feel superior to – those belonging to different cultures. Catholic Europe, which embraced the western and central parts of the continent (including the Poles, the Hungarians, the Czechs and the Slovaks), sought every opportunity to expand by converting and, often enough, by subjugating its neighbours. This included Eastern Europe's Orthodox adherents. If Orthodoxy seemed less aggressive, this was only because it was rarely in a position to be so, being menaced on numerous occasions by the threat or reality of invasion. Orthodoxy also lost most of the Middle East at an early stage because of the rapid spread of Islam. It had more enduring success in the Balkans, where much was done to Christianize the Slavs and, at the same time, lay the foundations of a Slavic literature. After AD 988, Orthodoxy was also accepted by the Rus, ancestors of the present-day Russians, Ukrainians and Belarus'ians (see Figure 2.3).

The rise of the Muscovite or Russian state from the late medieval period marked an important point in the history of Orthodox Europe (Obolensky 1950). The Russians proved staunch adherents of Orthodoxy. When an attempt was made to unite the Orthodox and Catholic Churches at the Council of Florence in 1439, Metropolitan Isidore of Russia, who had agreed to the merger, was denounced by Moscow as a heretic. The capture of Constantinople by the Turks in 1453 and the subsequent extinction of the Byzantine Empire were regarded as divine punishment for the heresy perpetrated at Florence. Soon the Russian tsars were being urged by some among the Orthodox to proclaim themselves successors to the imperial dignities of the Byzantine (and hence Roman) emperors on the grounds that they now ruled the only unconquered Orthodox realm. Whether or not the tsars officially embraced the so-called 'Third Rome' doctrine remains a matter of controversy. But it may help explain Tsar Ivan the Terrible's contemptuous response to the Catholic teachings of the Jesuit Antonio Possevino, who came to Moscow on a diplomatic mission in the early 1580s (Milyukov 1974; Shaw 1997).

Down to the reign of Peter the Great (1682–1725), Russians seemed more concerned with the Orthodox world than with the European one. Few appeared inclined to learn from the latter. Peter, however, conscious of the way that Europe had forged ahead in both economic and military terms, believed such attitudes to be a threat to Russian security. He therefore embarked on a policy of forced modernization (Sumner 1951; Anderson 1978; Anisimov 1993). It was a strategy bound to cause offence to Orthodox traditionalists, who regarded Russia as the centre of the Christian world and not as being on the periphery of an increasingly secular, European one. Peter's policies fomented a set of controversies which continue to resonate in Russia to this day. The central issue is: what kind of modernization should Russia pursue? Behind that lies the even more significant question: is Russia part of Europe, as the nineteenth-century Westernizers averred, or is it a civilization in its own right (the view of the nineteenth-century Slavophils and – to some extent – of the so-called Eurasianists today)?

One effect of Peter's policies was to reduce the Russian Church to virtually the status of a department of state. After his reign it would be difficult to argue that the Orthodox Church played an important role in politics. And yet the ghost of the Orthodox *ecumene* could not be laid entirely to rest. In the late eighteenth and the nineteenth centuries, for example, Russia used the claims of Orthodoxy to support the rights of the Christian subjects of the Turkish Empire and thus to advance its own interests in the Balkans (Jelavich 1983). Echoes of this policy are still to be heard in the 1990s, as witnessed, for example, by Russian support for the Serbs in the Yugoslav conflict, and Greek interest in the resuscitation of an Orthodox axis comprising Serbia, Romania, Russia and Greece as a counterweight to a perceived threat from militant Islam in the Balkans, Turkey and beyond (Georgiev and Tzenkov 1994).

Until the secularization of the concept, Europe, whether Catholic or Orthodox, was equated with Christendom. Islam clearly represented a threat to this Christian Commonwealth. One source of danger was finally eliminated in 1492 with the completion of the unification of Spain under its Catholic sovereigns, Ferdinand and Isabella. In the East, however, the confrontation with Islam lasted much longer. From the thirteenth century, the Russian principalities were subject to the rule of the Golden Horde, an Islamic successor to the Mongol Empire. Their liberation was heralded by the Russian victory at Kulikovo on the Don in 1380 but only secured by Ivan the Terrible's conquest of the khanate of Kazan, on the middle Volga, in 1552. This was followed by Russia's gradual expansion southwards across the steppe, culminating in the annexation of Crimea in 1783 and the advance towards the Caucasus, where the wars against the mainly Muslim mountain peoples lasted until the middle of the nineteenth century. In Central Europe, the Islamic Turkish tide, which had swept across the Balkans – taking Constantinople in its course – twice reached the gates of Vienna (in 1529 and 1683). There then began the slow decay of the Turkish Empire in Europe, culminating in the celebrated nineteenth-century 'Eastern Question'. Turkey's European empire was virtually eliminated by the first Balkan War of 1912–13.

The long confrontation between Christianity and the Muslim world continues to resonate through the eastern parts of Europe which still have considerable Muslim populations. One of its most dramatic expressions during the 1990s has been the frightful conflict in Bosnia-Herzegovina, involving mainly Catholic Croats, Orthodox Serbs and Bosnian Muslims. This threatened to stir up religious and political animosity among Christians and Muslims far from the scene of the actual fighting. Another instance has been the bloody war between Russia and Muslim independence-fighters in the Caucasian republic of Chechnya. Problems in the Russian republic of Tatarstan, among the Crimean Tatars in Ukraine, and elsewhere are a continuing reminder of one of the sadder aspects of the European heritage.

Imperialism and the state

From the point of view of the small nations, Pan-Slavism was a programme based on the brotherhood of equal Slav nations, including Catholics and Protestants as well as Orthodox. The Russian Pan-Slavs saw things differently.

(Seton-Watson 1967, p. 448)

Central and Eastern Europe is divided by more than religion. Compared to the western part of the continent, it is a veritable *mélange* of ethnic groups and cultures (see Figure 7.1). The Slavs (Russians, Ukrainians, Belarus'ians, Poles, Czechs, Slovaks, Slovenians, Croats, Serbs, Bulgarians), who speak related languages, give the appearance of a certain superficial unity. But these peoples are divided by their different histories, religions, cultures and much else. Other peoples, such as the Greeks, Romanians and Albanians, speak languages distantly related to Slav, falling within what is known as the Indo-European group. Conversely, the Hungarian language is quite unrelated to its near neighbours.

The region's complex ethnic and linguistic geography, which corresponds neither with political boundaries nor with any other straightforward lines of demarcation, is related to its chaotic history of settlement. The Great Migrations, which had helped overthrow the Roman Empire in the West, ended earlier there than in the East. Successive waves of peoples from Asia, Northern Europe and elsewhere descended on Central and Eastern Europe, displacing or intermingling with earlier settlers, and sometimes being absorbed by them. Consequently, despite the arguments of many nineteenth- and early twentieth-century nationalists, the idea that any straightforward relationship exists between language, territory and ethnicity is largely mythical, at least when applied to this area. It is worth noting in passing, however, that the ethnic geography is somewhat less complex today than it was before 1945. German colonists in particular, often settling by invitation of the local rulers, had played an important role in the region's settlement from the medieval period onwards. In the eighteenth and early nineteenth centuries, Germans even settled in the Ukrainian and Russian steppelands. The migration of most of these peoples back to Germany during and immediately after World War II (the Soviet Germans mainly being deported to Siberia) and other instances of 'ethnic cleansing' (most shockingly, the Nazi genocide of the Jews) have somewhat simplified the ethnic map.

Ethnicity, of course, is never simply a 'given'. Ethnic identity is influenced by state-building and other processes, and identities can change through time (Smith 1991). As explained in Chapter 1, a complex pattern of feudal loyalties and landholdings arose in Western Europe after the age of the Great Migrations, eventually consolidating, over the course of centuries, into a series of states (Giddens 1985). The medieval idea that the rulers of Catholic Europe were answerable to the Holy Roman Emperors (successors to Charlemagne) was largely a myth, except perhaps in Germany. The process of state formation was far more tortuous in the East, where, even by the early nineteenth century, the political geography was still dominated by large imperial states (Figure 4.2). This was partly because its peoples were so frequently the victims of invasion and empire-building. Mention has already been made of the Byzantine, Mongol and Turkish Empires. From the sixteenth century onwards, the Muscovite state began to transform itself into the Russian Empire and, from the middle of the following century, began to expand westwards at the expense of its neighbours. Meanwhile, the Habsburgs, who were Holy Roman Emperors with their capital in Vienna, began to build up their dominions to the east, in the process incorporating a good deal of Central Europe and the Balkans. After 1806, their territories became known as the Austrian Empire, and after 1867 as the Austro-Hungarian Empire. Further north, Prussia, which like Austria was essentially a German state, also began to expand to the east. Russia, Prussia and Austria divided Poland among themselves in the second half of the eighteenth century. In the following century these three empires (Prussia uniting Germany around itself in 1871) ruled over Central and Eastern Europe and tried to expand into the Balkans at the expense of the Turks. The three empires (as well as the Turkish Empire) finally came to grief in the First World War, but unfortunately this was not the end of empire-building in the region. The Nazi empire (1938–45) and the Soviet empire (1945–89/91) were the latest episodes of imperialism to afflict the region's peoples.

The modern states of Central and Eastern Europe, therefore, date only from the nineteenth century at the earliest (Figure 4.3). Several (notably Poland, Czechoslovakia, Hungary, Yugoslavia) appeared only in 1918, and the successor states to the USSR only in 1991. With such brief histories of state-building and lacking a legacy of mutual co-operation, these polities have proved extremely delicate. The post-Communist breakup of Yugoslavia and of Czechoslovakia are recent evidence of this fragility. At the conclusion of World War I the eminent British geographer Halford Mackinder pleaded in his book *Democratic ideals and reality* (1919) for the peacemakers of

Figure 4.2 Central and Eastern Europe, *c.* 1815

Versailles to establish a buffer zone of powerful, independent states to separate Germany and the Soviet Union. His hope was that the zone would prove strong enough to prevent a future aggressive alliance between Germany and the USSR. Mackinder's ideas were accepted only in part, for the peacemakers preferred to create states based on ethnic principles. During the 1930s, these states proved incapable of defending themselves and were easily swept aside by Nazi Germany. Today, as they emerge from yet another episode of imperialism, the Central and East European states are only too conscious of their past weakness and are haunted by memories of Nazism and Soviet expansionism. Consequently, most are seeking salvation in membership of NATO and the EU. It remains to be seen, however, whether this strategy can be reconciled with the loudly expressed desires to maintain their new-found sovereignty.

Figure 4.3   Central and Eastern Europe, *c.* 1914

Economic dimensions

> The strength or weakness of a society depends more on the level of its
> spiritual life than on its level of industrialization.
>
> (Solzhenitsyn 1991, p. 44)

Part of the appeal of a 'return to Europe' after 1989 lay in the belief that the
relative economic backwardness of Central and Eastern Europe was mainly
the product of the Communist command economy. The answer appeared to

lie in a rapid and decisive move to the market. As Alexander Solzhenitsyn (1991, p. 33) passionately demanded of his fellow Russians in the last days of Soviet power, 'Why should we cling to the centralized, ineffective and ideologically regulated economic system that has reduced the whole country to poverty?' Marketization therefore became the central goal of the economic policies of all the post-Communist states, a key element in their quest to rejoin the modern world.

There is, of course, much to be said for the charge that the command economy ultimately held back economic development in the region and brought numerous problems in its wake. In the years after 1945, the Stalinist economic system was imposed on all the members of the Soviet bloc (although Yugoslavia and, later, Albania began to seek their own routes to socialism). The centralized, planned economy and collectivized agriculture followed patterns previously established in the USSR, even though the Soviet-bloc countries had economies which varied among themselves: pre-war Bohemia (the western part of Czechoslovakia), for example, was relatively industrialized, as were the GDR and significant parts of Poland. By contrast, Hungary and much of the Balkans were more agrarian. The Stalinist approach led to a more than doubling of industrial output in many Soviet-bloc countries between 1948 and 1953 and a rise of some 50 per cent in the USSR. But these figures conceal the point that part of the growth came from a restoration of pre-war activity. In any case, Stalinist economic development caused much resentment on account of the coercive and often exploitative methods employed and the lack of attention paid to living standards. It was inevitable that 'de-Stalinization' in the USSR after 1953 would have repercussions in the rest of the Soviet bloc, most notably expressed in the series of uprisings of the early to mid-1950s. In their wake, several of the more ideologically inspired policies were modified and the different members of the CMEA were, to some extent, permitted to pursue their own independent policies (despite the fact that after 1954 the CMEA was given the task of co-ordinating the economic plans of all its members and integrating them with Soviet plans). Thus different countries experimented with economic reform, Hungary, for example, being allowed to introduce its own version of market socialism (known as the New Economic Mechanism). Again defying principles of socialist integration, Romania pursued autonomous policies in both economic and political realms. It is one of the ironies of Soviet-type Communism that a creed which preached internationalism actually achieved less economic integration among its client states than did the capitalist Member States of the then European Community. The limits to Soviet tolerance of deviance in the Communist camp were firmly established in 1968, however, when an experiment in socialist democracy in Czechoslovakia was suppressed by armed intervention.

Superficially, all the countries of the Soviet bloc made impressive economic gains down to the 1980s, closing the gap between themselves and their neighbours to the west. Thus industrial production rose six and a half times between 1955 and 1980 in Czechoslovakia and Hungary, eleven times in Poland, sixteen times in Bulgaria and twenty times in Romania (Longworth 1992). These data take no account, however, of the very low starting-points which characterized several of these countries, and they also ignore qualitative factors. Communist priorities emphasized heavy industries and particularly those sectors having a close affiliation with military requirements. As time went on, it became clear that the command economies remained fixed in what many social analysts describe as the 'Fordist' era and were failing to adapt to the new technologies and employment patterns increasingly characteristic of the capitalist world. Communist economic growth tended to be by 'extensive' (resource-demanding) rather than 'intensive' (resource-saving) methods. The former were expensive in terms of energy, raw materials and labour, and extremely harmful to the environment. Moreover, despite some changes in priorities following Stalin's death, the consumer sector and agriculture were relatively neglected and living standards lagged behind those of the West.

The reasons for the conservative nature of Communist development patterns have been analysed elsewhere (see Gregory and Stuart 1993; Ellman and Kontorovich 1992). Through time, rates of growth began to slow and increasingly failed to satisfy the aspirations of the citizenry. Policies pursued by many of the Communist countries, such as incurring ever greater debts on foreign markets, only stored up problems for the future. Not surprisingly, when Poles, East Germans, Estonians and Latvians compared their living standards with their capitalist neighbours, Soviet-style socialism began to seem less attractive than ever and the consciousness of failure could only grow.

There is no doubting the problems which the command economy brought in its wake, but a complete appreciation of the region's economic difficulties must also take account of a longer historical perspective. As we have seen in Chapter 2, the legacy of political and social turmoil and the discrepancies existing today between the economic development levels of western and eastern parts of Europe are in fact a continuation of deep-seated historical processes. Before World War II, their interaction resulted in many of the countries of Central and Eastern Europe still having largely agrarian peasant economies, while Germany was economically dominant (both before and after World War I). A number of long-term factors were responsible for this. As observed above, the age of the Great Migrations had lasted far longer in the East than in the West, one result being that population densities tended to remain much lower in the former. Again – as explained in Chapter 2 – feudalism and its attendant system of contracts and liberties did not develop in the East in the same way it did in the West, or took a different form there, while, notwithstanding the economic renaissance of the high Middle Ages, the growth of large towns and a self-confident mercantile class also tended to lag behind the more economically dynamic regions of first the Mediterranean and later North-West Europe. The nobility was frequently more numerous and more powerful in the East, to the detriment of the urban middle classes on the one hand and (outside Russia) of the state on the other. This tended to focus economic and political life on the countryside rather than on the town. The widespread adoption of serfdom in East-Central Europe and Russia in the early modern period, at a time when this institution was fast disappearing in the West, has been explained by some historians as a by-product of the rise of capitalist markets in the West. Other factors, however, were undoubtedly responsible, including – in Russia's case – military needs. But serfdom certainly reinforced the increasingly apparent differences between the two parts of the continent and may have impeded economic growth in the longer term.

Although their relative significance remains a matter for debate, many processes are thus implicated in the long-term relative backwardness of the East. One further historical factor, which helps to put the command economy into perspective, concerns the role of the state. From the early modern period, as a reaction to their economic difficulties, Central and Eastern European states tended to adopt policies to encourage economic growth and what, in the eighteenth century, was often referred to as economic 'rationality'. This, for example, was true of many German states, of the Habsburg Empire, and of Russia from the time of Peter the Great. The Habsburg rulers Maria Therese (1740–80) and Joseph II (1765–90) were among those noted for their 'enlightened' policies of social and economic reform. They sought to apply cameralist and physiocratic principles, which were early economic doctrines focusing on development and the enhancement of state revenues, in the hope of strengthening their empire. But they ultimately failed to overcome the regional particularism which was so much a feature of their realm. Nevertheless, this shows that the idea of state intervention in the economy has a long history in the region.

After 1989, however, the countries of the former Soviet bloc set their faces firmly against state control of the economy. The patent failure of Communist policy to satisfy their aspirations meant that they turned wholesale to the market as the only viable alternative and one that seemed to be associated with so much success in the West. While detailed policy has varied from one country to another, the broad outlines remain similar. Thus the state has withdrawn from its dominant economic role and firms have been required to adopt market principles as prices were freed and central planning and resource allocation replaced by the rules of supply and demand. The state, too, is now expected to pay its way and to control spending to achieve and maintain macro-economic stability. Economies are being opened up to the outside world to encourage domestic competition and allow inward investment. State-owned enterprises are gradually being privatized.

Needless to say, such policies, pursued to their logical conclusion, could cause enormous suffering in economies constructed to pursue goals which did not include the need to compete in the world market-place. Every transitional economy has passed, or is still passing, through a difficult period with steep falls in output, raging inflation and rising unemployment. This has occurred even when governments have continued to subsidize firms and have compromised in other areas of reform. Safety-net policies for those thrown out of work or on fixed incomes have proved inadequate in compensating for the problems which have ensued. The mechanical application of policies derived from liberal economic theory has also been severely criticized by those who point to the necessity of paying due heed to institutional and historical factors in designing reforms (Aligica 1997). The stresses and strains of the transition period have inevitably had their political repercussions, including – in some cases – a reversion to socialist- or Communist-inclined political parties. Nowhere has the issue of correct economic policy been more contentious than in Russia, where, as we have seen, the whole idea of a 'return to Europe' has provoked centuries of controversy.

One effect of marketization has been a reorientation of the trade relations of ex-Communist countries away from their former CMEA partners and towards the capitalist world. To a lesser extent, this is also true of trade relations among the states of the former USSR, which have faced particularly acute problems of transition. Their period of Communist control, which began in 1917, was much longer than that of the other ex-Communist states, while their economies were closely intertwined as a result of the Soviet mode of development. Indeed, it would be true to say that the USSR had no regional or republican economies as such, only separate economic sectors which spanned the territory as a whole (Bradshaw 1993). The establishment of the CIS in December 1991, which heralded the breakup of the Soviet Union, was meant to ease the co-ordination problems that would inevitably follow the split. In the event it has proved rather ineffective, not least because of widespread fears that Russia might use the organization to reassert its dominance over the other states. The three Baltic states (Estonia, Latvia and Lithuania), for example, simply refused to join, arguing that their annexation by the USSR in 1940 was illegal while their future lay with the West rather than the East. Together with Russia, they have pursued radical policies of economic reform, whereas Ukraine and especially Belarus' have proved more cautious. Different republics have therefore moved in different directions though not always in easily understandable ways. Thus Belarus', under the autocratic and crypto-Communist President Lukashenka, has been seeking closer union with Russia under the much more radical President Yeltsin.

Most of the countries outside the former USSR, together with the Baltic states, have been securing association agreements with the EU and hope to join in the foreseeable future (see Figure 5.1). Membership is a real possibility in the near future for several countries, but the EU is unlikely to look kindly upon those countries in which outstanding problems of economic transition have yet to be solved. Membership of NATO is also on the agenda, much to Russia's annoyance.

Peoples and places

> The national conflicts tearing the continent apart in the 1990s were the old chickens of Versailles once again coming home to roost.
>
> (Hobsbawm 1994, p. 31)

Hungarian resistance was the principal reason for the failure of Emperor Joseph II's policy of reforming and rationalizing his late-eighteenth-century empire. This proved to be a portent of the anti-imperial nationalism which engulfed the region in the nineteenth century. Joseph died in 1790 just a few months after the outbreak of the French Revolution, which did so much to spread ideas of liberty and national rights across the continent. What the French Revolution began was furthered by Napoleon, who not only took the new ideas into Central and Eastern Europe but also, by his military successes, helped to discredit the empires which held the peoples of the region in their sway. The stage was now set for a series of struggles which ultimately undermined the viability of the empires and helped pit their heterogeneous peoples against one another.

Scholars have classified nationalism into two main types: civic and ethnic. According to Smith (1991), the civic model of the nation presupposes the existence of common institutions, a single code of rights, and a well demarcated and bounded territory with which the nation's members identify and to which they feel they belong. This form of nationalism is regarded as typical of Western Europe, where both modern states and belief in the idea of 'the nation' originally arose together, reinforcing one another. It is also applicable to the United States and to other countries where different ethnic groups intermingled to form the basis of new nations. Ethnic nationalism, by contrast, is based on the idea of a single, ethnic identity, a community of birth and native culture. Unlike the civic nation, to which one may belong by virtue of living within that nation's territory, membership of the ethnic nation is a matter of blood. Civic nationalism is regarded by many scholars as having derived from the rationalism of the eighteenth-century Enlightenment, whereas ethnic nationalism emerged from organic notions of identity associated with the Romantic movement and, particularly, the teachings of J. G. Herder.

In reality, the two forms of nationalism are by no means mutually exclusive. In the United States, for example, neither Native Americans nor Afro-Americans were regarded as full members of the civic nation until relatively recently, while British nationalism has had problems embracing the Celtic peoples with their distinctive cultures and, more recently, immigrants from the former British colonies. Ethnic nationalism, however, seems to be most characteristic of situations like that in Central and Eastern Europe, where peoples have had to struggle for their liberty and distinctive identities against dominant imperial powers. In opposition to the 'modernizing' and 'rationalizing' projects of such imperial reformers as Emperor Joseph II, colonized peoples found it necessary to insist upon their separate identities based upon such distinctive markers as ethnic descent, language, customs and history. On such a basis, they could argue for their right to an independent state. In response to the ethnic nationalism, which threatened the integrity of their empires, nineteenth-century rulers commonly adopted repressive measures. Some, like the rulers of Austria and, to some extent, Russia, tried to avoid adopting an overtly nationalistic stance, thereby placing themselves above nationalism. In other cases rulers pursued aggressive assimilationist policies to the benefit of the dominant national culture. The Russian tsars, for example, veered between policies to promote ethnic Russian (*russkii*) culture and those emphasizing a more inclusive (*rossiiskii*) and tolerant approach to ethnic issues (Dixon 1996).

With the fall of the Central and East European imperial powers at the close of World War I, two contrasting policies were applied to the national question. In East-Central Europe and the Balkans, the Versailles peacemakers adopted the principle of 'self-determination' espoused by the American president, Woodrow Wilson. Hence, a structure of nation-states was created on the ruins of the former empires, each representing one people or a group of interrelated ethnicities. Inevitably, given the complexities of language and culture, the peacemakers found it impossible to draw boundaries that reconciled the interests of all parties involved. In the event, moreover, their decisions were only partially influenced by ethnic geography. Strategic issues, the vested interests of the victorious Allies, and the desire to weaken the former enemy states (Germany, Austria, Hungary and others) were also important considerations.

The Bolshevik victory in the Russian Revolution of 1917 and the ensuing civil war ensured that the Wilsonian model could not be applied in the Soviet Union. As internationalists rather than nationalists, the Bolsheviks aimed at establishing an international socialist society, which, it was believed, would eliminate the national antagonisms characteristic of capitalism. At the same time Lenin, the Bolshevik leader, was a pragmatist who realized the necessity of making some concessions to national feeling in the former Russian Empire, without going so far as to undermine the unity of the new Soviet state and thus the future of the Revolution. The eventual solution lay in the establishment of the USSR as a type of centralized federation, binding together a series of nominally equal republics and, within these, various autonomous units designed to represent and promote the national interests of minorities.

Neither of these solutions to the national question has stood the test of time. In East-Central Europe and the Balkans, national antagonisms were not pacified by the Wilsonian approach and the new states were unable to withstand either German or Soviet expansionism (Figure 4.1). While the many mutual antagonisms and conflicts were held in check during the Communist period, they did not disappear. Indeed, in some cases they were actually fostered by Communist leaders who were quite ready to exploit nationalist sentiment to deflect attention from their own failings or to solve other pressing problems (Glenny 1990). In the USSR, while the federal structure helped to ensure the persistence of national feelings (and may even have created them in some cases), those sentiments were frustrated by Soviet centralism and especially by the pro-Russian policies pursued under Stalin and, to some extent, by his successors. Gorbachev's mishandling of the nationalities issue after 1985 led to an upsurge of nationalism and to the breakup of the USSR, demonstrating once and for all the failure of Soviet policy to deal with this problem.

Ethnic nationalism's insistence on the importance of the ethnic community and on the existence of a historic territory or homeland, to which it alone has an exclusive right, creates numerous disadvantages, especially in a region so culturally diverse as Central and Eastern Europe. Each ethnic group or would-be nation has a tendency to try to encompass all its members, no matter how geographically scattered they may be. As discussed in Chapter 3, this leads to a determination to expand the state's frontiers, a phenomenon known as irredentism. Similarly, in defining the boundaries of the 'homeland', history is almost invariably invoked, and especially those historical periods in which the nation and state experienced their most expansive phases. Needless to say, the competing territorial claims to which such tendencies give rise are simply irreconcilable unless some compromise can be reached (Kolarz 1946). Ethnic nationalism also creates problems for minorities. If the state exists essentially to promote the interests of the dominant ethnic group and that group's rights override those of all others, then what of the minorities' rights? All too often in the past the solution has been sought in genocide, forced evacuation or forced assimilation, rather than in mutual tolerance, federation or power-sharing (Wilson 1996).

While the peoples of Central and Eastern Europe seem to have learned some lessons from the bitter events of the 1914–45 period, the ghosts of the past are still there to haunt them. The problems of the Hungarian minorities in Romania (Transylvania) and southern Slovakia, the Turks in Bulgaria, Greeks in Albania, Albanians in Macedonia and Serbia, to say nothing of the complex ethnic mixtures in Bosnia and Croatia, have been readily apparent since the fall of Communism. Attitudes towards the Jewish and Roma communities, which have no historic homelands in the region, still sometimes recall the racism of the past. In the former Soviet Union, the mass migrations of the pre-1917 and Soviet periods have left large numbers of people living in the 'wrong' state, now that the USSR is no more.

In the new post-Communist era, the states of Central and Eastern Europe all face the challenge of making a success of their independence without provoking either conflict with their neighbours or political instability within. Among other things, this means the construction of a national identity to which most of their citizens can subscribe without threatening the rights of minorities or of neighbouring states. For each state the challenge is somewhat different. The Poles, for example, can point to past glories and achievements, while striving to reconcile their Catholic and secular inclinations. The Hungarians must somehow find a way to combine the desire of many to respect and uphold the unique Hungarian culture with the need to progress as a modern European state. The Romanians face the task of building a modern society on the foundations of an authoritarian and still part-peasant structure. In the former Soviet Union, the problem of nation-building for most countries revolves around securing their independence from Russia. Thus Estonia and Latvia, with well entrenched national identities based upon their pre-war independence, must promote distinctive national traditions, meanwhile ensuring that the presence of large Russian minorities does not lead to political instability or a threat from Russia itself.

For Slavic Ukraine and Belarus', the problem is somewhat different. The history and culture of both countries intertwine with those of Russia and neither has a history of independent statehood. Ukraine's difficulties include a substantial Russian minority and regional differences between a more nationalistic west and a Russophone east. Attempts to build a secure national identity have included promoting the Ukrainian language and culture and providing a Ukrainian dimension to historical writing (for example, by advancing exclusive claims to the heritage of the medieval Kievan state and to a unique Cossack past) (Torbakov 1996). But these strategies obviously run the risk of alienating the Russophone regions. In Belarus', the lack of an apparently secure national identity has allowed the ruling group to seek reunion with Russia. Perhaps the most difficult task of nation-building faces the Russians, who must not only cope with a myriad of minorities within their frontiers and up to 25 million Russians outside them, but also come to terms with the collapse of their imperial dreams (Shlapentokh 1997). Whether Russia can transform itself into a truly post-imperial state, and stabilize its relations with its immediate and more distant neighbours, is a problem in which the world as well as Russia itself must share.

Conclusion

> History is the mighty force which enables the dead to dominate the living.

This aphorism, commonly attributed to Jules Michelet, succinctly summarizes the dilemma of the new Central and Eastern Europe. The collapse of Communism has opened a new era in the lives of the peoples of the region, an epoch in which unprecedented opportunities exist to make a success of independence and redefine relationships with neighbouring states, the western part of the continent and the wider world. It remains to be seen, however, whether, in seizing those opportunities, the peoples of the region will repeat the grievous mistakes of the past. In looking to the future, the region's peoples must undoubtedly be influenced by history. The hope must be that they will not become history's slaves. Part of the challenge which faces the peoples of the central and eastern parts of the continent is to define their place in Europe. This chapter has suggested a little of the complexity of

Europe's history and of the absurdity of claiming that one part of the continent is somehow more 'European' than another. The events of 1989–91 constitute a challenge to all the peoples of Europe, without exception. Adopting a narrow and exclusive definition of Europe cannot solve the problem of European identity, and is extremely unlikely to secure its stability in the future.

Notes

1 Of popular geographical texts of the past, Osborne's *East-Central Europe* (1967) excludes East Germany, Greece and the USSR; Mellor's *Eastern Europe* (1975) excludes Greece and the USSR, as do Turnock's *Eastern Europe* (1978), Singleton's *Background to Eastern Europe* (1965) and Dawson's edited volume, *Planning in Eastern Europe* (1987). Hoffman's edited text, *Eastern Europe: essays in geographical problems* (1971), includes Greece but not the USSR.
2 The sources in question are Sigismund von Herberstein, *Rerum Moscoviticarum commentarii* (1549); Giles Fletcher, *Of the Russe Commonwealth* (1591); and Adam Olearius, *Neue Beschreibung der muscowitischen und persischen Reyse* (1647).

References

Aligica, P. 1997: The institutionalists take on transition. *Transition* 7 March, 46–9.

Anderson, M. S. 1978: *Peter the Great*. London: Thames & Hudson.

Anisimov, E. V. 1993: *The reforms of Peter the Great: progress through coercion in Russia*. London: M. E. Sharpe.

Baron, S. H. 1991: European images of Muscovy. In Dukes, P. (ed.), *Russia and Europe*. London: Collins & Brown, 36–47.

Barraclough, G. 1957: *History in a changing world*. Oxford: Basil Blackwell.

Bradshaw, M. J. 1993: *The economic effects of Soviet dissolution*. London: Royal Institute of International Affairs.

Brechtefeld, J. 1996: *Mitteleuropa in German politics, 1848 to the present*. London: Macmillan.

Dawson, A. H. (ed.) 1987: *Planning in Eastern Europe*. London: Croom Helm.

Dixon, S. 1996: The Russians and the Russian question. In Smith, G. (ed.), *The nationalities question in the post-Soviet states*. Harlow: Longman, 47–74.

Droz, J. 1960: *L'Europe Centrale: évolution historique de l'idée de 'Mitteleuropa'*. Paris: Payot.

Ellman, M. and Kontorovich, V. (eds) 1992: *The disintegration of the Soviet economic system*. London: Routledge.

Georgiev, A. and Tzenkov, E. 1994: The troubled Balkans. In Miall, H. (ed.), *Redefining Europe: new patterns of conflict and co-operation*. London: Pinter, 48–66.

Giddens, A. 1985: *A contemporary critique of historical materialism*, vol. 2: *The nation-state and violence*. London: Polity Press.

Glenny, M. 1990: *The rebirth of history: Eastern Europe in the age of democracy*. London: Penguin.

Gregory, P. R. and Stuart, R. C. 1993: *Soviet and post-Soviet economic structure and performance*, 5th ed. New York: HarperCollins.

Hobsbawm, E. 1994: *Age of extremes: the short twentieth century, 1914–1991*. London: Michael Joseph.

Hoffman, G. (ed.) 1971: *Eastern Europe: essays in geographical problems*. London: Methuen.

Jelavich, B. 1983: Tsarist Russia and Balkan national liberation movements: a study in Great Power mythology. In Sussex, R. and Eade, J. C. (eds), *Culture and nationalism in nineteenth-century Eastern Europe*. Columbus, OH: Slavica Publishers.

Kolarz, W. 1946: *Myths and realities in Eastern Europe*. London: Lindsay Drummond.

Kumar, K. 1992: The 1989 revolutions and the idea of Europe. *Political Studies* 40, 439–61.

Likhachev, D. S. 1991: *Reflections on Russia*. Boulder, CO: Westview Press.

Longworth, P. 1992: *The making of Eastern Europe*. Basingstoke: Macmillan.

Mackinder, H. 1919: *Democratic ideals and reality: a study in the politics of reconstruction*. London: Constable.

Massey, D. 1995: The conceptualization of place. In Massey, D. and Jess, P. (eds), *A place in the world: places, cultures and globalization*. Oxford: Oxford University Press/The Open University, 45–86.

Mellor, R. E. H. 1975: *Eastern Europe*. London: Macmillan.

Meyendorff, J. 1981: *Byzantium and the rise of Russia: a study of Byzantino-Russian relations*. Cambridge: Cambridge University Press.

Meyer, H. C. 1955: *Mitteleuropa in German thought and action*. The Hague: Martinus Nijhoff.

Miall, H. 1994: Wider Europe, fortress Europe, fragmented Europe. In Miall, H. (ed.), *Redefining Europe: new patterns of conflict and co-operation*. London: Pinter, 1–15.

Milyukov, P. 1974: *Outlines of Russian culture*, vol. 3: *The origins of ideology*. Gulf Breeze: Academy International.

Naumann, F. 1966 (originally 1946): *Behemoth: the structure and practice of National Socialism, 1933–44*. New York: Harper & Row.

Obolensky, D. 1950: Russia's Byzantine heritage. *Oxford Slavonic Papers* 1, 37–63.

Osborne, R. H. 1967: *East-Central Europe*. London: Chatto & Windus.

Seton-Watson, H. 1967: *The Russian Empire, 1801–1917*. Oxford: Oxford University Press.

Shaw, D. J. B. 1997: Geopolitics, history and Russian national identity. In Bradshaw, M. J. (ed.), *Geography and transition in the post-Soviet republics*. Chichester: John Wiley, 31–41.

Shlapentokh, V. 1997: Creating the 'Russian dream' after Chechnya. *Transition* 21 February, 28–32.

Singleton, F. B. 1965: *Background to Eastern Europe*. Oxford: Pergamon.

Smith, A. D. 1991: *National identity*. London: Penguin.

Solzhenitsyn, A. 1991: *Rebuilding Russia*. London: Harvill.

Sumner, B. H. 1951: *Peter the Great and the emergence of Russia*. London: EUP.

Torbakov, I. 1996: Historiography and modern nation building. *Transition* 6 September, 9–13 and 64.

Turnock, D. 1978: *Eastern Europe*. Folkestone: Dawson.

Wilson, A. 1996: The post-Soviet states and the nationalities question. In Smith, G. (ed.), *The nationalities question in the post-Soviet states*. Harlow: Longman, 23–43.

Wolff, L. 1994: *Inventing Eastern Europe: the map of civilization on the mind of the Enlightenment*. Stanford, CT: Stanford University Press.

P A R T

III

*The nature of European integration
and the consequences of diversity*

|5|
The political geography of European integration

RICHARD GRANT

Introduction

The political geography of contemporary European institutions has been shaped by the concerted efforts of a European policy-making élite to achieve economic union, a grand project now being extended into manifestations of union that encompass far more than the economic arena alone. In this latter process, European Union (EU) institutions will have their powers extended and, as a result, European standards and laws will impinge ever more directly on the daily lives of some 370 million citizens. For most social groups in the EU, the reality of integration is about the growing amount of European law governing their activities. The aim of this chapter is to describe the evolution and the growth of powers of the most important EU institutions. A network framework is employed to describe the role and relations among EU institutions in making policy decisions. Within this framework, decision-making in the Council of Ministers, at the apex of the policy process, is highlighted. Recent attempts to bring the union closer to citizens are outlined. In the concluding section a set of observations are presented on the EU's 'democratic deficit' and the inability of EU policy-makers to establish policy legitimacy among the public.

The institutional form of the Union has been created over time. The first building-block for European institutions was laid down with the establishment of the European Coal and Steel Community (ECSC) by the Treaty of Paris in 1951. The treaty also set up a series of institutions which subsequently have been greatly modified and strengthened. These include the predecessor of the contemporary European Commission – the High Authority – as well as a Council of Ministers, a Court of Justice and a Common Assembly (later transformed into the present-day European Parliament). Following the

1957 Treaty of Rome, the European Economic Community (EEC) was formed in 1958. Originally, this comprised Belgium, France, the Federal Republic of (West) Germany, Italy, Luxembourg and the Netherlands. The United Kingdom, Ireland and Denmark joined in 1973, to be followed by Greece in 1981 and Spain and Portugal in 1986. The former German Democratic Republic (GDR or East Germany) was added in 1990, following the unification of Germany. Most recently, Austria, Finland and Sweden joined the Union in 1995 and further enlargement is imminent (Williams 1996; Figure 5.1). As discussed in Chapter 4, many of the countries of Central and Eastern Europe – together with Cyprus – are actively seeking membership (Table 5.1). The next enlargement could be as soon as 2001, although 2003 is more likely.

The EU is a unique international entity with a succession of defining characteristics. First, the limitations placed on national sovereignty make EU institutions distinctive. The fact that EU Member States have been willing and able to 'pool' their sovereignty and delegate sovereign powers to central

Table 5.1   European Union enlargement: criteria for membership and applicant states as of 1997

Criteria of application

A country must have:

1 Stability of institutions, guaranteeing democracy, the rule of law, human rights, and respect for – and protection of – minorities.

2 A functioning market economy.

3 Ability to take on obligations of membership, including adherence to the aims of political, economic and monetary union.

Applicant states

1 Cyprus – application has already received a favourable opinion from the European Commission

2 Applicant states listed in *Agenda 2000* (1997):
   (a) More favoured group: Czech Republic
                              Estonia
                              Hungary
                              Poland
                              Slovenia
   (b) Possible entrants: Bulgaria
                              Latvia
                              Lithuania
                              Romania
                              Slovakia

3 Turkey's application for membership has been rejected although it has a customs union with the EU which came into force on 31 December 1995.

Source: *Agenda 2000* (Commission of the European Communities 1997).

Figure 5.1 The countries of the European Union and its various enlargements

institutions in making common decisions is epoch-making. Because the powers and relations among EU institutions are delineated by co-operating over common policies with national governments, policy-makers must be cognizant of both the Community and the national implications of their decisions. Second, approximately 22 500 professionals (popularly known as 'Eurocrats') are employed to assist in policy-making (Commission of the European Communities 1995). Third, the EU has a large reserve of financial resources at its disposal. Fourth, the founding treaties (Paris and Rome) do not remain the only constituting agreements by which the organization functions. Co-operation among states and institutions has grown continuously (both formally and informally). Agreement was reached in 1986 to create a Single European Market (SEM) by 1992, an 'ambitious attempt to eliminate non-tariff barriers to trade and also to facilitate the mobility of the factors of production' (Chisholm 1995, p. 5). The Maastricht Treaty for Political, Economic and Monetary Union, signed in December 1991 (and subsequently amended at Amsterdam in 1997), rewrote the original Treaty of Rome and established the EU to succeed the EEC. Community policy-makers are planning to introduce a single European currency in 1999, and full European Monetary Union (EMU) (a European Central Bank and harmonization of macro-economic policies) in the early twenty-first century. Fifth, the EU institutional framework offers an alternative structure for national policy-makers. In mediating the global economy, European policy-makers operate simultaneously in Community, national and local policy environments. Finally, however, there is a major negative factor in that EU institutions and decision-making processes lack public legitimacy, having been created by policy élites, largely independent of the will of national populations.

Evolution of European institutions and their roles in policy-making

Western Europe's remarkable institutional transformation, from a collection of fractious states to an increasingly amalgamated union, was inspired by the desire to prevent another world war as well as by visions of economic advantage and union among people. According to one of the most prominent among the early architects of European unity, Jean Monnet, 'We are not forming coalitions between states, but union among people' (cited in Fontaine 1988, p. 6). European integration also received strong support from the United States as part of its Cold War strategy to strengthen Western Europe as a buffer to Soviet encroachment in the East. However, co-operation has been primarily economic. The early successes in reducing internal barriers to trade, erecting a common external tariff and the creation of a Common Agricultural Policy (CAP) all strengthened this particular emphasis, as has the subsequent development of the SEM. Conversely, initial attempts at political union failed. The European Defence Community and a European Political Community were discarded in 1954 as a result of opposition in the French Parliament. Only now, in the post-Maastricht era, is the EU tentatively moving towards a 'new European architecture' that reflects its role as a supranational government with jurisdiction (Williams 1996, pp. 27–9).

The division of tasks and responsibilities among the institutions of the EU shapes the balance of power within the Community (Moravcsik 1993). The most important policy institutions are the Brussels-based Commission, the Council of Ministers, the European Court of Justice and the European

Parliament. The Commission is an independent institution that transcends national interests and occupies a central place in the formulation and execution of EU policy. The ECSC and European Atomic Energy Agency (Euratom) had separate Commissions until they were merged in 1967. Today the Commission – and its bureaucracy – epitomizes the concept of 'Europe'. It has a monopoly on proposing economic legislation and is considered the 'boiler-room' in the process of integration (*The Economist* 1991). Commission staff are recruited by competitive examination from each of the 15 Member States by numerical quotas. Once appointed, they do not serve their individual countries; rather, their mission is to represent the Community. The Commission is headed by a President who is appointed by agreement among Member State governments. The current incumbent, Jacques Santer, is assisted by a cabinet of 19 Commissioners, who each have separate portfolios for policy areas (for example, Competition, Fisheries, Economic and Financial Affairs, and Transport). The Commissioners are nominated by the governments of the 15 Member States in consultation with the new Commission President every five years. The President and Commissioners are then subject to European parliamentary approval before they can take office. Two Commissioners are appointed from each of the larger Member States (Germany, Spain, France, Italy and the UK) and one each from the smaller Member States (Austria, Belgium, Denmark, Finland, Greece, Ireland, Luxembourg, the Netherlands, Portugal and Sweden). The Commission is answerable to the European Parliament, which retains the power to dismiss it as a whole in a vote of censure.[1]

The Council of Ministers is the decision-making authority within the Union. Most meetings take place in Brussels (except in April and June, when Luxembourg is the venue). The Council of Ministers is a forum for inter-governmental negotiation between Member State governments. The members of the Council of Ministers tend to favour national interests and assess Commission policy drafts from national and regional perspectives. The Council always meets behind closed doors, only the outcomes of its discussions being published afterwards. Its membership is not fixed but varies according to the approximately 25 policy areas that fall within its remit. For example, in a discussion of agriculture policy, the Ministers for Agriculture from each of the Member States constitute the body. Meetings of the Council of Ministers are presided over by a six-month presidency, rotating among all EU Member States. Most of the day-to-day judgements on EU policy are made by the Committee of Permanent Representatives (Coreper), whose members are appointed by national governments. Coreper members have an important gatekeeper function, representing EU views to national experts and national views to their counterparts and the Commission. Up to 1994 voting was relatively infrequent in the Council of Ministers and consensus remained the preferred path for the enactment of legislation (Dinan 1994).

Related to – but separate from – the Council of Ministers is the European Council, established in 1975. While not a Community institution, this is increasingly active in the policy process and policy-makers generally resort to it in order to resolve contentious issues – such as monetary union – that have proved too stubborn for the Council of Ministers. The European Council consists of government ministers and a European Commissioner, who meet at least twice a year at intergovernmental summits and often deliver important communiqués. There is a high degree of overlap between the Council of Ministers and the European Council. For instance, agreement on the Maastricht Treaty was reached at a European Council session that directly preceded a Council of Ministers meeting. In contrast to the Council of Ministers, all summit meetings are public relations exercises and are relatively transparent, with a huge media presence. The role of the European Council is to provide strategic direction by considering Community and Member State policies as an organic whole, rather than as separate and competing interests. In the 1990s, the European Council has become a crucial conduit for resolving policy differences between the Member States and for maintaining momentum on policy initiatives.

The European Parliament's role is mainly consultative on policy proposals and it undertakes timely debates on current issues. The Parliament, based in Strasbourg, is the guardian of the European interest and the defender of citizens' rights. Parliament members (MEPs) are selected in Community-wide elections for five-year terms. Each state is allocated a quota of MEPs, largely based on population size. Presently there are 626 MEPs, although parliamentary membership is set to rise after the next enlargement (*The Economist* 1997a). The Parliament has little legislative authority, although its powers have been increased by every treaty amendment over the past twenty years. Two important parliamentary powers shared with the Council of Ministers are its equal control over the Community's annual budget and its co-decision powers on all resolutions passed by majority voting. Within Parliament, members are organized by transnational political groupings. Individually, or collectively, European citizens have a right to petition the Parliament, and can seek redress of their grievances on matters that fall within the EU's sphere of responsibility. The Parliament also appoints an ombudsman to investigate allegations of maladministration brought by citizens.

The European Court of Justice (ECJ) is located in Luxembourg, far from the political fray of Brussels and Strasbourg. Each Member State appoints a judge to the ECJ and contributes to a small staff (approximately 600 in total), who provide research, translation and administrative support. For much of its existence, the ECJ has been the least-known of the EU's institutions. The Court's principal purpose is 'to ensure that in the interpretation and application of [the treaties] the law is observed' (Dinan 1994, p. 296). The ECJ has presided over 9000 cases since 1954 and delivered over 4000 judgments (Europa 1997). The original treaties (the Treaties of Paris and

Rome), the treaties of accession and the various treaty amendments constitute the Community's 'primary legislation', whereas the laws made by Community institutions in accordance with the treaties constitute the Community's 'secondary legislation' (Dinan 1994, p. 296). The ECJ's power is limited to economic union; it has no legal powers in the other two pillars of the Maastricht Treaty (security and defence, and home and judicial affairs). The ECJ has consistently ruled that EU law has supremacy over national law. Cases may be brought before the ECJ by an EU institution, a Member State or an individual. Importantly, under Article 173 of the Maastricht Treaty, the European Parliament does not have the right to bring action before the ECJ and is limited to protecting its privileges (Bermann *et al.* 1995). The Court may also be requested by the legal authorities of a Member State to give a preliminary ruling on a question of Community law in order for national courts to make a judgment so that the harmonization of policies can proceed. Because the EU's other institutions are often litigants in court cases, the ECJ has unique institutional relationships. The Commission is usually the ECJ's ally, although there are instances where the ECJ has ruled against it (Dinan 1994). A close working relationship is facilitated by the ECJ and the Commission sharing responsibility for ensuring that Community citizens comply with EU directives and regulations. By contrast, the Court's relations with the Council of Ministers are more tense, largely owing to differences over prioritizing EU or national interests.

The EU as an institutional network

The EU is the world's most successful example of institutionalized international policy co-ordination. It is best characterized as an institutional network that comprises complex linkages among institutions, governments, organizations and individuals. Although their dynamics are not always sufficiently understood, these linkages can be viewed as hanging together in complex configurations. Actors in a network have preferences for interaction with one another, rather than with outsiders. Successful interactions among network actors create a policy-making history and culture, in addition to incentives for the continuation of co-operation and the maintenance of relationships. For example, long-term partners typically have a greater incentive to exchange reliable information than do rivals.

Within the EU network, a 'European policy style' of decision-making has emerged – a cumulative process of accommodation whereby the participants refrain from unconditionally vetoing proposals, and seek instead to attain consensual agreement through compromise, in the process complying with institutional regulations (Mazey and Richardson 1995).[2] The European policy culture is facilitated by the propensity for working groups and committee meetings to conduct business; the ensuing complex system of committees, working groups and expert groups creates subsystems within the network. Some are dominated by national administrators, who play dual roles as representatives of their states and as European policy-makers, while others are controlled by 'Eurocrats'. Lobbyists attempt to influence the policy process at three levels: by inducing national governments to represent their position in Brussels; by persuading EU officials directly; and by enlisting the support of their affiliated EU-wide lobbying organizations (Grant 1993). However, the concentration of policy formulation in the hands of the bureaucracy means that certain types of groups (notably producer groups) are likely to be more successful than others, such as, for example, voluntary associations (Mazey and Richardson 1995). In particular, multinational corporations have some special advantages because of their technical expertise and their ability to co-ordinate lobbying across Member States.

Like other international organizations such as the General Agreement on Tariffs and Trade (GATT), the EU establishes common expectations, provides information and facilitates arm's-length intergovernmental negotiation. However, the EU's policy-making apparatus receives a much higher level of commitment from its constituent countries. Member States defer to the EU because the institutional policy network assists in protecting them against external uncertainties such as the volatility of global currency markets. Moreover, Member States surrender sovereignty to the EU because it has been a productive policy-making apparatus; very few sectors remain unaffected by EU laws and decision-making. National governments find the institutional structure preferable in so far as it strengthens, rather than weakens, their control over most domestic affairs, permitting them to achieve otherwise unattainable goals such as currency stability. The institutions of the EU also facilitate the efficiency of inter-state bargaining, their network reducing the transaction costs for governments by identifying issues, negotiating bargains, codifying agreements, and monitoring and enforcing compliance. The autonomy of national political leaders *vis-à-vis* domestic political groups can also be strengthened by EU institutions. In some instances, EU policies can be used to support the credibility and legitimacy of governments. Common Community policies can be used to bolster support and strengthen the domestic agenda-setting power. Furthermore, by referring to common EU policies, governments can reduce the risk presented by radical political agendas such as neo-Nazism. The existence of EU institutions increases the bargaining power of Member States with external actors, as they negotiate as a bloc rather than as individual countries. Such has been the practice since 1962 when the Commission first negotiated on behalf of all Community members at the GATT (Grant 1995).

Policy-making in the EU

The legislative process begins with a Commission proposal for legislation, a prerequisite for making EU law. In devising its proposals, the Commission has three objectives: first, to identify the European interest; second, to consult as widely as is necessary; and third, to respect the principle of subsidiarity that competence be exercised at the lowest level – as near to citizens as possible, thereby maximizing flexibility and local discretion (see below and also Chapter 6 for a more detailed discussion). The policy process is initiated by the Commission formally sending a proposal for legislation to the Council of Ministers and the European Parliament for consideration. The Commission retains the power to initiate legislation in most policy areas (exceptions are security and defence policy, justice and home affairs). Within the network, there is a clear centre for activities and decisions. Lines of authority permeate from the Council of Ministers; specific tasks such as negotiating the specifics of the EMU (where agreement could not be reached in the Council of Ministers) can be directed to a subsystem such as an intergovernmental summit. In theory the functioning of the network hierarchy involves the flow of authority in 18 movements; sometimes sideways and upwards, but mostly downwards (Figure 5.2).

Complicating the policy-making system are the three procedures in effect for the European Parliament, each of which allocates it a different role. The first is the traditional procedure, in which the Parliament is consulted only before a Commission proposal can be adopted by the Council of Ministers (the Parliament's advice can be disregarded). Formerly this consultation applied to all initiatives, but it is now limited to specific areas; for example, agricultural price reviews. The second is the co-operation procedure, which allows the Parliament to improve legislation by amendment. This procedure involves two readings in Parliament, giving members ample opportunity to review and amend the Commission proposals and the preliminary position of the Council of Ministers, which, nevertheless, can still decide to disregard the Parliament's opinion. This procedure applies to a large and growing number of policy areas such as regional development, research, the environment, social policy and overseas co-operation and development. In exceptional cases, the Parliament also can force the Council of Ministers to decide by unanimity to overrule the Parliament's position. Third, there is the 'co-decision' procedure, whereby Parliament has the power to reject a proposal. A conciliation committee, comprising an equal number of members of the Parliament and the Council of Ministers, then seeks a compromise on the proposal. If no agreement is reached in the conciliation committee, the Parliament does have a veto power, although an absolute majority is needed to reject a proposal. The co-decision procedure applies to a wide range of issues such as migration, consumer protection, education, culture and health. Furthermore, the Parliament's assent is required for important international agreements such as the accession of

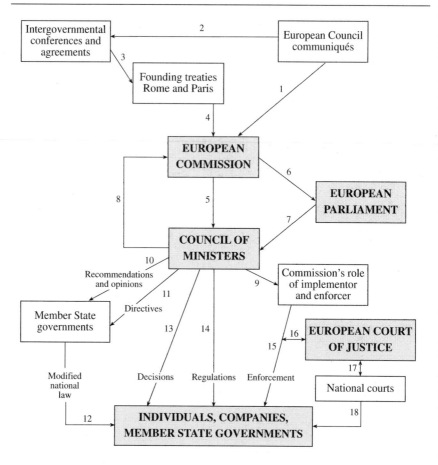

Figure 5.2　Decision-making in the European Union
Source:　Based on Swann (1996, p. 16)

new members, EU agreements with external countries, the organization and objectives of the structural funds, and the task and powers of the European Central Bank.

Following parliamentary participation in reviewing an issue (by one of these procedures), final decisions are made by the Council of Ministers. These can take a number of forms (regulations, directives, decisions and recommendations/opinions). Regulations are applied directly without any need for national measures to implement them. In general, directives that bind Member States to the objectives, but which leave their governments with the power to choose the form and means to be used, are preferred. Decisions are binding in all aspects upon those at whom they are specifically targeted. Recommendations and opinions are not binding but provide a rule of thumb. The ECJ can also become involved in making decisions binding through its interpretations of the law. Finally, Member State governments and authorities, individuals and companies are expected to comply with EU policies.

Decision-making in the Council of Ministers

The intricate network for policy-making in the EU is not widely understood by the public. Even less is known about decision-making in the Council of Ministers. Only publications with specialized readerships like the *Financial Times* and *The Economist* carry partial coverage of its discussions. The EU's attempt to improve public access to information on its activities – by updating its Internet site on daily activities and providing a phone hot-line for information – has not included any effort to make the activities of the Council of Ministers more transparent. In short, these remain veiled in secrecy, and members of the Council of Ministers have so far resisted any attempts to bring their activities into the public domain, partly because they are accountable to their national parliaments and citizens for positions taken in the Council's meetings. The secrecy of meetings provides Council members with a smokescreen for an unpopular decision. They can denounce other Member States (without means of verification) for decisions made by the Council of Ministers, or they can adopt a more pro-European policy stance over a national one and blame the outcome on the voting procedure. Another, albeit more positive reason for secrecy is that, because closed meetings are more conducive to efficient and productive deliberations, members are more likely to work towards a consensus with each other than they would perhaps do under public scrutiny.

None the less, conflicts between Member States over voting procedures in the Council of Ministers have exacerbated the problematical aspects of secrecy. The founding treaties originally envisaged two voting procedures as coequal: unanimity and majority voting. In the mid-1960s, however, France, under its then President, Charles de Gaulle, resisted any attempts to introduce more majority voting. De Gaulle believed that majority voting would have federalizing effects and subject France to the strictures of a 'foreign majority' (Teasdale 1996). The French strategy to block this change in voting procedure entailed French ministers boycotting the Council of Ministers in the second half of 1965 and resulted in 'an empty chair' at its meetings. A compromise was reached in 1966 when the Council of Ministers agreed to seek unanimous agreements (even when the Treaty of Rome provided for majority voting) on issues that individual states regarded as vital to their national interests. This solution became known as the 'Luxembourg Compromise'. Even though the compromise had no legal standing, its impact was to prove highly significant. It created a climate in which, in most policy areas, unanimous decision-making in the Council of Ministers continued to be the norm, even when no 'very important national interest' was formally asserted. As a consequence, decision-making on

Table 5.2   Present and possible future allocations of institutional powers
among EU Member States

	Percentage of EU population	Votes in Council of Ministers		Seats in EU Parliament		Number of EU commissioners	
		Now	Reweighted	Now	Reweighted	Now	Reweighted
Germany	22.0	10	19	99	136	2	4
UK	15.7	10	14	87	98	2	3
France	15.6	10	14	87	98	2	3
Italy	15.4	10	14	87	96	2	3
Spain	10.5	8	9	64	66	2	2
Netherlands	4.1	5	4	31	26	1	1
Greece	2.8	5	2	25	18	1	1
Belgium	2.7	5	2	25	17	1	1
Portugal	2.7	5	2	25	17	1	1
Sweden	2.4	4	2	22	15	1	1
Austria	2.2	4	2	21	14	1	0
Denmark	1.4	3	1	16	9	1	0
Finland	1.4	3	1	16	9	1	0
Ireland	1.4	3	1	15	6	1	0
Luxembourg	0.1	2	0	6	1	1	0
Total	100	87		626		20	

new policy issues slowed dramatically. The public, informed only of the outcomes of decisions made by the Council of Ministers, felt alienated from the process.

An attempt to clarify the Luxembourg understanding was reached at the European Council meeting of March 1994 at Ioannina, Greece. An agreement was reached – the Ioannina Compromise – that permitted qualified majority voting and disallowed a single-state political veto in an effort to improve decision-making efficiency. As a result, the Council of Ministers now decides unanimously only on major policy decisions as expressed in the treaty provisions (for example, taxation, industry, culture, regional and social issues and the framework programme for research and technology development). In most other policy areas, such as agriculture, fisheries, the internal market, environment and transport, decisions are made using a qualified majority threshold of 70 per cent of votes cast in favour within the Council of Ministers. In allocating voting rights, each Member State's votes are weighted in less than strict proportionality to the number of inhabitants) and votes are cast in a member-state bloc (Table 5.2).

In some cases, the majority in favour must also include at least 10 countries and – in certain circumstances – the Commission can decide that an issue is sufficiently important to warrant a unanimous vote. The Commission can prevent decision-making in the Council of Ministers by withdrawing a proposal if it is believed that agreement can be secured only with major amendments that are not in the interests of the Union as a whole (Commission of the

European Communities 1995). For example, in 1996 the Commission withdrew 50 legislative proposals that had been accepted by the Council of Ministers (*The Economist* 1997a). The Council attempts to reach the widest possible consensus before taking a vote. For instance, in 1994 only 14 per cent of legislation adopted was subject to a negative vote (Europa 1997). Research on decision-making procedures of the Council of Ministers suggests that, on average, 80 per cent of differences of opinion are resolved at the level of working groups and that Coreper has a strong influence on decision-making processes within the Community (Van Schendelen 1996). Unfortunately, we know even less about the functioning of Coreper than about the Council of Ministers.

Thus, despite attempts to clarify decision-making procedures, the lack of transparency is still the most undemocratic aspect of the EU. Furthermore, the Ioannina Compromise has not solved the voting issue. The UK favours the right of Member States to opt out on future policy decisions, and argues that future voting on any effort to deepen the Union should require unanimity(Blitz 1997). The implications are far-reaching. The UK prefers a 'multi-speed' Europe in which each Member State decides which part of the process to be involved in. In contrast, Germany and France favour a 'two-speed' EU with a leading group of Member States deciding to adopt a core range of policies through majority voting (Blitz 1997). Germany and France also visualize the lagging group eventually catching up so that, ultimately, all Member States will be at the same level.

Reform in the EU

The closed and secretive nature of policy-making in the EU, particularly in the Council of Ministers, has been strongly attacked by several Member States (led by the Netherlands and Sweden), journalists (in the UK led by the *Guardian*) and members of the European Parliament, particularly the Socialists (Van Schendelen 1996). The Commission's top priority in bringing the Union into the twenty-first century is to improve 'transparency and openness and in bridging the gap between the activities of the European institutions and ordinary citizens' (Commission of the European Communities 1995, p. 44). The same report, however, does not offer any real proposals on how this might be achieved. For the most part, the people who run European institutions are faceless. The average European-in-the-street is hard-pressed to name any European Commissioner, 'Eurocrat' or MEP, let alone have an opinion on how to reform EU institutions.

The European Parliament proposes that public access to European documents be increased and that the public should have access to drafts of proposals (European Documents 1995). Furthermore, it has argued that meetings of the Council of Ministers 'be held in public unless a specific and duly justified exception is decided by a two-thirds majority' (European

Documents 1995). Even then, these modest initial efforts at openness have met with some national resistance. For example, the Eurosceptic UK Conservative government (overwhelmingly ejected from power in 1997) decided not to promote the European Commission's hot-line on EU information (Southey 1997).[3]

The introduction of the principle of subsidiarity under Article 3 of the Maastricht Treaty was an attempt to bring decision-making closer to EU citizens. In essence, its application means that the EU cannot regulate when policies can be made more effectively at the local level. However, the definition of subsidiarity remains contested. The UK, for example, understands the principle as a vital safeguard for national sovereignty, one that prevents the EU from involving itself unduly in Member States' affairs. Conversely, in claiming that subsidiarity is an attempt to define clearly who does what, the European Commission argues that EU institutions will act only in circumstances in which policy objectives are better achieved by common EU action than by Member States acting independently. Furthermore, the practical implications of subsidiarity need further illumination. It is not clear how 'closeness' can be operationalized geographically. For example, in Ireland, closeness to citizens may mean Dublin, but not Galway or Waterford. A Council of Ministers agreement on the precise meaning of subsidiarity would give more substance to the concept, as would a body of EU case law on the subject, which inevitably will take some time to accrue.

The Committee of the Regions (COR), also established by the Maastricht Treaty, is another attempt to give regions and local authorities some EU policy influence. It also represents an effort to move beyond the binary relationship of the EU and Member States. The 222 members, directly elected regional or local politicians, must be consulted for policy-making in the following areas: trans-European networks; public health; education; youth; culture; and economic and social cohesion. The COR, however, as merely an advisory body to the Council of Ministers and the Commission, lacks legislative authority, nor does it have any direct dealings with the European Parliament. The low profile of the COR seems likely to continue. The combination of slowly increasing powers for the European Parliament and highly organized lobbies in Brussels appears to leave little room for the COR.

In order to promote more democracy among EU institutions, proposals are being made to reweigh votes and influence in the institutions. Currently, every EU institution is biased in favour of countries with small populations. For example, in the Council of Ministers, Germany, the UK and the Netherlands cannot block a qualified majority even though they account for 42 per cent of the EU's population. Conversely, Belgium, Denmark, Finland, Ireland, Luxembourg and Portugal can prevent a qualified majority, even though their combined population is only 12 per cent of the EU total. The problem may be accentuated by the relative or absolute small size of most of the potential new members from Central and Eastern Europe and the Mediterranean (excepting Poland and Turkey), which will exacerbate the bias if the same system prevails (Table 5.1). Deriving a formula to increase both democracy and decision-making efficiency is highly sensitive and proving difficult to achieve.

Table 5.2 shows the current political geography of voting power and compares the distribution to a reweighted distribution based on population size. In a system based on population size, Germany would be in a dominant position, with 19 votes in the Council of Ministers, 136 MEPs and four Commissioners. Luxembourg, by contrast, would lose all votes in the Council of Ministers, retain only one vote in the European Parliament and be unable to appoint a Commissioner. Thus to smaller countries, the principle of territoriality, whereby one country has one vote, is far more appealing. Other possibilities have been suggested to promote greater democracy. One proposed formula is based on double-majority voting, requiring a majority of Member States and their populations within the particular institution (Council of Ministers and the Parliament). Another possibility of increasing EU institutional democratization is provided by the bicameral system, which requires all EU law to be accepted by both the European Parliament and the Council of Ministers, thereby increasing the consultation powers of these institutions in legal issues (Holsti 1995).

The European Parliament has also put forward some innovative reform proposals to compensate for the democratic deficit, suggesting that its assembly form a partnership with national parliaments, which would strengthen the role of various governments in the EU policy network. This could be accomplished by joint meetings of European Parliamentary committees and their national equivalents to discuss major EU proposals with their ministers, prior to meetings of the Council of Ministers (Europe Documents 1995). In addition, broad European Parliamentary agreement has been reached on limiting the number of MEPs to 700 (European Documents 1995), their distribution both reflecting population size and ensuring the continuing adequate representation of small countries. It is inconsistent, however, that the Parliament, which is supposed to represent EU citizens, should be so preoccupied with Member State representation.

Other obstacles constrain the development of the Union. Two pillars of the Maastricht Treaty remain outside the scope of formal EU institutions. The responsibility for policy formulation in home and judicial and foreign and security affairs does not belong to the Commission, Member State sovereignty over these policy domains never having been challenged. Any effort to bring these domains into the EU policy apparatus has failed. For example, the Schengen Agreement, which allows for the complete removal of barriers to population movement within the EU (it was agreed in 1985 and came into operation in 1995), has not been implemented by all Member States. Partial co-ordination on these wider issues is so far limited to an intergovernmental framework, rather than an EU institutional framework. It seems likely that

these policy areas will remain the domain for intergovernmental conferences rather than Commission-led policy areas, and that the Council of
Ministers' role will be limited to declarations and recommendations rather
than decisions. One legal problem with this emerging procedure in home
and security affairs is that the ECJ has no legal jurisdiction over intergovernmental activity and decisions outside of the formal institutions. Because
intergovernmental decisions remain outside EU judicial review, there is no
institutional body to determine whether members are correct in their interpretation of decisions, nor to protect individuals affected by activities of the
Union beyond economic union. Another consequence of this arrangement is
that the Commission's role in proposing new Community legislation is slow-
ing: only 12 new Commission legislative proposals were initiated in 1996.
According to one senior official, the Commission's role is now 'only in subordinate legislation, especially in food safety and farming' (*The Economist*
1997b, p. 56).

Conclusion: towards the creation of public legitimacy?

One of the unanswered questions is whether the Union, as an integrated
polity under construction, can attract consent from EU citizens. Until
recently, the legitimacy of the project has attracted little attention. In the
early days, the public did not perceive itself as being directly affected by
Community decisions, and was thus unlikely to protest against these. As a
result, a 'permissive consensus' developed for accepting, although not necessarily embracing, Community policies. However, following the Maastricht
Treaty, increasing EU institutional competence and the trend towards more
majority voting, the need to establish policy legitimacy among the public has
become vital to the survival of the EU experiment.

Legitimacy as a concept is founded on the premise of popular sovereignty,
which holds that the people can be the only legitimate source of power, since
they represent ultimate authority. The universality of this belief is enshrined
in the United Nations Declaration of Human Rights. Legitimacy, which
entails a sense of public obligation to a system of authority, is not just about
establishing more democratic mechanisms or writing more EU law. It also
requires the public to embrace a vision that typically – in the case of the
nation-state – is based on the myth of the imagined community (Anderson
1991). As discussed in Chapter 1, this embodies symbolic values through
which people share an idea of origin, continuity, historical memories, collective remembrance, common heritage and traditions, as well as common destiny. The creation of national popular sovereignty has involved a binding
political process by which one vision becomes hegemonic (Obradovic 1996).

The EU entirely lacks such an overarching origin-myth to underpin its legitimacy, and it may prove impossible to create and rationalize one *ex post facto* because of the plethora of extant national identities. Attempts at creating a community by providing more information, telephone hot-lines and web sites (such as Europa[4]) are scarcely sufficient, nor do they impact on the vast majority of the EU's population.

Even the rule of EU law cannot be viewed as a sufficient legitimizing force, because of the public's low awareness of the legal milieu and especially its lack of knowledge of the ECJ. The Council of Ministers may also be a poor legitimizing vehicle, as Member State governments like to evade responsibility when EU decisions are difficult to deliver domestically. Consequently, they often refer to decisions made by some generalized 'Other', of which that national government is no part. Nor can the European Parliament fulfil a legitimizing function. It is not constituted to be representative of the European electorate, nor has it the powers of a true parliament. The European Commission is viewed as being very distant in geographical and perceptual space from most citizens.

Presently, there is the uneasy understanding that multiple loyalties can coexist in the EU. This rests on the assumption that public legitimacy to the national state in 15 different countries can be subsumed under the umbrella of a stronger EU identity. A nested hierarchy of local, regional and national identities is supposed to fit neatly together. However, this thesis of multiple loyalties is based on a theory of divided sovereignty that has never been adequately formulated and examined. For instance, we do not know if sovereignty can be shared without being lost, or if a European identity can be established at all. It requires something that is a great deal more than common passports and a European flag. The Union currently lacks the tangibility and intelligibility that would enable it to capture the imagination, and therefore gain the voluntary support, of its citizens. To achieve otherwise, the ethos of European integration must be placed in public discourse and Europeans will have to rally behind and identify with bold new steps to integration if the EU experiment is ever to attain public legitimacy. This task will be immensely difficult and time-consuming but remains one that must be placed at the top of the EU agenda.

Notes

1 In January 1997, for the first time, the European Parliament threatened to force a constitutional crisis by censuring the Commission over incompetence in handling BSE – 'mad cow' disease. It was reported that the Commission withheld information from the European Parliament (Barber and Southey 1997).

2 There are of course exceptions to the norm. For example, France vetoed the UK's application for membership for many years. Furthermore, the UK threatened to sabotage all EU business if France did not concede in the Council of Ministers on

moderating its demands, thereby permitting the EU Commissioner to have more freedom in negotiating agriculture in the Uruguay Round of trade talks (Brock 1993).

3 Of the 122 000 callers to the hot-line only 1000 have been British (Southey 1997).

4 http: //europa.eu.int

References

Anderson, B. 1991: *Imagined communities: reflections on the origins and spread of nationalism*, revised ed. London: Verso.

Barber, L. and Southey, C. 1997: MEPs may censure Brussels on BSE. *Financial Times* 24 January, 2.

Bermann, G. A., Goebel, R. J., Davey, W. J. and Fox, E. M. 1995: *1995 supplement to cases and materials on European Community law*. St Paul, MN: West Publishing.

Blitz, J. 1997: Major acts over EU policy deadlock. *Financial Times* 13 January, 6.

Brock, G. 1993: Hurd bolsters threat to sabotage EC business over GATT. *The Times* (London) 9 November, 1.

Chisholm, M. 1995: *Britain on the edge of Europe*. London: Routledge.

Commission of the European Communities 1995: *The European Commission, 1995–2000*. Brussels/Luxembourg: CEC.

Commission of the European Communities 1997: *Agenda 2000*. Com (97) 2000. Brussels/Luxembourg: CEC.

Dinan, D. 1994: *Ever closer union? An introduction to the European Community*. Boulder, CO: Lynne Rienner.

The Economist 1991: A view from the Brussels boiler-room. 1 June, 48.

The Economist 1997a: European Parliament. Looking for legitimacy. 11 January, 49–50.

The Economist 1997b: European Commission. The big squeeze. 8 February, 55–6.

Europa 1997: Internet site at www.Europa.eu.int

European Documents 1995: The European Parliament's stance in the 1996 IGC. No. 1936/37, 25 May.

Fontaine, P. 1988: *Jean Monnet: a Grand Design for Europe*. Luxembourg: Office of Official Publications of the European Communities.

Grant, R. 1993: Against the grain: agricultural trade policies of the US, the European Community and Japan at the GATT. *Political Geography* 12, 247–62.

Grant, R. 1995: From Blair House to the farm house: negotiating agriculture in the European Union. Pew Case Studies in International Affairs, Case no. 167. Georgetown: Institute for the Study of Diplomacy.

Holsti, M. O. 1995: The balance between small and large: effects of a double-majority system on voting power in the European Union. *International Studies Quarterly* 39, 351–70.

Mazey, S. and Richardson, J. 1995: Promiscuous policy-making: the European policy style? In Rhodes, C. and Mazey, S. (eds), *The state of the European Union*, vol. 3. Boulder, CO: University of Colorado Press, 337–59.

Moravcsik, A. 1993: Preferences and power in the European Community: a liberal intergovernmental approach. *Journal of Common Market Studies* 31, 473–524.

Obradovic, D. 1996: Policy legitimacy in the European Union. *Journal of Common Market Studies* 34, 191–221.

Southey, C. 1997: UK left in the dark about EU rights. *Financial Times* 16 January, 2.

Swann, D. 1996: *European economic integration: the Common Market, European Union and beyond.* Cheltenham: Elgar.

Teasdale, A. L. 1996: The politics of majority voting in Europe. *Political Quarterly* 67, 101–15.

Van Schendelen, M. P. C. M. 1996: The Council decides: does the Council decide? *Journal of Common Market Studies* 34, 531–48.

Williams, R. H. 1996: *European Union spatial policy and planning.* Liverpool: Paul Chapman.

| 6 |

Convergence, cohesion and regionalism: contradictory trends in the new Europe

MARK HART

Introduction

The origins and idealistic motivations for European integration in the 1950s have lost some of their resonance during the intervening four decades. The original objectives to prevent another war in Europe and create a bloc against the Communist East, while countering the economic power of the United States, have, more recently, been overshadowed by the emphasis placed on economic integration. Now, however, Europe is being driven by a political agenda which seeks to move beyond the SEM towards greater political union – perhaps even a supranational 'United States of Europe'. The 1990s are proving to be an important watershed in the history of the EU. As explained in the preceding chapter, the programme to complete the internal market, which was launched in 1985, ended in December 1992, abolishing many of the remaining non-tariff barriers to the free movement of goods, services, labour and capital. The Maastricht Treaties for Political, Economic and Monetary Union, signed in December 1991, committed the EU to the introduction of a single currency by no later than 1999. Following a brief overview of the contemporary political and economic challenges confronting the EU, this chapter provides a detailed examination of the complex relationship between geographical, economic and political change. The aims of this analysis are twofold.

First, the nature of the emerging spatial structures within the 'New Europe' is examined, particular attention being given to the concepts of convergence and cohesion. Socio-economic convergence involves an equalization of basic incomes through higher GDP growth, competitiveness and employment, while cohesion policies are concerned with improving the quality of European citizenship through measures that combine the promotion of solidarity, mutual support and social inclusion with sustainability of economic growth (Commission of the European Communities 1996). The chapter offers an assessment of the ability of EU Member States to achieve cohesion as a basic condition for further economic and social progress. As Hall and Van der Wee (1995, p. 14) argue, 'Cohesion [can] no longer be viewed as an accompanying measure to the completion of the internal market, as in the past, but as an integral component of the Community's political, economic and social development.'

Throughout the 1990s, considerable resources have been allocated to structural policies that promote cohesion. These have been further enhanced during the period 1994–9 with, for example, just under ECU 30 billion being allocated in 1996. The mid-term review of progress since 1994 has yet to be published (it was due in late 1997) and therefore the official Commission assessment of the impact of Structural Funds (including the Cohesion Fund) remains unknown. (The Cohesion Fund is not strictly one of the Structural Funds, but was introduced as part of the broader financial package accompanying the 1993 reforms of those funds.) However, great care must be exercised in assessing the impact of resources allocated to promote cohesion. There can be no simple relationship between any change in the relative performance of particular regions, and the effects of either EU interventions or the completion of the Single Market. Any discussion of the effects of EU policies upon Member States must be viewed within the context of the broader processes of economic change, which operate at a range of spatial scales.

The second major aim of the chapter is to examine the tensions and contradictions between the politically inspired project of European integration, and the rise of regionalism against a background of continuing uneven geographical development. The discussion will focus on the many threats to cohesion emerging in the watershed decade of the 1990s. The important question to address here, as already mentioned, is the extent to which the Commission's celebration of 'Europe of the regions' actually runs counter to the parallel project of achieving social and economic cohesion. According to the Commission, 'Cohesion and diversity are not conflicting objectives, but can be mutually reinforcing' (Commission of the European Communities 1996, p. 15). These are fine sentiments but they depend on exactly what is meant by diversity. The regionalist, or even separatist, movements found in virtually every nation-state within the EU may yet constitute a very real threat to cohesion if integrative federalist principles are not the shared political objective.

Contemporary political and economic challenges for the EU

That the EU is undergoing a process of great transition is without question. However, that transformation is not a smooth one, the path towards greater economic and monetary integration being strewn with a multitude of internal and external obstacles. By late 1996, for example, the timetable for monetary union was in disarray, with the partial breakdown of the Exchange Rate Mechanism (ERM), coupled with the emergence of increasing opposition from some Member States (most vociferously the UK). Rising unemployment in the reunified Germany and the defeat of right-wing governments in the UK and France in the spring of 1997 cast further doubts on the ability to meet the original deadline for European Monetary Union (EMU).

Outside the EU, the political and economic collapse of Communism in the nations of Central and Eastern Europe and their attempts to convert to a market economy and representative democratic system have unquestionably presented the West, and the EU in particular, with one of the greatest challenges of the late twentieth century, and indeed the next century. As the Union moves towards greater economic integration, the rapidly changing context further east casts doubt on many of the predicted scenarios concerning the path towards economic and social cohesion. The unification of Germany and the general westward reorientation of the countries of Central and Eastern Europe has created tensions in the 'European economic space' which demand immediate responses. These changes have added many of the nations of Central and Eastern Europe to the rump of European Free Trade Association nations applying for EU membership. As we have seen in Chapter 5, the 1997 EU of 15 Member States might increase to 26 countries during the early twenty-first century. This rapidly enlarging 'European economic space' will pose fundamental questions concerning the nature of economic integration that can be sought and attained at this pan-European level. For example, the currently 'agreed' agenda for EMU, and the search for monetary stability among the 15 Member States, would seem totally inappropriate in the context of current economic developments in Poland and Russia.

General agreement exists that the move towards greater European integration will have a profound impact on the current 15 Member States of the EU and their peoples. However, the precise nature of that impact remains the subject of intense political and academic debate. One area of notable apprehension, essentially the subject matter of this chapter, concerns the relationship between the move to greater integration, and existing spatial patterns and structures of economic and social well-being across Europe. To what extent can the long-standing pattern of regional disparities within the EU be ameliorated with the advent of economic integration? Many commentators and experts (e.g. Cecchini 1988; Delors 1989) have either paid scant attention to the spatial effects, or simply assumed that the trickle-down effects of European economic growth associated with the creation of the SEM would automatically lead to regional convergence. Fortunately, such simplistic reasoning has now been replaced with a more considered investigation of the regional impact of economic integration (Bachtler and Clement 1992; Bachtler 1995; Leyshon and Thrift 1995; Quevit 1995). In broad terms, these studies conclude that as a result of the completion of the internal market, the prospects for the regions are not encouraging in that an increased level of economic activity is being concentrated in a smaller number of places, where cost reductions and scale savings can be realized. As a result of such analyses the decision was quickly taken to double the resources available under Structural Fund interventions (Quevit 1995). Despite this being the appropriate response on the grounds of equity, however, the question remains as to whether there has been a more favourable economic performance in the regions concerned. We return to this issue later in a more detailed discussion of the convergence question within the EU.

The publication by the European Commission in July 1997 of *Agenda 2000* provided further evidence that the process of enlargement will have a profound impact on the future of EU finances and policies (Commission of the European Communities 1997). The document contains outline proposals for changes to the current arrangements for the operation of the CAP and the Structural Funds (European Regional Development Fund – ERDF, and the European Social Fund – ESF) in the 15 Member States. The pressure for change stems both from the results of the internal reviews of the efficiency and impact of these policies, and from the impending membership of certain Eastern European countries to the EU with their associated demands for financial assistance. One anticipated outcome of these proposals will be a rationalization of the six Objectives currently acting as guidelines in the allocation of Structural Funds (for instance, Objective 1, which promotes the development and structural adjustment of regions whose development is lagging behind; Objective 2, which covers the regions or parts of regions seriously affected by industrial decline). The precise implications for the regions of the present 15 Member States have not yet been assessed, but it is difficult to anticipate anything other than a change in their priority status as other, more deprived regions within an enlarged EU become eligible for assistance.

Alongside the recognition of the potential spatial unevenness of the effects of greater economic integration, it is apparent also that the emerging spatial structures within the EU, or new industrial spaces, are linked intrinsically to the variety of economic forces currently at work in each of the Member States and their regions. The globalization of production and financial markets and rapid technological progress have led to far-reaching changes in national and regional economies, patterns of employment and the organization of work. What is clear from the process of globalization is that, despite a well defined set of regional economic networks (old and new) within the EU, there has been a parallel movement towards the organization of enterprises and financial institutions on an international scale. One of the key concepts associated with these processes is that of increased *flexibility* of production techniques and technologies, industrial structures, capital, work practices and the labour market, trends that have defined the nature of economic change during the past twenty years.

Whether this increased flexibility within the international economy may be interpreted as evidence to support the emergence of a new regime of accumulation remains a matter of considerable debate (Piore and Sabel 1984; Allen and Massey 1989; Benko and Dunford 1991; Lipietz 1993; Dunford and Perrons 1994). Has there really been a break with the past? The central issue in the debate is the extent to which one accepts the claim that Fordism (see Chapter 2) can reconstitute itself by incorporating increased flexibility in the production process and in the organization of the workplace.

Flexible production processes would appear to be closely associated with the creation or revitalization of regional economies as, according to Sabel (1989, p. 22), these areas, or industrial districts, 'escape ruinous price competition with low-wage mass producers by using flexible machines and skilled workers to make semi-custom goods that command an affordable premium in the market.' Among the best-known European examples of such industrial districts is the 'Third Italy', which was heralded as one of the regional 'success' stories of the 1980s, although the 1990s have witnessed a reversal of past growth performance (Bellini and Pasquini 1996). The characteristic feature of the Third Italy was the emergence of powerful economic networks of mainly small enterprises, spatially concentrated in the Emilia-Romagna region in the north-east of the country.

Alongside flexibility, the issue of competitiveness has come to dominate debates on industrial development and policy in recent years. Francis (1992) identifies two broad approaches in its conceptualization. The first is concerned with the analysis and explanation of competitiveness at the level of the individual firm or industry, while the second investigates the link between competitiveness and national and regional economic performance. The competitive position of regions is seen by many observers as a key element in patterns of trade and national economic performance (Porter 1990; Krugman 1996). For Porter (1996), this competitive position is developed by regionally concentrated clusters of firms and industries, connected by specialized buyer–supplier relationships, or related by technologies or skills. Further, as Krugman (1991, p. 70) has argued, the economic benefits to firms of locating in close proximity stems 'from the standard Marshallian Trinity of labour market pooling, supply of intermediate goods and knowledge spillovers'.

The Commission's 1993 White Paper, *Growth, competitiveness and employment*, underlined the importance of the concept of competitiveness to the cohesion process (Commission of the Economic Communities 1993). As a consequence, initiatives designed to improve regional competitiveness have been placed at the centre of national and European industrial and regional policies. However, a fierce critique has emerged recently of the undue emphasis that has been placed upon competitiveness, especially at the level of the regional economy (Group of Lisbon 1995; Krugman 1996). The excessive concentration placed on competitiveness in a world of increasing globalization is leading to increased socio-economic inequalities and unacceptable levels of unaccountable power in multinational corporations. What are the mechanisms which result in these intensified uneven spatial structures? According to the Group of Lisbon (1995, p. 120),

> the ideology of competitiveness and the corresponding process of delinking between the most developed and the poorest cities, regions and countries of the world constitute severe obstacles to the emergence of regionally integrated units in the less-developed regions. There is no real alternative, however, for these categories of regions than regional integration within the context of a co-operative global framework.

In other words, competitiveness instils a 'winner takes all approach', which is destructive to any attempt to move towards greater economic and social cohesion.

The Single European Act and the Maastricht Treaties identified the regions as key actors in a stronger and more integrated EU, but the question remains whether regionalism, and its obvious expression in autonomous regional government, can actually serve the broader project of economic and social cohesion. The importance of the role of decentralized governance structures – such as regional governments and development agencies – in the success of internationally competitive regional networks constructed around the principles of flexible production processes has been the subject of much discussion (Benko and Dunford 1991; Dunford and Hudson 1996). To return to the example of the Third Italy, it has been suggested that its success was heavily dependent upon the process of institution-building, which in turn had its roots in politics and the construction of political alliances in the region (Best 1990). More generally, Best argues that successful industrial development is the result of a 'new competition', based on new production and organizational concepts within firms (including 'just in time' production methods, and computer-aided design and manufacture), and the means by which these interact with external conditions, such as systems of industrial relations, to determine and influence regional competitiveness and development.

None the less, despite some overlap between the rise of regionalism and the emergence of strong regional economic networks, an obvious tension remains in the notion of a 'Europe of the regions'. On the one hand, the continent as a whole is moving towards economic and political integration, while on the other, there has been a resurgence of regional identities within a number of nation-states, sometimes accompanied by a transfer of power to the decentralized authorities. At a basic level there would appear to be a conflict between the logical outcomes of these two processes. The rise of regionalism is seeking to bring power closer to the people of the region, while greater integration is, on the basis of experiences to date, designed to produce a transfer of power to new centres, which by definition are further away from the people of the regions. Instead of advocating the benefits of a 'Europe of the regions' it would be more appropriate, perhaps, to talk in terms of a 'Europe against the regions'.

Regional disparities

The enlargement of the Community from six to 15 countries has been accompanied by an increase in the scale of socio-economic differences between Member States and their regions. The standard indicator of national and regional economic performance is GDP. At the national level those countries significantly above (i.e. 10 per cent or more) the Union average in GDP per capita in 1995 include Belgium, Denmark, Luxembourg and Austria. Germany, France, Italy and the Netherlands were above average, the UK was around average and both Sweden and Finland were slightly below average. That leaves four Member States – Greece, Ireland, Portugal and Spain, commonly referred to as the 'Cohesion Four' – which had GDP per capita figures of between 40 and 10 per cent below the EU average (Commission of the European Communities 1996). This gap is considerably wider now than at the time of the first enlargement of the Community, in 1973 (Commission of the European Communities 1996). The EU exhibits a distinctive core–periphery pattern in regional GDP per capita disparities, with the more prosperous regions located close to the geographical centre of the Union (Figure 6.1). Other indicators of regional economic and social well-being, such as unemployment and net migration rates, clearly support this broad core–periphery pattern of EU regional disparities.

The major social and economic challenge confronting the EU must surely be the problem of persistently high levels of unemployment (Figure 6.2). National unemployment rates in 1995 varied considerably, from less than 5 per cent in Luxembourg and Austria to 15 per cent or more in Spain (the highest in the EU) and Finland. Unemployment rates in France and Italy, at nearly 12 per cent, were also above the EU average. At the regional level, using 1994 data (the latest date for which disaggregated regional data are available at NUTS Level 2 – EUROSTAT 1995), the variations were even more acute, with many areas in southern Italy, Spain and Finland experiencing unemployment rates almost double the EUR12 average for that year of 11.4 per cent. Examples include Calabria, 21.1 per cent; Campania, 19.8 per cent; Andalusia, 32.4 per cent; Estremadura, 30.3 per cent; Comunidad Valenciana, 23.9 per cent; Pohjois-Suomi, 20.8 per cent; and Ita-Suomi, 20.4 per cent.

Disaggregated by gender and age, the geographical pattern of unemployment becomes even more complex. In April 1994, 10.2 per cent of men were out of work compared to 13 per cent of women. This trend was not consistent across the EU, with certain national and regional labour markets, notably in the UK and Sweden, being characterized by a female unemployment rate much lower than that for their male counterparts. In the rest of the EU, the gap between female and male unemployment rates varied enormously, the greatest differentials being found in Italy, where in regions such as Campania the unemployment rate for women was almost twice that for men. Youth unemployment also continues to pose major problems for Member State

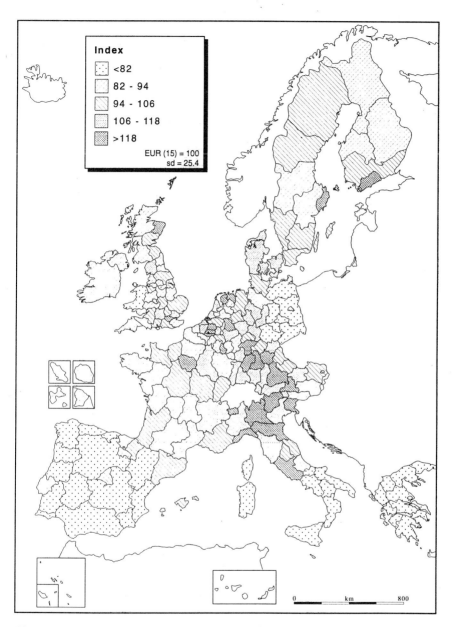

Figure 6.1 European Union: GDP per capita by region (purchasing power standards), 1993
Source: Based on CEC (1996, p. 22)

governments. Unemployment rates among the under-25-year-olds in the majority of Spanish, southern Italian and Finnish regions followed the pattern for total unemployment and – at some 40 per cent or even higher – were well above the EUR12 average of 21.6 per cent. However, above-average unemployment rates for this age group were also experienced in parts of Belgium, in large parts of France and in Ireland (EUROSTAT 1995).

One attempt to translate these socio-economic indicators into a more generalized spatial structure was presented in Figure 2.1 and is one that has become familiar in recent years. Europe's regional core, or vital axis, it is argued, runs from London and south-east England through the Paris basin, Benelux, southern Germany, Austria, Switzerland to northern Italy and south-eastern France. The centre of gravity of this axis is located in former West Germany. Set against the summary evidence in Figures 6.1 and 6.2 it would appear that there is considerable validity in the stylized spatial structure presented in Figure 2.1. What is clear from such a spatial structure is that the periphery of Europe can no longer be presented as the Mediterranean countries and Ireland, but also includes large areas of western France, England, northern Germany and Denmark.

Employing the REGIO database for NUTS 2 regions, Dunford and Hudson (1996) have undertaken a more detailed examination of regional disparities throughout the EU than space permits in this chapter. One of the most interesting conclusions to emerge from this analysis, which is pertinent to the cohesion debate, is the degree of spatial polarization evident within each Member State. This can be characterized in a number of ways. First, an increasing divide exists between a 'hierarchical network metropolis' (Dublin, Barcelona and Paris are good examples), where there are concentrations of skilled labour, high-level service activities and new technologies, and the rest of the national territory. In other words, high-level, high-status jobs are concentrated in metropolitan areas along Europe's 'vital axis', while peripheral areas make do with low-level jobs. Second, these regional and inter-urban differences are overlain by a finer-grained set of intra-urban inequalities, which exist equally in both core and less developed areas. These arise out of a combination of rising unemployment and the demands of high-growth sectors in these urban areas for routine consumer services and low-level manufacturing goods. As a result, a demand exists for 'insecure' work that can be characterized as part-time, short-term and informal. In certain national contexts, such as the UK, these trends have been intensified by political actions in areas such as regional policy, tax and welfare. As Dunford and Hudson (1996, p. 28) conclude, 'The result of these economic and political trends is an intensification of socio-spatial segregation and a differentiation and polarization of lifestyles within and between cities and regions.'

How best can these manifestations of regional economic disparity in the EU be explained? One of the most important determinants of regional economic performance is the interaction between regional employment rates and productivity differentials. An important study of this dynamic (Dunford

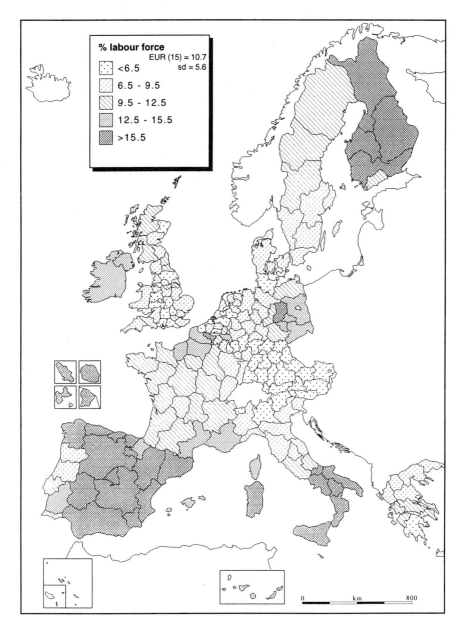

Figure 6.2 European Union: unemployment rates by region, 1995 (Greece and French Dominions, 1994)
Source: Based on CEC (1996, p. 26)

1996) demonstrates that the ability of a region to mobilize its human resources (the employment rate) contributes to regional disparities in GDP. Simply, the important point to grasp here is that high productivity rates and high employment rates go hand in hand in generating regional economic growth. Setting this principle against the evidence on productivity and employment rates throughout the EU, Dunford demonstrates that the strongest regions can indeed be characterized by the combination of high productivity and employment rates. To use the UK as an illustration, low levels of productivity demonstrate that, while a large proportion of the population work, they are not doing so very productively.

The importance of high rates of employment to the process of regional economic growth sets a challenging agenda both for the Commission, and its project of economic and social cohesion, and for the governments of Spain, Italy and Finland, which appear to face the greatest problems in attempts to raise employment rates. On the theme of economic and social cohesion in the EU, it is important to reach some sort of assessment of whether, over the lifespan of the Community, it is possible to determine if convergence or divergence has been taking place and at what spatial scale. From the above discussion it would appear that the core–periphery spatial structures presented for the EU of the 1990s differ very little from those identified over twenty years ago at the time of the first enlargement in 1973. What has actually changed? We now turn to examine the evidence on regional performance over time.

The convergence debate

The ability of certain regions to improve their position, relative to the wider national or Europe-wide contexts, has only served to reinforce the view that the map of inequality should not be regarded as a static one. As Dunford (1993, p. 742) argues, 'Its shape and intensity have changed, as networked metropolitan economies have reinforced their position, as formerly dominant economies have undergone relative decline and as new areas of growth have emerged.' But does this changing map lend weight to the view of orthodox trade and regional growth models which underlie the convergence thesis?

The importance of the cohesion objectives, and indeed the impact of the Cohesion Fund itself to the harmonious development of the EU, as set out in the Maastricht Treaty, has led to numerous studies which seek to establish whether economic and social disparities between Member States, regions and social groups have narrowed over time. The search for an answer to this question has fuelled a heated debate as to whether growth and economic integration in the EU will lead to a narrowing or widening of regional disparities. This 'convergence–divergence' controversy has its foundations in quite diametrically opposed theoretical camps within economics and its associated disciplines.

The neo-classical theory of *international* relations holds that economic integration, through the increased movements of goods, services and the production factors of labour and capital, tends to equalize factor returns and hence incomes. It is argued that trade in goods alone, irrespective of the freedom of factor movement, will achieve equal factor prices. At the international level, therefore, the economic development process will lead to convergence of income levels. By contrast, the neo-classical account of *inter-regional* relations stresses the mobility of capital and labour rather than the trade of goods and, as a result, argues that economic growth is faster in more peripheral regions than in core regions, a trend which, in turn, fuels the convergence process. In addition to the importance of capital mobility and labour migration to these neo-classical convergence theories of regional growth, there are three other equilibrating mechanisms. These are, first, technology diffusion from the more techno-logically advanced regions to their lagging counterparts; second, capital accumulation, whereby regions with high productivity and income levels per capita find that growth is constrained owing to the onset of diminishing returns, and therefore lagging regions, starting from a lower base, can grow more rapidly; third, free trade, which allows regions to concentrate on the production of goods and services in which they have a comparative advantage and redress the problem of surplus labour through increased export activity.

The alternative school of thought emphasizes mechanisms that lead to a greater divergence in the economic trajectory of regional economies over time. Proponents of this particular school have a long tradition in Europe and have developed the early work of Myrdal (1957) and Hirschmann (1958). The modern cumulative growth theories stress the role of Verdoorn's law, which holds that regions that develop early reap additional productivity gains through scale economies and specialization of production (Kaldor 1970; Dixon and Thirlwall 1975). In the 1990s, these views have been developed further by 'new growth economics', which stressed the importance of tech-nological change and skills acquisition as the key cause of cumulative growth (Crafts 1992). Advanced (core) regions perpetuate the gap by concentrating on the development of high-value new products, while the less developed regions are dependent upon more mature production systems and products (Molle and Boeckhout 1995).

It is frankly impossible to reconcile these two theoretical perspectives on the convergence–divergence debate, with each side interpreting the recent trends in regional economic development in the EU to support their particu-lar viewpoint. The problem lies in the dilemma that, over the past three decades, the process of regional economic development has provided evi-dence of both divergence *and* convergence. This has led some commentators to talk in terms of consecutive periods of divergence and convergence (see, for example, Barro and Sala-i-Martin 1991). They argue that periods of widening disparities are a function of short-term cyclical shocks (national or regional), which take time to be absorbed before the overriding trend towards convergence eventually returns.

Others, however, prefer to explain the trends by using a framework which, in effect, acknowledges the possibility of a simultaneous process of convergence and divergence (Boltho and Holtham 1992). In other words, some regions are more successful than others over similar time periods at harnessing the essential requirements for growth. These more successful regions benefit from a combination of factors, which include, for example, market access, human capital, technological change, international competitiveness, economies of scale, public infrastructure and institutional efficiency. The recent study by Dunford and Hudson (1996) illustrates very well the importance of the structure and dynamics of the governance system, regional innovation systems, internationalization, co-operation and networking, and institutional performance in the more successful regions of the EU during the 1990s.

Having briefly set out the opposing theoretical views on the convergence debate, I now turn to the task of reviewing the empirical evidence on the trends in regional disparities within the EU. As Armstrong (1995a, p. 259) states,

> While the existing pattern of regional disparities within the EU suggests that forces in the past may have systematically favoured the more central regions, this does not necessarily mean that current growth is cumulative. If the poorer peripheral regions are catching up with the more prosperous core of the EU, but the process is a slow one, then the neoclassical explanation may be compatible with the evidence.

At a national level, data on GDP per capita (measured in purchasing power standards – PPS – to allow for differences in the prices of goods and services in different areas) indicate that the convergence observed between Member States during the 1960s and early 1970s was sharply reversed in the late 1970s and early 1980s (Dunford 1993; Dunford and Hudson 1996). The early convergence period coincided with a period of quite unprecedented fast economic growth, while the widening took place against a background of much slower economic growth. As Dunford and Perrons (1994) are quick to point out, however, the causal link between these trends is not unidirectional, in that, for example, divergence actually contributed to slower growth through increased unemployment and lower incomes.

Recent evidence has indicated that, over the 1950–85 period, regions within the EU have experienced convergent growth in GDP per capita, albeit at a very slow rate of 2 per cent per annum (Barro and Sala-i-Martin 1991). This study further concludes that the rate of convergence among regions *within* Member States was similar to the rate of convergence *between* Member States. One of the major weaknesses of this study was its restriction to regions in the former West Germany, France, the UK, Belgium, Denmark, the Netherlands and Italy. Armstrong (1995b) revisited the dataset for the slightly longer 1950–90 period and extended the geographical coverage of the analysis to include Ireland, Portugal and Luxembourg as single spatial entities, together with some regions in Spain and Greece. His study supports the conclusions of Barro and Sala-i-Martin (1991) and confirms a process of convergence across the Member States, albeit at a significantly lower rate of 1 per cent per annum. This can best be explained by the inclusion in the analysis of the disadvantaged Southern European countries. The results also confirm that rates of convergence, particularly within each Member State, have fallen since their peak in the 1960s. Overall, therefore, the general conclusion to emerge is that convergence is under way but at a very slow pace, while, simultaneously, wide disparities in economic and social well-being continue to exist at the regional level.

Employing a different methodology and a shorter time period (1977–91), Dunford and Hudson (1996) demonstrate that, since the early 1970s and for almost all of the 1980s (until 1989), a widening of the degree of inequality has been taking place at regional level within each of the EUR12 countries. This would tend to confirm the earlier finding of Armstrong (1995b) that within-country disparities have been widening. As noted earlier, however, what is interesting is not simply that already strong regional economies are getting stronger and weaker ones declining, but that 'new' growth regions are emerging which complicate even further an already complex spatial structure. Examples include Friuli–Venezia Giulia and Lazio, both in Italy.

One of the questions which these empirical studies seem unable to answer, however, is the extent to which the factors set out in the neo-classical models are responsible for the convergent growth observed over the three decades to 1990. A detailed study of the emerging successful regions, such as those referred to above, would offer opportunities to investigate the processes of growth in greater depth and to move beyond the range of factors specified within the neo-classical and cumulative models of regional growth. An analysis of this form would permit consideration of the adequacy of new economic growth theory in the context of regional development. Unfortunately, despite the increasing evidence to the contrary, the position of the neo-classical growth models appears unchallenged at the centre of the policy formulation process within the Commission.

Another question which emerges from the evolving spatial structure is the net effect that national programmes of regional policy assistance and/or EU intervention policies (e.g. Structural Funds) have had on the growth process. That is, without these interventions what might have been the rate at which the inequality gap closed in the 1950–90 period? Further, it must be accepted that the impact of initiatives set in motion by the Maastricht Treaty upon the pattern of regional disparity remains largely unknown. One reason for this is that the nature of the economic union itself is not well defined in terms of macro-economic policy. Apart from the encouragement to adhere to the 'stability package' for EMU, there appears to be no overall co-ordinated economic policy.

In the *First report on economic and social cohesion*, the Commission itself now accepts that in the 1990s the patterns of economic and social cohesion *within* most Member States experienced a reverse (Commission of the European Communities 1996). Against this background of widening disparities, the most extensive of which still exists at the level of the EU itself, the Commission remains determined to maintain and strengthen the EU's structural policies for promoting greater economic and social cohesion. However, nothing is contained within that *First report* to indicate that this is any more than a hope as distinct from a clear and co-ordinated strategic framework for action. This criticism is grounded on the premise that until the process of regional economic growth is recognized to be more complex than the simple equilibrating mechanisms of neo-classical thinking, the net effect of the package of cohesion policies will be less than optimum.

Cohesion and regionalism

The previous section began to highlight some of the many problems associated with the convergence process in the EU. We now return to the theme of a 'Europe of the regions' in order to examine in greater detail the potential threats, which may serve to interrupt the cohesion process stemming from the rise of regionalism and the redefinition of the nation-state. The key question here is whether regionalism and the associated institutionalization of regional government are in the best interests of European integration and, in particular, the promotion of economic and social development. The advocates of regionalism would argue that it helps promote democracy, administrative efficiency/accountability and participation in the policy process of regional interests. As discussed earlier, many commentators place a great deal of significance on the existence of decentralized governmental institutions in stimulating economic growth, although very few studies have sought to identify the precise ways in which decentralized governments actually achieve this goal.

In recent years, several European countries have undertaken a process of administrative decentralization which have created subnational tiers of government with associated new powers. Within the UK, for example, the establishment of Government Offices in the Regions (GORs) under the Conservative government in power from 1979 to 1997 is one indication of such trends. Most EU Member States now contain some expression of regionalist or, more rarely, separatist movements. Examples include the Scottish Nationalist Party, the Lega Lombarda in Italy, the separatist groups in Corsica and the Basque 'nationalist' movement in Spain. Increasingly, such groups or movements are looking to the EU as a means of 'opting out' of their own national political structures, with which they have little affinity. Indeed, many regional authorities have created their own representations to European institutions, which serve to highlight the interests of the regions they are representing.

Nevertheless, it has to be stated that, overall, European states are not at all well advanced in terms of having regional or federal structures.

In an attempt to assess the impact of regional government activity on the process of regional economic development, Leonardi and Garmise (1992) undertook a study of subnational political structures and politicians and related the results to the regional pattern of economic performance over the period 1981–7. The results generally support the argument that the strongest regions (as measured using the REGIO database on GDP), found in Germany, have the most developed regional institutions with the most widespread powers. A closer examination of the Italian data revealed some of the most dynamic regions in the EU, all characterized by regional governments that had been growing in institutional competence since the early 1970s. The corollary of this pattern was that the weakest regions, located in Greece, Ireland and Portugal, have regional bodies but in name only. In the UK, a number of regions were declining rapidly during the 1980s, a trend that coincided with a move towards greater centralization of power from subnational institutions. The implications, therefore, are that decentralized and competent forms of regional government can make a positive contribution to economic growth. From this discussion one can argue that a great deal of overlap exists between the consequences of strong regional institutions (i.e. enhanced economic performance) and the objectives of the Commission to promote greater economic and social cohesion.

Two further concepts are relevant to this discussion on regionalism and the clamour for democratic regional institutions. They arise out of what Teague (1995, p. 150) calls the 'project to recast the EU along integrative federalist principles'. The first is subsidiarity and the second is the issue of fiscal transfers. As we have seen in Chapter 5, subsidiarity, in its widest definition, should be interpreted as a set of guiding principles which govern the shared responsibilities of the individual and society as a whole, and encourage individuals and groups to take an active part in society (Teague 1995). In return, individuals enjoy both freedom and protection against poverty with the guarantee of a minimum standard of living – in other words, the Social Chapter. The Commission defines subsidiarity as follows (Curwen 1995, p. 204):

> in areas which do not fall within its exclusive competence, the Union shall take action, in accordance with the principle of subsidiarity, only and insofar as the objectives of the proposed action cannot be sufficiently achieved by the Member States and can therefore, by reason of the scale or effects of the proposed action, be better achieved by the Union.

The arguments over the British opt-out of the Social Chapter illustrate very graphically the problems surrounding the concept of subsidiarity. The UK government felt that the EU should only do things which the UK cannot do at all, an interpretation that precludes any attempts by the Commission to introduce new social policy legislation.

At a very broad level, the concept of subsidiarity is embedded within discussions on the impact of the rise of regionalism on the territorial aspects of European governance. In simple terms, the debate at the core of the 'Europe of the regions' thesis concerns the extent to which regions can become the dominant unit of government below the Commission institutions, 'thus superseding the national states, which will gradually atrophy as the principle of subsidiarity renders them obsolete' (Sharpe 1993, p. 33).

More generally, subsidiarity has become associated – in the UK and Ireland, at least – with the idea of facilitating the decentralization of administrative decision-making. This is certainly consistent with the views of those who seek greater regional autonomy. However, two caveats arise. The first and more important is the availability of adequate resources, human and financial, to permit the implementation of the decisions taken at local level. Second, it must be accepted that certain administrative functions of government are best retained at the centre as they benefit from specialization, economies of scale and other positive externalities. Consequently, a range of policies exist – including such areas as transport and the environment – which the Commission would prefer to control from the centre. Therein lies one key source of tension between the Commission and the regionalist and federalist camps.

Yet another area of tension concerns the concept of fiscal transfer (interregional resource transfers such as those which exist in nation-states and federal systems) and its ability to act as a binding agent in the cohesion process. This factor operates between geographical areas and communities in order to ensure that the differentials in incomes and well-being are ameliorated. Within the EU, however, there is a transparent difficulty with the concept of fiscal transfer in that, as Mackay (1995, p. 229) states, 'solidarity, unlike trade, has problems in escaping national boundaries'. In other words, although the process of economic convergence is in part dependent upon fiscal transfers between the different Member States, it demands a clear political will to ensure that these transfers actually take place. Therein lies the difficulty, as political will is almost always likely to be stronger within national boundaries. It has been claimed, for example, that the neglect of fiscal transfer may halt the integration process, or even dissolve the Union.

In effect, the kinds of fiscal transfer mechanisms found in many other federal countries (such as the United States, Canada and Australia), which ensure that poorer regions benefit disproportionately from federal expenditures, are largely absent within the EU. The Union lacks many of the tax-raising powers so common in federal systems elsewhere, which exhibit high ratios of consolidated central government taxes to total taxes. For example, in 1988, this ratio was 76.5 per cent in the United States, while the corresponding figure for the EU was 2.6 per cent (Armstrong 1995c). In these circumstances it is not surprising that the EU has had little choice but to rely upon regional policy to reduce regional disparities, rather than a system of fiscal transfers from richer to poorer areas.

Regionalism can also interact to frustrate the fiscal transfer process and, ultimately, the convergence process, by insisting on complete autonomy over the manner in which the additional resources are deployed. As Mackay (1995, p. 230) concludes, 'fiscal transfers compensate for the loss of independence, but their counterpart is a mature political structure'. So while fiscal transfers are seen as an important ingredient in the convergence process, the existence of strong regionalist movements may delay their introduction.

Finally, this debate on the tensions between the project of European integration and the demands of the regionalists needs to recognize the distinctions and relationships between *regionalism*, representing the political and economic mobilization within a specific territorial unit, and the *regionalization* of European policy, which has provided an environment within which the notion of a 'Europe of the regions' has been encouraged (Keating and Jones 1995; Christiansen 1996; Farrows 1997). It is argued that, even though the rise of regionalism in recent years is now being reinterpreted and presented as a reaction to European economic integration, the growth of the EC's regional policy since the late 1960s, and particularly since 1988 with the reform of the Structural and Cohesion Funds, had already provided a catalyst for the development of regionalism. For example, the introduction of such guiding principles as additionality, programming and partnership to the operation of the Structural and Cohesion Funds provided opportunities for local and non-state actors to participate in the policy process at a level that was not previously possible. Furthermore, Article 10 of the ERDF emphasizes the value of inter-regional co-operation which transcends state boundaries, and the growth of these informal networks, as representing an important source of influence for the regions.

Since the signing of the Maastricht agreement, a clear political and institutional dimension has been added to the regional debate, which seeks to control more formally the developments at subnational level. As indicated above, the principle of subsidiarity is a potential source of greater subnational involvement in an emerging framework of European governance, although the effects of the operation of the principle are not yet clear. Finally, the establishment of the Committee of the Regions seeks to institutionalize regional interests across the EU and provides an opportunity for a greater regional involvement in the policy process. However, an early review of its activities and influence would suggest that it has some way to go before it is firmly identified with the regionalist movement in the EU (Farrows 1997). At present it seems to be carrying out the functions of a consultative or advisory body with no real power in the policy process. The problem would appear to be twofold. First, the process of European integration has not yet reached a stage sufficiently mature to allow real transfers of power to regional levels of government at the expense of national governments. Second, the complexity and diversity of development at regional level constitutes an almost impossible agenda for the Committee of the Regions.

Conclusion

The chapter began by outlining the nature of the challenges confronting the EU at the end of the twentieth century and expressed concern over the ability of the process of European integration to resolve the twin problems of regional economic inequalities and social exclusion. The widening and deepening of the EU is creating economic, social and political tensions, which may yet prove to be even more damaging to the prospects for regional convergence. Unfortunately, the discussion has provided no clear assessment about the way in which the spatial structure in the EU will evolve in response to these internal and external pressures. We have been heavily reliant upon identifying the key regional trends over the previous three decades and, in so doing, attempted to understand more effectively the processes determining regional economic growth performance.

In simple terms, the convergence process is a constantly adjusting trade-off between the flows of capital to the disadvantaged regions and the flows of labour to existing core regions. Convergence of GDP per capita can occur through the operation of either mechanism or a combination of them both. The unknown in all this is the nature and scale of the effects of these flows upon the performance of individual firms, industries and regions. From the evidence and views presented in this chapter, it would be difficult to disagree with the sentiments of Shepley and Wilmot (1995, p. 61) when they stated, 'We know that regional centres of industrial production, once established, become very hard to shift.'

Certainly, some convergence has taken place but only at a very gradual pace, and greatly assisted in the 1960s and early 1970s by unprecedented levels of economic growth. The most worrying trend, however, has been the growing inequality gap between regions within each Member State. At a time of great uncertainty the persistence of this process of divergence at national level is a major problem for the EU. The regional implications of EMU are as yet not fully understood, but they would be unlikely to benefit disproportionately the lagging regions of the EU on the grounds that these start the transition towards EMU with unfavourable initial conditions. The Central and Eastern European shock has yet to impact fully on the resource allocations under the Structural and Cohesion Funds and, when it does, the within-country inequality gap may widen quite sharply among the current Member States of the EU. The *Agenda 2000* proposals provide the first clear indications of the possible impact of enlargement on EU finances and policies.

What we can say, however, is that the rise of regionalism will continue apace and that, although this would appear compatible with the cohesion process, substantial tensions still need to be recognized. Regions distinguished by strong networks of competent and efficient regional institutions have been more successful at harnessing the appropriate resources necessary in producing economic and social development than have regions within more centralized nation-states. Consequently, regional autonomy is a key element in the cohesion process. The moves towards greater integration within the EU, however, are not necessarily in step with these trends. For example, the objective of creating EMU by the end of 1999 carries with it the need to transfer power to new centres far removed from the regions' control. Further, the implementation of the twin issues of subsidiarity and fiscal transfer, which theoretically provide the 'teeth' to the actions of regional government and institutions, poses serious questions about the viability – and indeed sincerity – of the centre's celebration of a 'Europe of the regions'.

References

Allen, J. and Massey, D. 1989: *The economy in question*. London: Sage.

Armstrong, H. W. 1995a: The regional policy of the European Union. In Healey, N. M. (ed.), *The economics of the new Europe: from Community to Union*. Routledge: London, 255–76.

Armstrong, H. W. 1995b: Convergence among regions of the European Union, 1950–90. *Papers in Regional Science* 74, 143–52.

Armstrong, H. W. 1995c: European Union regional policy: sleepwalking to a crisis. *International Regional Science Review* 19, 193–210.

Bachtler, J. 1995: Policy agenda for the decade. In Hardy, S., Hart, M., Albrechts, L. and Katos, A. (eds), *An enlarged Europe: regions in competition?* London: Jessica Kingsley, 313–24.

Bachtler, J. and Clement, K. 1992: 1992 and regional development. *Regional Studies* 24, 305–6.

Barro, R. J. and Sala-i-Martin, X. 1991: Convergence across states and regions. *Brookings Papers of Economic Activity* 1, 107–82.

Bellini, N. and Pasquini, F. 1996: The case of ERVET in Emilia-Romagna: towards a second generation Regional Development Agency. Paper presented to the Regional Development Agencies in Europe Conference, Aalborg, Denmark, September 1996.

Benko, G. and Dunford, M. (eds) 1991: *Industrial change and regional development*. London: Belhaven.

Best, M. H. 1990: *The new competition: institutions of industrial restructuring*. Cambridge: Polity Press.

Boltho, A. and Holtham, G. 1992: The assessment: new approaches to economic growth. *Oxford Review of Economic Policy* 8, 1–14.

Cecchini, P. 1988: *The European challenge: 1992, the benefits of the Single European Market*. Aldershot: Wildwood House.

Christiansen, T. 1996: Reconstructing space: from territorial politics to European multi-level governance. In Jorgensen, K. E. (ed.), *Reflective approaches to European governance*. London: Macmillan, 78–93.

Commission of the European Communities 1993: *Growth, competitiveness and employment: the challenges and ways forward into the twenty-first century*. Brussels/Luxembourg: Commission of the European Communities.

Commission of the European Communities 1996: *First report on economic and social cohesion 1996*. Brussels/Luxembourg: Commission of the European Communities.

Commission of the European Communities 1997: *Agenda 2000*. Com (97) 2000. Brussels/Luxembourg: Commission of the European Communities.

Crafts, N. 1992: Productivity growth reconsidered. *Economic Policy* 15, 388–426.

Curwen, P. 1995: The economics of social responsibility in the European Union. In Healey, N. M. (ed.), *The economics of the New Europe: from Community to Union.* London: Routledge, 188–205.

Delors, J. 1989: Regional implications of economic and monetary union. *Report on Economic and Monetary Union in the European Community.* Collection of papers submitted to the Committee for the Study of Economic and Monetary Union: Brussels: Commission of the European Community.

Dixon, R. J. and Thirlwall, A. P. 1975: A model of regional growth rate differentials along Kaldorian lines. *Oxford Economic Papers* 27, 201–14.

Dunford, M. 1993: Regional disparities in the European Community: evidence from the REGIO databank. *Regional Studies* 27, 727–43.

Dunford, M. 1996: Disparities in employment, productivity and output in the EU: the roles of labour market governance and welfare regimes. *Regional Studies* 30, 339–58.

Dunford, M. and Hudson, R. 1996: *Successful European regions: Northern Ireland learning from others.* Belfast: Northern Ireland Economic Council, Research Monograph 3.

Dunford, M. and Perrons, D. 1994: Regional inequality, regimes of accumulation and economic integration in contemporary Europe. *Transactions of the Institute of British Geographers* NS 19, 163–82.

EUROSTAT 1995: *Regions: statistical yearbook.* Luxembourg: Office of the Official Publications of the European Communities.

Farrows, M. 1997: *The Committee of the Regions: regionalising the Union or pacifying the regionalists?* Limerick: Working Paper No. 3, Centre for European Studies, University of Limerick.

Francis, A. 1992: The process of national industrial regeneration and competitiveness. *Strategic Management Journal* 13, 61–78.

Group of Lisbon 1995: *Limits to competition.* Cambridge, MA: MIT Press.

Hall, R. and Van der Wee, M. 1995: The regions in an enlarged Europe. In Hardy, S., Hart, M., Albrechts, L. and Katos, A. (eds), *An enlarged Europe: regions in competition?* London: Jessica Kingsley, 8–21.

Hirschmann, A. O. 1958: The strategy of economic development. New Haven, CT: Yale University Press.

Kaldor, N. 1970: The case for regional policy. *Scottish Journal of Political Economy* 17, 337–48.

Keating, M. and Jones, B. 1995: *The European Union and the regions.* Oxford: Clarendon Press.

Krugman, P. 1991: *Geography and trade.* Cambridge, MA: MIT Press.

Krugman, P. 1996: Urban concentration: the role of increasing returns and transport costs. *International Regional Science Review* 19, 5–30.

Leonardi, R. and Garmise, S. 1992: Conclusions: sub-national elites and the European Community. *Regional Politics and Policy* 2, 247–74.

Leyshon, A. and Thrift, N. 1995: European financial integration: the search for an 'island of monetary stability' in the sea of global financial turbulence. In Hardy, S., Hart, M., Albrechts, L. and Katos, A. (eds), *An enlarged Europe: regions in competition?* London: Jessica Kingsley, 109–44.

Lipietz, A. 1993: *Towards a new economic order: post-Fordism, ecology and democracy.* Cambridge: Polity Press.

Mackay, R. R. 1995: Non-market forces, the nation-state and the European Union. *Papers in Regional Science* 74, 209–32.

Molle, W. and Boeckhout, S. 1995: Economic disparity under conditions of integration: a long-term view of the European case. *Papers in Regional Science* 74, 105–24.

Myrdal, G. 1957. *Economic theory and underdeveloped regions.* London: Duckworth.

Piore, M. and Sabel, C. 1984: *The second industrial divide: possibilities for prosperity.* New York: Basic Books.

Porter, M. E. 1990: *The competitive advantage of nations.* New York: The Free Press.

Porter, M. E. 1996: Competitive advantage, agglomeration economies, and regional policy. *International Regional Science Review* 19, 85–90.

Quevit, M. 1995: The regional impact of the internal market: a comparative analysis of traditional industrial regions and lagging regions. In Hardy, S., Hart, M., Albrechts, L. and Katos, A. (eds), *An enlarged Europe: regions in competition?* London: Jessica Kingsley, 55–69.

Sabel, C. F. 1989: Flexible specialisation and the re-emergence of regional economies. In Hirst, P. and Zeitlin, J. (eds), *Reversing industrial decline?: industrial structure and policy in Britain and her competitors.* Oxford: Berg, 17–70.

Sharpe, L. J. (ed.) 1993: *The rise of meso-government in Europe.* London: Sage.

Shepley, S. and Wilmot, J. 1995: Core vs periphery. In Amin, A. and Tomaney, J. (eds), *Behind the myth of the European Union: prospects for cohesion.* London: Routledge, 51–82.

Teague, P. 1995: Europe of the Regions and the future of national systems of industrial relations. In Amin, A. and Tomaney, J. (eds), *Behind the myth of the European Union: prospects for cohesion.* London: Routledge, 149–73.

|7|

Room to talk in a house of faith: on language and religion

Colin H. Williams

Introduction

An inherent ambiguity lies at the heart of the debate on the social and cultural future of Europe. Both individuals and a multitude of ethnocultural groups increasingly interact within decentralized and diverse frameworks, while the political system which guarantees such autonomy is becoming ever more centralized within the supranational structures of an expanding EU and its associated organizations. The traditional method of reducing ethnocultural diversity by closing or redrawing borders no longer applies, as the geography of bounded space gives way – at least in theory – to a geography of communication flows, and a new strategy for mutual coexistence demands the deterritorialization of identity. The sad truth for many in Europe today, however, is that the direction and speed of this transition are precisely the point at issue. Although the most satisfactory method of ensuring cultural autonomy is to allow individuals to determine group membership for themselves, this dilutes the geographical concentration of ethnic groups and renders many of them vulnerable within a multicultural framework; ambiguity, tension and conflict are the inevitable consequences (Lijphart 1995).

In addressing the contradictory and spatially complex nature of multicultural Europe (Figure 7.1), this chapter focuses on language and, to a lesser extent, religion, as components of a European identity that offer a profound insight into the range of possibilities which the contemporary world allows. At one open and receptive pole of a European spectrum are the key players of international politics and commerce, together with their advisers, civil servants, lobbyists and journalists, each of whom brings into play a diverse range of language skills, whether active or passive, and a different set of assumptions as to the appropriate role of cross-cultural communicative competence.

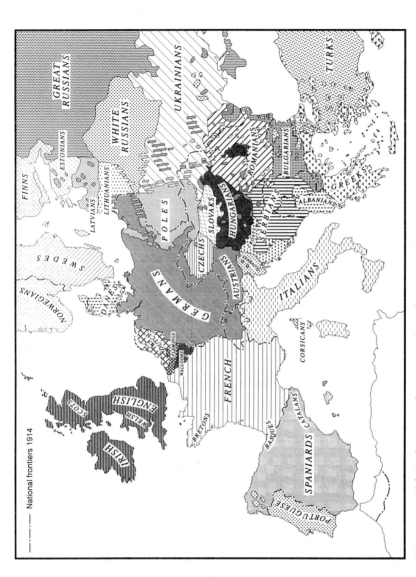

Figure 7.1 Ethnic divisions of Europe in the early nineteenth century
Source: Based on *The Times Atlas of World History* (1986, p. 214)

National frontiers 1914

Close behind, occupying a rather more independent and fluid position, are the educated youth of Europe, whose command of 'European pidgin English' or some other 'language of wider communication' allows them access to international culture, sport, entertainment and education. Occupying the centre of the spectrum is the post-1945 generation whose knowledge of two or more languages may be rather limited, but who nevertheless espouse a healthy attitude to Europe-wide culture, cuisine, films, travel and personal contact. Towards the other end of the spectrum – largely closed to external influences – are different types of relatively deprived groups. One is the monolingual majority of long-established nation-states, who may conceive of themselves as participants in European integration, but at a more prosaic level than that of their children. Finally, we have a number of social groups, many of them minorities, who by virtue of their exclusion from decision-making and wealth-creation potential are denied full access to participatory democracy and social progress. The chapter examines the ways in which language identification and religious affiliation influence access to mainstream European development. In so doing, it addresses the central question as to how we might construct a European identity which is based upon the recognition of cultural diversity as a key element of social and political life.

Historically, language and religion have often been treated as the critical markers of a distinct cultural identity, a useful shorthand in which to describe a complex reality. Such close correspondence can no longer be entertained, for linguistic identification and religious affiliation are increasingly divorced from each other. As discussed in Chapters 1 and 2, the cultural inheritance of a Catholic, Latin civilization gradually gave way in post-medieval Europe to a system of nation-states. Conflict was inherent in the political, cultural and economic processes, which underpinned the emergence of the territorial bureaucratic nation-state as the prime locus of political activity. This was particularly so in terms of language and religion, where both the past abrogation of rights and the continued refusal of states to grant social demands promoted ethnic conflict and fragmentation. Why is a common language so often seen as essential to 'nation-building' or state development, and if conflict is such a predictable outcome, why not opt for linguistic and cultural pluralism as a dominant ideology? The answer is surely that language is power – to confer privilege, deny opportunity, construct a new social order and radically modify an inherited past which is not conducive to the pursuit of hegemonic aims. Language choice is thus a battleground for contending discourses, ideologies and interpretations of the multiethnic experience.

The argument advanced in this chapter is that it is both advisable, and feasible, to construct a political framework in Europe which acknowledges the positive virtues of cultural pluralism on the basis of equality as a necessary prerequisite for democracy and freedom of action in an increasingly multicultural world order. This raises issues concerned with the distribution of power in society and the encouragement of democratic participation by previously beleaguered interest groups. A *sine qua non* of their recognition is mutual respect, and structures of freedom which guarantee the conditions for cultural reproduction. As the former Yugoslavia graphically illustrates, open conflict and warfare are too pressing a reality for many beleaguered groups, while also serving to remind us – as we have seen in Chapter 3 – that current periods of peace and tolerance almost always derive from previous periods of resistance and struggle.

The construction of Europe-wide institutions has to deal with the diversity and tensions resulting from an international political system designed to suit the vagaries of nineteenth- and early-twentieth-century statecraft, and the ensuing lack of congruence between multicultural citizenries and the sense of order prescribed by national conceptions of the modern state (Williams 1989). Although there are indications that the neo-liberal conception of the unitary state, based upon representative democracy, is yielding to the logic of the enabling state with its focus on participatory democracy, this transition is either illusionary or faltering in many parts of Europe. The management of cultural diversity as a permanent feature of the international social order is among the most taxing of political issues facing modern Europe (see Chapter 9).

Religion, language and diversity: the socio-cultural renegotiation of Europe

The foundation of any modern democracy lies in the ability of citizens to derive maximum security and satisfaction from their contribution to the common wealth of society. Historically, several societies allowed for instrumental pluralism as a societal norm because a broad measure of freedom from state interference provided the necessary breathing space for the peaceful coexistence of citizens. Habermas (1996) argues that such space permitted citizens of widely diverging cultural identities to be simultaneously members of, and strangers in, their own country. The historical reproduction of dissenting cultures was, in part, a function of relatively weak economic-structural assimilation, often compounded by geographical distancing from the cores of the emerging nation-states. Their maintenance today in the face of much stronger pressures for inclusion is all the more difficult. Increased secularization in the West and the enforced totalitarian conformity that described Central and Eastern Europe prior to 1989 have greatly damaged the primacy of Catholicism, Protestantism and Orthodoxy. When we add the catastrophic effects of the Holocaust on Jewish community life, then it is little wonder that organized religion is a markedly less salient part of mainstream society (see also Chapter 4). The only exceptions, which appear to have witnessed real growth, are the faiths of non-European migrants and their descendants, such as Sikhs and Hindus in Great Britain, together with Islamic believers throughout much of Europe (Gerholm and Lithman 1990) and a plethora of evangelical variants of fundamentalist Protestantism. However, although organized

Christian religion may have declined absolutely, ample evidence exists to suggest that much of the habit of obedience to their religion displayed by earlier generations was culturally determined, and not the outcome of an individual conviction to join a particular religious community. Thus despite the absolute decline in the number of adherents, religious faith will continue to contribute to European social life and its diversity, and not just as the tabloid banner headline where Catholic is pitched against Protestant in Northern Ireland, or where members of the Muslim nation within Bosnia-Herzegovina are set against their Orthodox and Catholic neighbours (Zametica 1992).

To turn to language, if it is accepted that a lack of congruence exists between the formal political system and the cultural inheritance of its constituent citizens, this is rendered even more complex by the friction that exists between attempts to maintain linguistic diversity and the increasing linguistic standardization apparent throughout the world. This contradiction is important because tensions related to religion and language often serve as mobilizing factors for conflict within which other grievances are then pursued. At the global scale, it is estimated that some 6170 living languages (exclusive of dialects) are contained within the 185 or so sovereign states – a number that rises to around 200 if dependencies and semi-autonomous polities are included (Mackey 1991). Less than 100 of these *c.* 6000 languages are 'official', since 120 states have adopted English, French, Spanish or Arabic as their official language, while some 50 states have their own indigenous official language (15 per cent have two or more). If a further 45 regional languages are added, it remains the case that only about 1.5 per cent of the world's total spoken languages are formally recognized. The situation is even more polarized in that only 1 per cent of the world's languages are used by more than half a million speakers and only 10 per cent by more than 100 000. Hundreds of languages have no adolescent speakers at all and thus we are continuously losing parts of our global linguistic diversity (Williams 1995).

If it is accepted that most minority ethnic–linguistic groups are also relatively underdeveloped economically and politically, it is clear that questions of language, culture and identity are not merely supplementary to the more routine socio-economic concerns of development. Rather, they may constitute the very essence of a subordinated group's relationship with the state in whose name the dominant group exercises power and control. It is not surprising that any threat to the immediate territory of a subordinated group is interpreted as a challenge to culture and group survival. Place and territory are critical in the process of control and development, and their appropriation by external agencies has a long history in Europe, related to the extension of state hegemony and projects of state-building and nation-formation, processes which are all essentially contestations over space.

The incorporation of ethnically differentiated territories was a necessary precursor to the creation of the territorial–bureaucratic state that came to dominate the political geography of Europe between the eighteenth and twentieth centuries. European state development was often over-centralized around national capital cores, without any corresponding attention to the interests of 'minorities', except, of course, the need to subject them to political and strategic integration. This has led, as for example in Spain and France, to charges of core discrimination, peripheral marginalization and the denial of group rights and cultural reproduction. For some this is a necessary product of global development, but should state formation involve the denial of local and regional distinctiveness? Must we perforce sacrifice cultural autonomy – at whatever scale – to promote the hegemony of particular political–economic structures? Multicultural societies necessitate choices but, given the competitive nature of cultures in contact, these options also promote conflicts and tensions, one person's choice being another's denial of opportunity.

In Western Europe, the twin processes of nation-building and state integration engaged language and religion as agencies of political unification. Thus since medieval times, religious competition, for example, was regulated by a layered system of legal securities. This involved the principle of coexistence in religiously mixed imperial cities and the realization of the *cuius regio eius religio* (implying that the formal religion of the territory was determined by the personal religious affiliation of the ruler), following the uneasy agreement between Catholics and Protestants which ensued from the Religious Peace of Augsburg in 1555. The law gave rights to overlords to determine the faith of their territories and not to the subjects, who were theoretically free to emigrate. The principle was hailed as the most important element of the rise of the secular territorial state, one that together with the *ius reformandi*, the law of reform (which was also tied to territorial privileges), formed the legal–geographical basis for the structural transition from Holy Roman Empire to Enlightenment state system. Nevertheless, most dissenting remnants in the secular territorial states, whether Catholic or Protestant, did not survive intact. They could survive 'only under cover, in isolation, and for short periods of time. Stringent control made it impossible to establish parties which did not conform to the official creed. Persecution and expulsion followed immediately upon discovery' (Klein 1978, p. 57).

None the less, although state-enforced linguistic and educational policies can create a common nationality in time, religious adherence is less amenable to state intervention, despite centuries of legally enforced observance, frequently at the point of the sword rather than the cross, star or crescent. Nowhere is this more evident than in East-Central Europe, which has long been characterized by its religious plurality. As Davies (1996, p. 504) observes, in a world of

> growing religious intolerance, Poland-Lithuania occupied a place apart. A vast territory . . . it contained a mosaic of the Catholic, Orthodox, Judaic, and Muslim faiths even before Lutheranism claimed the cities of Polish Prussia or Calvinism a sizeable section of the nobility.

Magosci (1993, p. 48) emphasizes the region's pivotal role within European religious history when he states that it was at the center of the dividing line between the Catholic and the Orthodox worlds; it experienced the first serious challenges to the unity within these two Christian worlds (Bogomilism in religiously mixed Bosnia during the thirteenth and fourteenth centuries and Hussitism in Catholic Bohemia-Moravia in the fifteenth century); and virtually its entire southern half had by the sixteenth century come under the rule of the Ottomans, who implanted the Muslim faith throughout their expanded domain. With such a tradition of religious pluralism, it is not surprising that the lands of East Central Europe, in particular its northern 'non-Ottoman' half, proved to be fertile ground for the spread of the Protestant Reformation.

Our present conception of the rights of subjects and citizens has evolved such that the principle of freedom from state direction or oppression in religious or linguistic matters has given way to a demand for freedom to be represented on the basis of equality within society as the determining essence of the participative state. Nowhere are these rights so fiercely conjoined and attacked as in the question of ethno-linguistic identities in the modern state. The new politics of recognition in Europe belatedly represents an attempt to compensate for the earlier systematic exclusion of many minority groups from the decision-making structures of society. This is not to deny earlier such attempts to specify the rights and obligations of minority cultures, but these prerogatives were often granted on the assumption that no permanent change would result to the state from such reforms. It was understood that the clarification of the nature and meaning of minority rights would not overly interfere with mainstream political business and economic development. However, the current transition from representative democracy to participatory democracy, at least within parts of the EU, requires the decoupling of the state majority from its hegemonic position. As Habermas (1996, p. 289) argues,

> Hidden behind such a facade of cultural homogeneity, there would at best appear the oppressive maintenance of a hegemonic majority culture. If, however, different cultural, ethnic and religious subcultures are to co-exist and interact on equal terms within the same political community, the majority culture must give up its historical prerogative to define the official terms of that *generalized* political culture, which is to be shared by all citizens, regardless of where they come from and how they live. The majority culture must be decoupled from a political culture all can be expected to join.

The nature of civil rights: inclusion versus exclusion

Thus those who are constructing the new Europe must search for a binding substitute for state nationalism. Geographically fixed identities, based upon real or putative representations of cultural homogeneity, have to yield to more fluid, heterogeneous forms of social interaction. Our conceptions of human rights are now being formulated in an increasingly comprehensive manner to include elements which earlier theorists would have considered to have lain outside the proper remit of the citizen–state relationship (Close 1995). This relationship is central to the analysis, since democracy avers that citizens are entitled to certain minimum rights, chiefly those of participation in, and protection by, the state. The changing nature of the state, however, both as ideology and practice, has encouraged a more pluralist view of its responsibilities.

The conventional view, characteristic of many Western societies until the early post-World War II period, held that the state should not discriminate against, or in favour of, particular subgroups, however these might be defined. This view, the individual rights approach, is often justified by majoritarian principles of equality of all before the law, and is implemented through policies of equal opportunity for socio-economic advancement based upon merit and application. In reality, however, in most multi-faith societies, the state persistently discriminated, by law, against religious and other minorities, be they Jews, Catholics, Protestants or Romanies. However, the admittedly patchy improvement in the treatment of minorities, and the resultant constructive dialogue between representatives of the various interest groups and governmental agencies at all levels in the political hierarchy of contemporary Europe, obviously bodes well for the medium-term future enactment of minority rights.

An alternative view, the group rights approach, has found increasing favour of late, for it recognizes that there are permanent entities within society whose potential and expectations cannot be met by reference to the recognition of individual rights alone. In the main, such recognition is offered grudgingly, reflecting a minimalist stance which seeks to extend the tradition of individual rights into a multicultural context. Such extensions tend to obscure the key issue of group tension, namely the ability of the minority to preserve and, if possible, develop its own group characteristics and desires in the face of state-inspired assimilation.

Two sorts of argument are posed to counter the 'special pleading' of constituent differentiated groups in the contemporary world (Williams 1993a). The first contends that the prime duty of the democratic state is to treat all its citizens equally, regardless of racial, national, ethnic or linguistic origin. Conversely, it can be argued that minorities should not require 'extra rights' if the necessary democratic guarantees are in place. Both interpretations are understandable in a political context which stresses the role of the reformed state as the agent and co-ordinator of radical change. Thus the growth of democratic representative power in Central Europe enables the state to pose as the guardian of civic rights, while individualism strives to triumph over collectivism in terms of human rights, social justice and economic productivity.

The challenge to liberal democracies is both real and very pressing, but what does equal representation mean if public institutions do not recognize particular identities, and allow only for general or universal recognition of shared interests based upon civil and political liberties, education, health care and economic participation? Public recognition of the worth of constituent cultures, as *permanent* entities in society, is what is at stake in liberal democracies. For many engaged in the politics of their group's survival, this is precisely what democracy should guarantee in practice as well as in principle. When it refuses to engage in the politics of recognition, liberal democracy appears arrogant and denies the life-enhancing spirit upon which all forms of democracy are based. It accentuates fragmentation and anomie within society, ultimately leading to various forms of disengagement from public life and community responsibility.

> Democratic citizenship develops its force of social integration, that is to say it generates solidarity between strangers, if it can be recognized and appreciated as the very mechanism by which the legal and material infra-structure of actually preferred forms of life is secured.
>
> (Habermas 1996, p. 290)

The real difficulty for such interpretations lies in maintaining the active participation of all citizens in the ensuing political process. It is far too tempting for many to yield responsibility and to opt out of formal politics, joining instead informal pressure groups or single-issue movements, meanwhile leaving proponents of the community drained of their energies to mobilize and agitate on behalf of all. A concern for an active participatory democracy is surely relevant in most developed societies, where talk of 'the hollow state' and of 'the democratic deficit' reveal the shallowness of the general public's trust in professional politicians (Williams 1994). There is, moreover, an urgent need to establish democratic credentials in the more 'liberal' post-Communist societies after forty or so years of state totalitarianism. But the pristine democratic principles of coequality, majoritarian tolerance and freedom under the law to reproduce individual or collective identity have not necessarily guaranteed or satisfied minority aspirations in Western European societies. In Central and Eastern Europe, there is even less consensus about the nature of mass society, let alone the legitimacy of selected minorities in multiethnic societies (Williams 1997). Social justice is not necessarily served by a compliant reliance upon a constitutional majority, whether it acts in a benign manner or otherwise. Indeed, the contemporary situation in Central and Eastern Europe is enigmatic, many previously warring factions apparently converging on a new conformity. This threatens, however, to be every bit as stultifying as the old system, as new forms of radical dissent are marginalized. We may be set for a new round of language-related conflict as geostrategic considerations clash with the emancipatory demands of mobilized minorities.

This situation arises because reconstructing societies face two opposing notions of justice. While one holds that justice is the apportioning of rewards to groups on the basis of proportionality, the other suggests that justice should consider the established rights of individuals, regardless of national origin, language, religion or any other diacritical cultural marker (Glazer 1977). In an individualistic society, the majority would favour merit as a guiding principle of selection. The various constituent minorities would counter, however – and with some justification – that this merely reproduces their position, marginalized and disparaged as permanent dependencies. If the state adopts a diffusion perspective of ethnic change, viewing group identities as malleable and group membership as a purely private affair, it will conceive of group rights as a barrier to minority assimilation and as a basis for maintaining permanent divisions within the state (Glazer 1977). Conversely, if the state conceives of its constituent cultural groups as forming part of an established ethnically plural society, then it must legislate and act to demarcate the rights of each group.

None the less, a profound difference exists between the stated policies of governments towards minorities, as enunciated, for example, at international conventions, and the actual treatment of differentiated citizens at the local level. It is unduly facile to rest content with either individualist or collectivist paradigms of language contact management. Liberalism is not a neutral ideology. The liberal democratic state is much more than a referee for the warring factions contained within its bosom. Thus rather than presume that one universal solution exists to the question of managing ethnic pluralism in Europe and elsewhere, it is more instructive to draw attention to the sheer variety of assumptions about the nature of majority–minority relations inherent in models of ethnic integration. One of the most important is a scheme developed by Kallen (1995) (Figure 7.2).

This conceptualization, which informs the remainder of the discussion in the chapter, envisages a fourfold model of integration. The first is the 'melting pot', with its emphasis on integration and the creation of one nation/one people. In theory, this system should produce a non-discriminatory national identity. Second, the 'mosaic' refers to the cultural pluralism of a nation comprising many peoples and cultures. Presently more applicable, perhaps, to the extra-European world, this model leads to hyphenated ethnic-national identities in societies that do not depend on institutionalized discrimination. It is possible to envisage, therefore, a future Europe made up of Irish-Europeans, German-Europeans and so on, the situation that still pertains in the United States. Third, dominant conformity effectively describes the formation of nation-state Europe, a dominant culture and people framing the state in its own image and discriminating institutionally against minorities. Finally, paternalism exists in the relationships between colonizer and colonized, a dominant nation, culture or people imposing its values on subordinated minorities. This model effectively describes the history of Ireland or Poland and, although no longer applicable, its legacy can still be detected in the national constructions of such states.

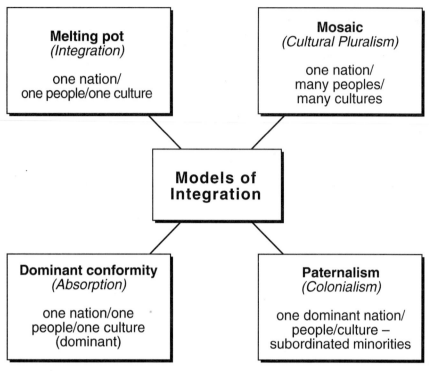

Figure 7.2 Models of ethnic integration
Source: Adapted from Kallen (1995)

The politics of equal respect in multicultural societies

Kallen's model encapsulates the dichotomy between exclusive and inclusive definitions of civil rights. Contemporary Europe is facing a period of re-adjustment, following a series of structural transformations which, cumulatively, have seen a greater recognition being accorded to its indigenous minorities. But what sort of multiculturalism are we discussing and how does it seek to promote the politics of mutual respect? Let us consider the difficulties of operating an ideology of multiculturalism within a model of liberal society which purports to recognize more than the mere survival of cultures by acknowledging their permanent worth.

A major difficulty in the postmodern period concerns the threat posed to communitarian democracy by the growth of special-interest groups. The recognition of religious and linguistic rights has encouraged the belief that increased cultural pluralism is a reflection of increased mutual tolerance. Yet tolerance *per se* does not necessarily follow, for it depends upon which groups have received recognition for what purpose and in which contexts. There is at least one further complication. While some autochthonous (or indigenous) minorities are being treated according to policies which assume pluralism as the basis of equality, other non-indigenous minorities are being treated according to policies predicated on pluralism as the basis of non-equality. How do these two trends impact the one upon the other with regard to a liberalism of rights in language and religious matters?

Taylor (1992, p. 61) argues that the politics of equal respect embodied within the discourse of recognition does not serve us as well as we imagine. Instead, there is 'a form of the politics of equal respect, as enshrined in a liberalism of rights, that is inhospitable to difference'. This 'insists on uniform application of the rules defining these rights, without exception, and . . . is suspicious of collective goals'. Taylor calls it 'inhospitable' because the construct cannot 'accommodate what the members of distinct societies really aspire to, which is survival'. This is a collective goal, which will inevitably demand some variation in the kinds of law that are deemed permissible from one cultural context to another.

Clearly, the dread hand of homogenization can lie heavy on any attempt to maintain diversity through an appeal to universal considerations of human dignity and worth. Yet the filter process by which some decide the relative worth of others is far too imperfect. At its root lies an inherent paradox between the demands for equality and the forces for efficiency. As Taylor (1992, p. 71) contends,

> The peremptory demand for favourable judgements of worth is paradoxically – perhaps one should say tragically – homogenising. For it implies that we already have the standards to make such judgements. By implicitly invoking our standards to judge all civilisation and cultures, the politics of difference can end up making everyone the same.

Taylor believes that the demand for equal recognition is unacceptable. Rather we require a humbler approach which does not imply the pejorativization of all other cultures. This, of course, presumes that cultural diversity is not only a growing feature of most societies, but also a positive and worthy feature. It has value both at the individual level of recognizing human worth and also at a societal level, where it legitimizes access to political power and frees participation in the democratic process. It follows that the ways of managing cultural diversity will vary according to why we think it has value.

A great stumbling-block, however, in the recognition of equal worth is the operation of the market and bureaucratic state, which, in Taylor's view, 'tends to strengthen the enframings that favour an atomist and instrumentalist stance to the world and others' (1991, p. 11). Community solidarity and public participation in the decision-making process are weakened, and religious or ethnolinguistic groupings, intent upon their own survival, may be drawn closer into their own subcultures rather than forming a distinct part of the whole society, a process that accentuates the fragmented nature of mass society:

This fragmentation comes about partly through a weakening of the bonds of sympathy, partly in a self-feeding way, through the failure of democratic initiative itself. Because the more fragmented a democratic electorate is in this sense, the more they transfer their political energies to promoting their partial groupings, and the less possible it is to mobilise democratic majorities around a commonly understood programme.

(Taylor 1991, p. 113)

Fragmentation and anomie seem to be winning the day for far too many previously committed citizens. And yet, in principle, they all possess the right of free association, free speech and political representation, the hallmarks of a modern democracy. Religious and linguistic distinctiveness rests both on special-interest recognition and on the maintenance of a common social order, which both affirms ordinary life and recognizes its *limits*. This is an especially urgent consideration in South-Eastern Europe and on the margins of Central Europe. In order to sustain religious and socio-linguistic pluralism here, as elsewhere, a redefinition of the essence of political engagement is required. Elshtain (1994, p. 79) argues that the affirmation of ordinary life has not been a primary passion of political philosophy, but – in order to retrieve that ordinariness – we require a new socio-political project that tames and limits 'the demands of sovereignty – both sovereign state and sovereign self'. The politics Elshtain has in mind 'shifts the focus of political loyalty and identity from sacrifice and control to responsibility' (1994, p. 79). In questioning whether a post-sovereign state politics is possible, Elshtain cites the writings of the President of the Czech Republic, Václav Havel, and urges us to forge civic identities in such a way that blood sacrifice is not so pervasive a demand and possibility. Havel's move from sacrifice to responsibility is best summarized in *Disturbing the peace* (1990), in which he argues that

We are going through a great departure from God which has no parallel in history. As far as I know, we are living in the middle of the first atheistic civilization. This departure has its own complex intellectual and cultural causes: it is related to the development of science, technology, and human knowledge, and to the whole modern upsurge of interest in the human intellect and the human spirit. I feel that this arrogant anthropocentrism of modern man, who is convinced that he can know everything and bring everything under his control, is somewhere in the background of the present crisis. It seems to me that if the world is to change for the better it must start with a change in human consciousness, in the very humanness of modern man. Man must in some way come to his senses.

(Havel 1990, p. 11)

The politics of self-control does not mean, as some have implied, the removal of all limits on the human condition. Taylor offers a warning and a correction to some aspects of postmodernist thought when he writes:

it is a thinking which moves from the undeniable fact that the aspiration
to take responsibility involved in the displacement of *some* limits, to the
conclusion that this aspiration inherently amounts to a rejection of *all*
limits. In other words, it confounds the original aspiration and its aberrant form. This is a bad mistake, indeed, a potentially catastrophic one
for us who cannot but share the modern identity.

(Taylor 1994, p. 233)

Language policy and conflict in Europe

We now move on to consider the ways in which these socio-political forces are
played out in linguistic terms and to assess the ensuing consequences for social
policy and citizenship in the new Europe. As a mechanism for behaviour
modification, language policy and planning depend largely on four attributes
identified by Stewart (1968). These are the degrees of, respectively, standardization, autonomy, historicity and vitality. These characteristics of language
freedom are critical in helping planners to evaluate existing language functions
and to harness the dynamic cultural interactions that characterize many multilingual societies. Stewart (1968, pp. 540–1) proposes a typology of language,
which recognizes the multiplicity of linguistic functions that can exist even
within ostensibly one nation/one people/one culture states:

- official languages;
- provincial languages (such as regional languages);
- languages of wider communication (LWCs), which are used within a multilingual nation to cross ethnic boundaries;
- international languages, which are LWCs used between nations;
- capital languages (the means of communication near a national capital);
- group languages (often vernaculars);
- educational languages (used as the media of education);
- school-subject languages (those taken as second languages);
- literary languages (for example, Latin or Sanskrit);
- religious languages (such as Islamic Arabic).

Most European societies have retained the language policies which were critical to the nineteenth-century project of constructing the territorial–bureaucratic nation-state. Consequently they face severe difficulties in matching their
inherited institutional agencies and organizational structures to the reality of
serving the legitimate demands of increasingly multilingual populations.
Historically, we can identify four types of language policy implicated in the
processes of European state formation. The first and most common – as found
in France, Spain and Britain – reinforced political and cultural autonomy by
giving primacy to one indigenous language and thus enforcing it, and no other,
as the language of government, administration, law, education and commerce.
In so doing, a number of goals were achieved simultaneously. Among these was

the search for national integrity, the legitimization of the new regime and its state apparatus, the re-establishment of indigenous social organization, the reduction of dependence upon external organizations and influences, and the incorporation of all the citizenry in a wide range of para-public social domains. Foreign languages were reserved for the very specialized functions of higher education, international diplomacy and commerce.

A second type of language planning characterizes those situations in which the 'national' goal has been to maintain cultural pluralism, largely that the state might survive through containing its inherent tensions. Under this system, language was used to define regional associations rather than state or national citizenship. It is best exemplified in Europe by Switzerland's decentralization, which incorporates cantonal unilingualism within a multilingual federal system, and the rigidly enforced division of Belgium between its Walloon and Flemish-speaking populations.

The third form of language planning has occurred when a recognized minority was granted some degree of geographical distinction, based upon the territoriality principle of language rights. In Finland, for example, high levels of language contact along the west coast, marked religious uniformity, a strong tradition of centralized government and the unifying effects of long-term external threats led to a recognition that the minority Swedish-speaking population should be accorded official status. Following the 1922 Language Law, all communes were classified as unilingual Finnish, unilingual Swedish, or bilingual if they contained a linguistic minority of either group of 10 per cent or more; these classifications were revised after each decennial census to take account of changing linguistic geography (McRae 1997). In comparative terms, the Finnish system of official bilingualism is characterized by a gross disparity in numbers, asymmetry in language contact and instability over time, leading McRae to conclude that such institutional arrangements for language accommodation have been functional in terms of conflict moderation and management, but less so for language stability.

A fourth option, the modernization of an indigenous tongue, is diagnostic of societies disengaging from colonial relationships and the cultural hegemony of a dominant state. This form of language planning was a key goal of the nationalist intelligentsia in Hungary, Ireland, Finland and Norway prior to independence, and remains critical to the political programme of nationalists and regionalists in Catalonia, Euskadi (the Basque Country), Brittany and Wales today.

Clearly, language planning in a multilingual society is not a precise instrument and is as capable of being manipulated as is any other aspect of state policy. Nevertheless, it is an essential feature of the economic and political restructuring of many states. The key issues are:

- Who decides – and on what basis – that such bi- or multilingualism is to be constructed?
- Which languages are chosen?

- Who benefits by acts of state-sponsored social and identity formation?
- How is language-related conflict managed and reduced?

Two conditions are necessary for competition to arise between language groups. First, the languages must share a common contact space, however defined, be it geographic, political, social, economic, cultural or religious in nature. Second, the relationship between these two languages must become the symbolic stakes of the competition, which takes place on the level of the shared space.

Laponce (1987, p. 266) has advanced the following four propositions about languages in contact:

- 'languages tend to form homogenous spatial groupings';
- 'when languages come into contact they tend either to specialize their functions or to stratify';
- 'the specialization and the stratification of languages is determined by the socially dominant group';
- 'the social dominance of a language is a function of . . . the number of its speakers and . . . the political and social stratification of the linguistic groups in contact'.

How do minority-interest groups influence the state structure so that it concedes certain rights, which are not requested by the majority? Concessions always follow contact and conflict, and may involve the use of a previously disallowed language within public administration and the legal system, or a religious-based education provision, or differentiated access to the media. Such reforms may be predicated on the basis of a personality or a territoriality principle or some expedient admixture of both (Nelde *et al.* 1992).

Research in contact linguistics demonstrates that conflict cannot be resolved by means of a universal model for conflict reduction. On the contrary, procedures must be considered that are adaptable to each situation. Nelde *et al.* (1992) have argued that measures of linguistic planning, such as the principle of territoriality, are not in themselves sufficient to avoid conflicts, but they can soften the repercussions in the socio-economic, cultural and linguistic life of multilingual populations. They believe that many conflicts can be partially neutralized if the following conditions are observed:

- the territoriality principle should be limited to a few key areas like administration and education;
- the institutional multilingualism that emerges should lead to the creation of independent unilingual networks which grant equal opportunity of communication to minority and majority speakers. These networks should also exclude linguistic discrimination connected with speakers of the prestige language;
- measures of linguistic planning should not be based exclusively on linguistic censuses carried out by the respective governments. Rather, they must

genuinely take account of the situational and contextual characteristics of the linguistic groups;
- minority linguistic groups in a multilingual country should not be judged primarily on quantitative grounds. On the contrary, they should be awarded more rights and possibilities of development than would be due to them on the basis of their numbers and their proportion to the majority.

Nelde *et al.* believe that according such equality to minorities by assuring them of more rights could result in fewer people adopting an intransigent ideological position, thereby lessening the infusion of emotionalism so often apparent in linguistic conflicts. One obvious conclusion is that unless far more attention is paid to the rights of lesser-used-language speakers, more conflict will ensue. Another is the potential loss of creativity and spontaneity mediated through one's own language(s), thus contributing to a quenching of the human spirit and a reduction in what some have termed the ecology of 'linguodiversity' (Williams 1991; Skutnabb-Kangas 1997). This latter reason alone may prove convincing to many. But need linguistic decline sound the death-knell for particular ethnic identities in Europe? As Edwards (1994) and Williams (1991) have argued, no necessary correspondence exists between linguistic reproduction and ethnic identity. Indeed, cultural activities and symbolic manifestations of ethnicity often continue long after a group's language declines. None the less, increased interdependence at the European level implies more harmonization for the already advantaged groups. In making multiculturalism more accessible, long-quiescent minorities will be rebuffed as they seek to institutionalize their cultures. It is not necessarily a tale of decline and rejection, however, for opportunities do exist to influence state and provincial legislatures and metropolitan political systems.

Linguistic hegemony

The project of European unity, combined with other macro-processes such as globalization, poses a threat to conventional territorial relationships and simultaneously opens up new forms of inter-regional interaction such as cable television and global multi-service networks. However, choosing between the promotion of one or many languages in the educational domain and public agencies of new states is becoming less of a 'free choice', as the increasing burdens of economic, social and cultural development crowd in on the limited resources available for language planning and its implementation. The increasing globalization of economic, political and cultural relationships is one major constraint in language choice. But there is also the counter-trend of regional diversity, which emphasizes the value of cultural diversity and the worth of each specific language, not least as a primary marker of identity. Concern over endangered languages in Europe has produced a re-examination of the relationship between culture, development and political identity (Williams 1993b).

Ethno-linguistic minorities have reacted to these twin impulses by searching for Europe-wide economies of scale in broadcasting, information networking, education and public administration, meanwhile establishing their own EU networks and entering into new alliances to influence decision-making bodies. They believe that by appealing to the superstructural organizations of the EU for legitimacy and equality of group rights, they will force individual states and the Community to recognize their claims for political/social autonomy within clearly identifiable territorial/social domains.

Logically, if globalization and interdependence can enhance the productive capacity of majority, nation-state interests, they can also be harnessed to develop the interests of groups speaking the lesser-used languages. The EU has harmonized state and community policies so as to strengthen its majority language regimes. But the wider question of the relative standing of official languages makes political representatives wary of further complicating administrative politics by addressing the needs of approximately 55 million citizens who have a mother tongue that is not the main official language of the state which they inhabit. Historically, the recognition of linguistic minority demands is a very recent phenomenon. Since 1983, the European Commission has supported action to protect and promote regional and minority languages and cultures within the EU. In 1996 some ECU 4 million was allocated to European socio-cultural schemes. Equally significant is a raft of recent legislation and declarations upholding the rights of minorities to use their languages in several domains (Williams 1993b; *Declaració de Barcelona* 1996; Nelde 1997).

The most recent expansion of the EU in 1995 and its imminent further enlargement (see Table 5.1) have increased the difficulties in translating multicultural communication and guaranteeing access to information and hence power for all groups. The real geolinguistic challenge is to safeguard the interests of all the non-state language groups, especially those most threatened with imminent extinction. Additional issues include the adaptation by speakers of lesser-used languages to the opportunities afforded by changes in global–local networks, and the growth of information networks accessed by language-related skills. Accessibility to – or denial of – these opportunities is crucial to the politics of regional cultural representation and to mobilizing the regional economies discussed in Chapter 6.

As a result of imperialism, neo-colonialism, the modernization of technology and the globalization of information flows, an air of acute inevitability surrounds the universalization of hegemonic languages, primarily English, French and Spanish. Transnational languages have been significant elements of imperial rule since classical times, and greatly influenced the spread of capitalism and the modern world system. English and French are also crucial to the development of former colonies, providing the basis on which these countries relate to the former metropolitan powers and to each other through the Commonwealth and *la francophonie* (Gordon 1978; Williams 1996).

At the heart of the debate on European identity lies a consideration of the role of hegemonic languages as both symbol and instrument of integration. English – the premier language for international commerce and discourse – is used by over 1500 million people world-wide as an official language, of whom some 320 million have it as a home language (Gunnermark and Kenrick 1985; Crystal 1987). Should English be encouraged as the official language enabling most Europeans to communicate with each other? Or is it desirable to attempt to slow down its inevitable global spread? Critics claim that the spread of English perpetuates an unequal relationship between 'developed' and 'developing' societies. While access to information and power demands fluency, it also requires institutional structures, economic resources and power relationships. Tollefson (1990, p. 84) reminds us that

> in order to gain access to English-language resources, nations must develop the necessary institutions, such as research and development offices, 'think tanks', research universities, and corporations, as well as ties to institutions that control scientific and technological information. From the perspective of 'modernising' countries, the process of modernisation entails opening their institutions to direct influence and control by countries that dominate scientific and technical information . . . the result is an unequal relationship.

The functions of English are nearly always described in positive terms. Language, and the ideology it conveys, is thus part of the legitimization of positions within the global division of labour. However, attempts to separate English as a European bridge language or from its British and North American value systems are misguided, for English should not be interpreted as if it were primarily a *tabula rasa*. Any claim that English is now a neutral, pragmatic tool for global development is disingenuous, being

> part of the rationalisation process whereby the unequal power relations between English and other languages are explained and legitimated. It fits into the familiar linguistic pattern of the dominant language creating an external image of itself, other languages being devalued and the relationship between the two rationalised in favour of the dominant language. This applies to each type of argument, whether persuasion, bargaining, or threats are used, all of which serve to reproduce English linguistic hegemony.
>
> (Phillipson 1992, p. 288)

Such points emphasize that conflict is inherent in language issues and helps explain why multicultural education, far from encouraging the positive aspects of cultural pluralism, has hitherto been characterized by mutual antagonism, begrudging reforms and ghettoization. Spatial segregation and social isolation have become mutually reinforcing patterns in far too many communities.

There are signs, however, within contemporary Europe that official languages are coming to terms with the realities of integration and globalization. Of the major languages, German is likely to spread as a result of the revitalization of Central and Eastern European commerce. It is Spanish, however, which seems set to be a major international, trans-continental bridge language. In global terms, if Spanish can be seen as a resource for socio-cultural growth as well as commercial gain, there could be a win–win situation rather than a zero-sum stand-off between supporters of English and those of Hispanic-based bilingualism, especially in the Americas (Baker 1996).

Modernized indigenous languages, such as Irish or Catalan, are likely to grow but not displace hegemonic languages, especially within the civil service, or technological and commercial sectors. Renewed languages such as Welsh, Frisian or Basque have penetrated into new domains, such as local administration, education and the media. In contrast, threatened minority languages such as Romany or Skolt Lapp will be further marginalized. Because of their high fertility rates, some groups are experiencing linguistic reproduction rates greater than 1, and their prospects for survival look promising, especially in constructing an infrastructure for domain extension in education, government and broadcasting. Conversely, however, some autochthonous language speakers are rapidly losing their control in traditional core areas as a result of out-migration, capital-intensive economic development and increased mobility. The fact that speakers of lesser-used languages have become highly politicized in the past twenty years should not divert attention away from their fears of cultural attrition. So much of their collective hopes and aspirations rest on the construction of an appropriate political and socio-economic infrastructure.

An additional consideration is that many migrants and refugees who settled in another European country, often as a consequence of the two world wars, have lost the native language of their forebears. To them markers such as diet, music or the visual arts have replaced language as the primary linkages with their wider cultural communities (Rystad 1990). Fundamental questions surround the symbolic bases of their culture and the degree to which one may characterize residual elements as either authentic or expressing an integral identity. This is a major feature of Europe's cultural heritage and will prove a testing ground for more sensitive and flexible applications of any policy of multiculturalism. It is a problem most acutely expressed in the recent rediscovery among many residents of Central Europe of their German heritage and of their potential relationship with a unified and reinvigorated German-dominated *Mitteleuropa* policy.

Perhaps the greatest challenge facing framers of European identity comes from non-European migrants and their descendants, especially as globalizing perspectives will reinforce the need for link languages other than English in this realm. Initially this will result from private and commercial-oriented demands, but as the total size and significance of link languages – especially Islamic-related variants – grows, there will be pressure to reform public agencies and the educational system, particularly in France, the UK, Germany, Belgium and the Netherlands. The increased presence of non-nationals within European states will add to the alienated feelings of many recent migrants that they do not belong by right to any particular state. As Miles argues, the ideological notion of a 'migrant' becomes embedded in the reproduction of hegemony, especially in religious life:

> Once 'we' all know who 'the immigrants' are, and once the state has indicated that their presence is undesirable and that their numbers should be controlled or reduced, all uncritical and unreflexive use of the category legitimates the official definition and the related conspiracy of silence about all other immigrants whom the state does not 'see' and who are excluded from public view.
>
> (Miles 1993, p. 207)

As a result of all these processes, there is a growing demand for a pan-European educational policy in which 'isolated instruction in separate foreign languages should be replaced by instruction for multilingual communication. This requires training in the ability to quickly shift actively or passively from one language to another' (Posner 1991, p. 134). This general maxim is laudable, although it would need to accommodate the requirements of speakers of lesser-used languages such as Breton, Catalan or Friulian, together with major languages – most notably Arabic – which also have a religious function.

Does advocacy of a common educational approach imply that ultimately the EU will become the first postmodern, post-sovereign multicultural political system of the twenty-first century? For the optimist, this emphasis on accommodation, openness and diversity is an expression of a highly developed pluralist society, which demands mutual respect for and tolerance of its constituent cultures. To the pessimist, however, such openness is a recipe for continued strife, inter-regional dislocation, inefficient government, and the artificial reproduction of often misleading cultural identities.

Conclusion

Clearly, the state is deeply implicated in the direction of change with regard to multicultural policy. As society becomes more plural, and social mobility increases, greater tensions occur between the functional provision of bilingual public services and the formal organization of territorial-based authorities charged with such provision. These conflicts are exacerbated by immigration into fragile language areas, which leads to the public contestation of language-related issues as each new domain is penetrated by the intrusive language group. Such sentiments are hard to gainsay. The difficulty lies in determining what proportion of the public purse is to be expended upon satisfying the legitimate demands of this policy. Issues of principle, ideology and policy are frequently no more than thinly disguised disputes over levels of resource expenditure. One arm of government is involved in extending the remit of pluralism while another is reining in the fiscal obligations to so act. Either way, dependent cultures are tied inexorably to the largesse of the state. Governments are obliged to maintain their support for many multicultural projects, albeit simultaneously signalling their intent to withdraw public finances and welcome private-sector funding. Either way, the languages and cultures of visible minorities are in danger of being expropriated by external forces, while cultural dependency is being increased. As they become better organized, however, astute minority groups will press for greater recognition of their cultural rights, seeking the individual choice and empowerment to decide their lifestyle and future prospects as participative citizens. From this perspective, multiculturalism is a set of institutional opportunities for individual and group advancement in a competitive environment. In other words, it becomes a platform for social progress.

The politics of mutual respect presuppose a historically well entrenched democratic order. The watchwords of the open society are redistributive social justice, participatory democracy and mutual tolerance. Majorities must always seek to temper their individual rights in the light of their collective impact on minorities. We must also address the implications of this balancing act, if we are to honour the full range of multilingual expectations and needs in Europe's major cities and densely populated regions. This in turn presupposes a political–juridical framework adequate for ensuring that increased cultural contact will not lead to escalating conflict. Cultural communities are best represented when the state guarantees individual freedom of association and protection. 'Cultural rights protect autonomy. They do this inasmuch as they look to guarantee the stability of the cultural environment within which the individual is able to exercise the capacity to make meaningful choices' (Kukathas 1995, p. 241). Thus the urgent task of geographers, in association with others, is to locate the place of such cultural environments as constituents of European integration, and to identify the impact of policy on all citizens, regardless of the contested values that are ascribed to them.

References

Baker, C. 1996: *Foundations of bilingual education and bilingualism*, 2nd ed. Clevedon, Avon: Multilingual Matters.

Close, P. 1995: *Citizenship, Europe and change*. Basingstoke: Macmillan.

Crystal, D. 1987: *The Cambridge encyclopedia of language*. Cambridge: Cambridge University Press.

Davies, N. 1996: *Europe: a history*. Oxford: Oxford University Press.

Declaració de Barcelona 1996: *Declaració universal de drets lingüístics*. Barcelona: International PEN and CIEMEN.

Edwards, J. 1994: *Multilingualism*. London: Longman.

Elshtain, J. B. 1994: The risks and responsibilities of affirming ordinary life. In Tully, J. (ed.), *Philosophy in an age of pluralism*. Cambridge: Cambridge University Press, 67–80.

Gerholm, T. and Lithman, Y. G. 1990: *The new Islamic presence in Western Europe.* London: Mansell.

Glazer, N. 1977: Individual rights against group rights. In Kamenka, E. (ed.), *Human rights.* London: Edward Arnold, 115–36.

Gordon, D. G. 1978: *The French language and national identity.* The Hague: Mouton.

Gunnermark, E. and Kenrick, D. 1985: *A geolinguistic handbook.* Gothenburg: Gunnermark.

Habermas, J. 1996: The European nation-state: its achievements and its limits. In Balakrishnan, G. and Anderson, B. (eds), *Mapping the nation.* London: Verso, 281–94.

Havel, V. 1990: *Disturbing the peace.* London: Faber & Faber.

Kallen, E. 1995: *Ethnicity and human rights in Canada.* Don Mills: Oxford University Press.

Klein, T. 1978: Minorities in Central Europe. In Hepburn, A. C. (ed.), *Minorities in history.* London: Edward Arnold, 31–50.

Kukathas, C. 1995: Are there any cultural rights? In Kymlicka, W. (ed.), *The rights of minority cultures.* Oxford: Oxford University Press, 228–55.

Laponce, J. A. 1987: *Languages and their territories.* Toronto: University of Toronto Press.

Lijphart, A. 1995: Self-determination versus pre-determination of ethnic minorities in power sharing systems. In Kymlicka, W. (ed.), *The rights of minority cultures.* Oxford: Oxford University Press, 275–87.

Mackey, W. 1991: Language diversity, language policy and the sovereign state. *History of European Ideas* 13, 51–61.

McRae, K. 1997: Language policy and language contact: reflections on Finland. In Wölck, W. and de Houwer, A. (eds), *Recent studies in contact linguistics. Plurilingua* XVII. Bonn: Dümmler, 218–26.

Magosci, P. R. 1993: *An historical atlas of East-Central Europe.* Toronto: University of Toronto Press.

Miles, R. 1993: *Racism after 'race relations'.* London: Routledge.

Nelde, P. 1997: On the evaluation of language policy. In Generalitat de Catalunya (ed.), *Proceedings of the European conference on language planning.* Barcelona: Department de Cultura, 285–92.

Nelde, P., Labrie, N. and Williams, C. H. 1992: The principles of territoriality and personality in the solution of linguistic conflicts. *Journal of Multilingual and Multicultural Development* 13, 387–406.

Phillipson, R. 1992: *Linguistic imperialism.* Oxford: Oxford University Press.

Posner, R. 1991: Society, civilization, mentality: prolegomena to a language policy for Europe. In Coulmas, F. (ed.), *A language policy for the European Community.* Berlin: Mouton, 121–37.

Rystad, G. 1990: *The uprooted: forced migrants as international problem in the post-war era.* Lund: Lund University Press.

Skutnabb-Kangas, T. 1997: Language rights as conflict prevention. In Wölck, W. and de Houwer, A. (eds), *Recent studies in contact linguistics. Plurilingua* XVII. Bonn: Dümmler, 312–24.

Stewart, W. A. 1968: A sociolinguistic typology for describing national multilingualism. In Fishman, J. (ed.), *Language problems of developing nations.* London: John Wiley, 503–53.

Taylor, C. 1991: *The malaise of modernity*. Concord, MA: Anansi.

Taylor, C. 1992: *Multiculturalism and 'the politics of recognition'*. Princeton, NJ: Princeton University Press.

Taylor, C. 1994: Reply and re-articulation. In Tully, J. (ed.), *Philosophy in an age of pluralism*. Cambridge: Cambridge University Press, 213–57.

Times Atlas of World History 1986: London: Times Books.

Tollefson, J. W. 1990: *Planning language: planning inequality*. London: Longman.

Williams, C. H. 1989: The question of national congruence. In Johnston, R. J. and Taylor, P. (eds), *A world in crisis?* Oxford: Blackwell, 229–65.

Williams, C. H. (ed.) 1991: *Linguistic minorities, society and territory*. Clevedon, Avon: Multilingual Matters.

Williams, C. H. (ed.) 1993a: *The political geography of the new world order*. Chichester: John Wiley.

Williams, C. H. 1993b: The European Community's lesser used languages. *Rivista Geografica Italiana* 100, 531–64.

Williams, C. H. 1994: *Called unto liberty: on language and nationalism*. Clevedon, Avon: Multilingual Matters.

Williams, C. H. 1995: Global language divisions. In Unwin, T. (ed.), *Atlas of world development*. Chichester: John Wiley, 296–8.

Williams, C. H. 1996: Citizenship and minority cultures: virile participants or dependent supplicants? In Lapierre, A., Smart, P. and Savard, P. (eds), *Language, culture and values in Canada at the dawn of the 21st Century*. Ottawa: International Council for Canadian Studies, Carleton University Press, 155–84.

Williams, C. H. 1997: Language rights for all citizens of Europe? In Wölck, W. and de Houwer, A. (eds), *Recent studies in contact linguistics*. Plurilingua XVII. Bonn: Dümmler, 430–41.

Zametica, J. 1992: *The Yugoslav conflict*. London: Adelphi Papers of the International Institute of Strategic Studies, 270.

P A R T
IV

Identity and the renegotiation of the meanings of European place

8

European landscape and identity

JOHN AGNEW

Introduction

Fifteen hundred years ago most of Southern and Western Europe was still part of the Roman Empire. Later, other regions of what we now know as Europe were Christianized and incorporated into certain legal, social and military practices emanating originally from a Roman base, however much they changed in passing eastwards and northwards. 'Europe' has long had this connotation as a unified cultural realm, as much as a geographical expression referring to the land mass to the west of the Urals or the river Don or wherever. None the less, this seeming unity is belied by the fact that however much a common past can be fathomed from roots in ancient Rome, and the Christianity which came out of it, much of the history of modern Europe has been of territorial division and attempts at claiming the mantle of a 'new Rome' on the part of the various national states into which the continent was increasingly divided between the fifteenth and nineteenth centuries (Wilson and van der Dussen 1995; Pocock 1997).

Although the ideal of Europe is currently undergoing something of a revival in the guise of expanding and/or deepening the EU, the reality of Europe has long been of making differences between Europeans on the basis of certain common inheritances that have been given distinctive casts in different places. So, even if a given landscape vista can be thought of as having a certain 'Europeanness' to it, compared – say – to North American or African vistas, much stronger influences come from the national, regional and local contexts in which particular vistas are embedded.

Stories of landscape and identity, simple and complex

Dominant images of landscapes for outsiders and nationalizing intellectuals have been national ones. Often these are quite specific vistas turned into typifications of a 'national landscape' as a whole. Quaint thatched cottages in pastoral settings (England), cypress trees topping a hill that has been grazed and ploughed for an eternity (Italy), dense village settlements surrounded by equally dense forests (Germany), and high-hedged fields with occasional stone villages (France) constitute some of the stock images of European rural landscapes conveyed in landscape painting, tourist brochures, school textbooks and orchestral music. Ideas of distinctive national pasts are conjured up for both 'natives' and 'foreigners' by these landscape images. As discussed in Chapter 1, these 'representative landscapes' constitute visual encapsulations of a group's occupation of a particular territory and the memory of a shared past that this conveys (see, for example, Graham 1994, p. 258). They can also be thought of as one way in which the social history and distinctiveness of a group of people is objectified through reference (however idealized) to the physical settings of the everyday lives of a people to whom we 'belong', but most of whom we never meet. Yet these landscape images are both partial and recent. Not only do they come from particular localities within the boundaries of their respective nation-states (respectively, southern England, Tuscany, Brandenburg and Normandy), but their visualization as somehow representative of a national heritage is a modern invention, dating at the earliest to the nineteenth century. The history of these landscape images, therefore, parallels the history of the imprinting of certain national identities on to the states of modern Europe.

The agents of every modern state aspire to have their state represented *materially* in the everyday lives of their subjects and citizens. The persisting power of the state depends on it. Everywhere anyone might look would then reinforce the identity between state and citizen by associating the iconic inheritance of a national past with the present state and its objectives. Yet this association is harder to achieve than might at first appear. In cases such as the English, where the past can be readily portrayed as monolithic and uniform, consensus about a national past with unbroken continuity to 'time immemorial' suggests that a comfortable – even casual – association is easily accomplished. But nowhere else in Europe is landscape 'so freighted as legacy. Nowhere else does the very term suggest not simply scenery and *genres de vie*, but quintessential national virtues' (Lowenthal 1991, p. 213). Even in England, however, not all is as it seems. The visual cliché of sheep grazing in a meadow, with hedgerows separating the fields and neat villages nestling in tidy valleys, dates from the time in the nineteenth century when the landscape paintings of Constable and others gained popularity among the taste-making élite (Rose 1995). Nevertheless, the 'invented' ideal of a created and ordered landscape, with deep roots in a past in which everyone also knew their place within the landscape (and the ordered society it represents), has become an important element in English national identity, irrespective of its fabulous roots in the 1800s.

Elsewhere in Europe, capturing popular landscape images to associate with national identities or inventing new ones has been much more difficult.

The apparently straightforward English case is therefore potentially misleading. It suggests a simple historical correlation between the rise of a national state on the one hand, and a singular landscape imagery on the other, however insecure this may now be in the face of economic decline, north–south differences, a revival of Celtic nationalisms that challenges the presumptions of the English to represent something they alone now call 'British', and immigration of culturally distinctive groups unwilling to abandon their own separate identities (Daniels 1993). National-state formation elsewhere in Europe took a very different direction from that of England, although this is not to say that everywhere else it was the same. Two aspects of the difference are vital.

The first was the complex history of local and urban loyalties in many parts of Europe, particularly those unified in the later nineteenth century as Italy and Germany. In these contexts, there was often a long history of city independence and local patriotism, with little or none of the early commercialization of agriculture and industrialization that swept English rural dwellers into national labour markets and national social class identities at the very same time a state-building élite was strengthening and extending national institutions. The image of a bucolic past tapped the nostalgia of those experiencing the disruptions of industrialization, reminding them that all had not changed. Such landscapes could still be found, even if no longer experienced on a day-to-day basis. Later industrialization often also involved less disruption of ties to place. In particular, as electricity replaced steam-power, industries moved to areas of existing population concentration rather than, as in the case of the English coalfields, requiring that people move to where the industry was.

A second aspect was the external orientation of the English state and economy. The English merchants, industrialists and travellers, who were increasingly dominant within the evolving world-economy of the nineteenth century, were often nostalgic for what they had left behind when they travelled abroad. In their need to compare what they saw with a datum or steady point of view, many of them came to idealize an England in their mind's eye that was largely the product of a merging of their own experience and the renderings of England in paintings and other visual representations. This produced a unified vision that was much harder to achieve in those contexts where influential people travelled less and thus had less need of a single, stable vision.

The idea of a national landscape, however, and also that of national identity, is more complex than the English case might make it appear. A national identity involves a widely shared memory of a common past for people who have never seen or talked to one another in the flesh. This sense of belonging depends as much on forgetting as on remembering, the past being reconstructed as a trajectory to the national present in order to guarantee a common future (Gillis 1994). National histories, monuments (war memorials, heroic statues), commemorations (anniversaries and parades), sites of institutionalized memories (museums, libraries and other archives) and representative landscapes are among the important instruments for ordering the national past. They give national identity a materiality it would otherwise lack. But such milieux of memory must needs coexist with other memories and their identities. National identity does not sweep all others away. As discussed in Chapter 1, some local identities, such as the French *pays* or the German attachment to *heimat* (or home place), while remaining distinct, also feed into a wider national identity (Applegate 1990). Some diasporic groups, however, such as Scottish Hebrideans or many recent immigrants into Europe, retain local or religious rather than the national identities with which they are usually identified by outsiders (Agnew 1996).

The 'sacralization' of the nation-state has never been total; even within totalitarian states, sites of religious and local celebration have had their place. The totalization of national identities has faced a number of barriers. One is the difficulty of what to select from the past to identify and emphasize as distinctive and peculiar. National pasts are fraught with conflicts over dynastic claims, boundary disputes, religious pogroms and the meaning of historic events and personalities. The selection of objects for emphasis, therefore, is also fraught with potential conflict (see Chapters 1 and 9). Whose national past is it, anyway? As a result, national identities can have multiple definitions and are constantly in flux. Allied to the question of what to select from the past must be the presence of the self-conscious conviction that a bounded national space with a considerable degree of internal cultural homogeneity actually exists. A readily available 'Other' against whom to define one's national identity is a requirement for doing this (Conversi 1995). Without this widely shared conviction there is little that is 'national' for the identity to express. A second barrier to the totalization of national identities is the existence of alternative identities (such as class, religion, ethnicity and region) that do not always flow into – or easily coexist with – clear and coherent national identities. As the world-economy has become more integrated, and political boundaries have lost some of their force in regulating economic and cultural flows, these alternative identities have become increasingly potent (Gillis 1994). A third factor has been the re-emergence of powerful transnational identities, usually associated with religious beliefs, but also related to groups adhering to dominant ideologies such as neo-liberalism and its agents, namely international banks, multinational firms and large-scale regulatory organizations (the European Commission and the International Monetary Fund, to name but two). Finally, the idea of 'identity' is itself an intellectual imposition, implying that there are singular, stable and essential divisions with which people identify (Handler 1994). Yet there are societies in which conceptions of the self or person do not require the spatially bounded reference groups and physically bounded individuals, which much academic discussion of identity presupposes. Identifying with a *particular* group need not be a necessity of life. Nor, therefore, need national identity.

Italian unification and the problem of a landscape ideal

Italy stands as a persuasive case-study in examining the connection between landscape and identity. It was at the centre of the 'revolution' of the Renaissance in which visual representation became a vital part of the modern means of communicating the meanings and significance of religious and political messages. It is also a country in which the process of state formation was long delayed by the existence of alternative foci of material life (in particular, city-based economies), and local cultural identities alternative to that of the 'Italian'. Indeed, as a late-unifying state with much internal heterogeneity, it may be at the opposite pole to the English case, in so far as creating a match between a representative landscape and an Italian national identity was a difficult and obviously 'artificial' process from the outset. It thus draws attention to the process of linkage between identity and landscape in more complex ways than can the much-examined English case.

The Italian state, unifying the physically fragmented peninsula and islands, was formed only in the second half of the nineteenth century. Although this tardiness had numerous causes, among the most significant was the existence of strong municipal, city-state and regional-state governments (particularly in the north), which held off the forces pushing the country towards unification. Most importantly, in the late Middle Ages and during the Renaissance, at the same time that the great Western monarchies were consolidating territorial states in England, Spain and France, the politics of northern and central Italy was characterized by a fragmented mosaic of city-states and localized jurisdictions of a variety of types, from principalities to republics. It was the

> extraordinary energy and growing capacity of urban centres [that] led paradoxically to the early elimination from central and northern Italy's political firmament of any *superior* – king, emperor, or prince. The cities transformed themselves precociously into city-states with corresponding territorial dimensions and political functions.
>
> (Chittolini 1994, p. 28)

This is not to say that they did not try to turn themselves into territorial states. Rather, it is that they failed to do so. As their economic strength faded in the eighteenth century, with Europe's centre of political–economic gravity moving north-westwards, the Italian mini-states proved easy prey to the expansionist ambitions of Austria and Spain. Even with foreign domination, however – and this was a major stimulus to the development of Italian nationalism before unification – the capacity of Italian cities to penetrate into adjacent territories was relatively undiminished, although no one city was capable of winning control over the others. When nineteenth-century European politics opened up the possibility of a national state for Italy, the initiative came from a regional state, Savoy-Piedmont, whose social and political structure was different, and of less glorious tradition, from that of the city-based states. Inversely, the rapid fall of the city-based states signals the absence of effective power to sustain them, even though their long survival testified for centuries to the vitality of medieval urban civilization.

(Chittolini 1994, p. 40)

It was from northern Italy and, initially at least, by northern Italians that Italy was made (Figure 8.1). It was the traditions of the city-states and the Europeanness of the Savoyard regime which gave unified Italy its monarchy while providing the 'new' Italy with its mythic resources. The south, and the

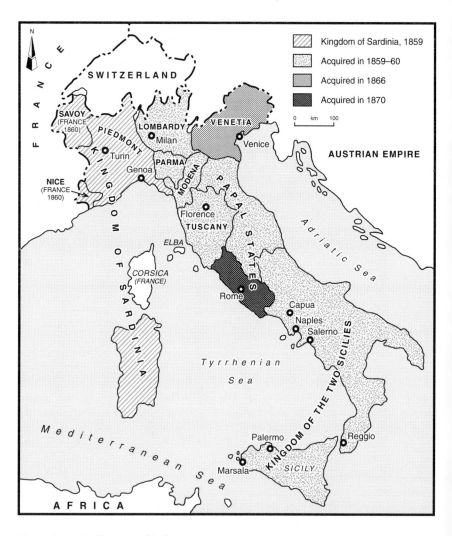

Figure 8.1   Unification of Italy

zones the Austrians had controlled in the north, had been 'won' from foreign domination, which, particularly in the south, was now seen as having created a society that was doubly disadvantaged. These regions were geographically marginal to Europe and politically marginal to the 'high' Italy of Renaissance city-states, from which the new territorial state could be seen as having descended.

Foreign political–constitutional models, particularly those provided by England, France and the new Germany, were also important to the nationalizing intellectuals who established themselves in Rome after the final annexation of that city to the new state in 1870. As the Italian historian Silvio Lanaro (1989, p. 212) notes, cynically, 'admiration for foreigners has been the principal ingredient of Italian nationalism'. Acceptance by other Europeans as a rising Great Power became a particularly important element in national policy that was to last until 1945. This meant taking very seriously what foreigners found exceptional in Italy. The new state could then build on foundations that would lead to respect from others. It was to ancient Rome, both Republican and Imperial, and to certain Renaissance landscape ideals, articulated by foreign visitors to Italy as well as by local savants, that the visionaries of the new state turned. Both of these represented powerful images that would serve double duty: to mobilize the disparate populations of the new state behind it, and to impress outsiders with the revival of a glorious past, only now in an Italian rather than a Roman or a Renaissance form.

Florence and the Tuscan landscape ideal

To turn first to the Renaissance inspiration, the Italian Risorgimento (revival through unification) was largely concerned with re-establishing Italy as a centre of European civilization, as 'it' had been during the Renaissance. Florence, of course, had been the pre-eminent centre of the Renaissance, although long since consigned to the role of *città d'arte* or storehouse for all that Italy had been. It was in Florence in the 1850s that a group of landscape painters set about putting their talents into service for the new state. The so-called Macchiaioli painters (from the various meanings of *macchia*: 'spot', 'sketch', 'dense underbrush') set about defining a representative landscape for Italy. Not surprisingly, they saw Tuscany – the region with which they were most familiar – as the prototypical Italian setting. Not only did it have impeccable Renaissance connotations but it corresponded also with the foreign (particularly English) Romantic attachment to much of northern and central Italy, as expressed, for example, by the early-nineteenth-century generation of poets and writers (Churchill 1980; Johnston 1987). The city of Dante, Michelangelo and Machiavelli was the appropriate centre for a national revival. The Macchiaioli used their Renaissance forebears and European contemporaries (particularly English painters) as their guides, expressing their nationalism through a search for images that could be used to tie the noble past to the developing present:

They searched the riverflats along the Arno, the orchards and farms of the suburbs of Florence, the hill pastures around Pistoia, and the wild Maremma region (with its thick *macchie* of scrub pine and under-brush) for motifs appropriate to their fresh viewpoint. Their topo-graphical specificity and personal response were totally integrated in what might be called a *macchia-scape* – the landscape that retained the sincerity of vision they admired in the Tuscan artists of the Quattrocento, but that also conveyed the modernity and nationalism of contemporary Italian life.

(Boime 1986, p. 33)

Like the Risorgimento itself, the Macchiaioli idea had both Italian and European dimensions. For all their other differences, leaders of the move-ments for Italian unification such as Giuseppe Mazzini and Count Camillo Cavour of Piedmont wanted to bring Italy 'up to date', and to a social and political equality with the rest of Europe. The Macchiaioli were also both nationally and internationally oriented. In Boime's words (1986, p. 36), 'by asserting Italian individuality they hoped to contribute to a release of energies needed to make Italy a great nation, able to assume a role in the affairs of Europe'. Their 'sketch tradition' drew directly on Renaissance prototypes but was linked also by the Macchiaioli to the French Barbizon school, just as the Risorgimento appealed for legitimacy to the French Revolution and the two Napoleons. But the self-attached label of the school also had subversive over-tones, if Boime's (1986, pp. 36–7) analysis has merit. Some of the painters were notorious punsters and self-defined 'outlaws'. The word *macchia* can mean 'hiding out in the woods' (*fare alla macchia*), 'living as an outlaw' (*vivere alla macchia*) or 'publishing illegally' (*stampare alla macchia*). In Florentine dialect, *macchia* has the additional meaning of 'child of the woods,' signifying someone without parents, marginal and disinherited. The wild and wooded Maremma region of southern Tuscany corresponded to this aspect of the Macchiaioli vision, associating the painters with the secret societies (such as the Carbonari) that had stimulated the first efforts at Risorgimento. The Carbonari took their name from the charcoal burners of the forest who laboured in secret away from the gaze of the authorities.

The 'bible' of the Macchiaioli movement, Telemaco Signorini's *Cari-caturisti e caricaturati al Caffè Michelangiolo* (Caricaturists and the carica-tured at the Caffè Michelangiolo), first published in 1893 but based on arti-cles written in 1866–7, re-creates the atmosphere of the popular café in Florence where the Macchiaioli congregated. There were ten 'core' mem-bers: Giuseppe Abbati, Cristiano Banti, Odoardo Borrani, Adriano Cecioni, Vincenzo Cabianca, Vito D'Ancona, Giovanni Fattori, Silvestro Lega, Raffaelo Sernesi and Telemaco Signorini himself. Underlying the book's gos-sip about who said what to whom is a narrative linking the history of the Macchiaioli to that of the Risorgimento. One connecting influence was Romanticism, even though the Macchiaioli were relentlessly realist in their artistic representations. As early as 1813 Laurence Sterne's Romantic mas-terpiece, *Sentimental journey*, was translated into Italian (Tuscan) by Ugo Foscolo, himself a famous poet and writer. Walter Scott's novels were also translated and widely read by literati in mid-nineteenth-century Italy. The great Italian patriot Garibaldi was later often compared to Scott's hero Rob Roy, from the 1817 novel of the same name. In 1827, Alessandro Manzoni came to Florence to purge his allegorical novel of Italian unification, *I promessi sposi* (The betrothed), of its Lombard expressions and rewrite it in the Tuscan (Italian) dialect. In accepting nations as natural forms whose lit-erary canons should reflect this fact, all these projects marked a break with the formal and pedantic works that preceded them. They also appealed to a certain naturalism that finds an important source of the 'spirit' of particular nations in physical landscapes. Manzoni, for example, was fond of saying that he came to Florence to bathe his masterpiece 'in the waters of the river Arno', as if the rewriting required his own presence in the physical sur-roundings, sounds and smells of Florence. Incidentally, this also gave a tremendous boost to the cause of the 'Tuscanizers', those who wanted to establish the Tuscan dialect (because of its historic connection to such great *Italian* writers as Dante) as the national language of the new state (Penman 1972, p. 11).

Another element in Signorini's story of the Macchiaioli is a critique of pre-vious landscape painters, who are seen as preferring foreign (particularly French) scenes to Italian ones. In particular, Signorini praises the autobiogra-phy *I miei ricordi* (Things I remember), of an older historical novelist and painter, Massimo D'Azeglio, which was published in 1867. Although a social conservative, D'Azeglio (famous for his aphorism after unification, 'We have made Italy, now we must make the Italians') argued strongly for a patriotic landscape ideal. This is what attracted the Macchiaioli to him. He celebrated the indigenous (Tuscan) landscape and shared their cultural aims and was heavily critical of his own generation of painters:

> We love independence and nationalism, we love Italy; further, the land-scape painters all chant together 'Rome or death', but when they take up their brushes the only thing they don't paint is Italy. The magnificent Italian landscape, the glorious light, the rich hues of the sky over our heads and the earth we tread; no one considers these things worthy of being painted. Go to exhibitions and what do we see? A scene from the north of France, imitation of so and so; a seascape at Étretat or Honfleur, imitation of someone else; a heath in Flanders; a wood at Fontainebleau, copied from God knows who! . . . They prefer a nature without a soul, without character, weak and tempered like a muted vio-lin. For this they renounce Italy, her sky and the beauty which once brought so many enemies into our land, but today, thank God, brings only friends who never tire of acclaiming it.
>
> (D'Azeglio [1867] 1966, p. 164)

Much of Signorini's story, however, is taken up with relating the history of the café and its patrons to the vicissitudes of the campaign to unify Italy. Even the recollections of friends killed in the wars or the ideals of the Risorgimento conjure up landscape images when at certain moments he recalls them. Especially:

> during a beautiful autumn morning, or on a balmy spring day, or in a winter mist, or amid the sultry passions and strident song of the harvest-time crickets, when it happens that I climb alone the smiling hills of memories which crown our city [Florence]; or stroll along the fields and gardens populated with farmhouses and villas, along the banks of the Mugnone or the Arno, the Mensola or the Affrico, and come upon a small grassy area, off to the side and in the shade; then, having put down my old paint box, the faithful custodian of my personal impressions, inseparable companion of my distant voyages and nearby excursions, I lie down on my back next to it, and gazing intently at the profound blue of the heavens, I return with my thoughts to the past, now having become more significant to me than the future! . . . And my entire past unfolds, not only its mad joys and its daring undertakings, but also its profound sadnesses and its infinite vexations.
>
> (Signorini [1893] 1952, pp. 186–7)

What Signorini finally reveals, therefore, is the deep relationship that existed for the Macchiaioli between landscape and the development of the Risorgimento (Figures 8.2, 8.3 and 8.4). But it is not just any landscape. The landscape impressions are those of Signorini's native Tuscany; of the river Arno, the hills surrounding Florence, the share-cropping peasants who are part of the landscapes in which they appear. The outstanding memories of his

Figure 8.2   Silvestro Lega: *Paese con contadini* (Landscape with peasants), *c.* 1871. Private collection, Montecatini
Source: I macchiaioli nelle collezioni pistoiesi e le evidenze culturali dell'epoca (p. 68)

Figure 8.3 Rafaello Sernesi: *Radura nel bosco* (Forest glade), *c*. 1862–3. Private collection, Montecatini
Source: Tonelli and Hart (1986, p. 138)

life return to him when he recalls the sites depicted by the Macchiaioli. The passage ends on the sad note with which many Italian patriots greeted the way in which Italian unification evolved: dependent on conquest and external (non)intervention more than popular uprising and revolt. Compensation is found in the private moments when art merged with life in the depiction of landscapes that expressed one's ideals and aspirations (Boime 1986).

This culture of the Risorgimento, so closely associated with Florence and Tuscany, was not to endure. Even as Florence became (temporarily) the capital of the new Italy in 1865, and was beginning to assert its position as a national cultural centre, the Macchiaioli started to lose their cohesiveness and common commitments (Boime 1993). Tuscany was not a smaller version of the whole of Italy and Tuscan history was not national history; Florence was not to be the permanent capital of the country. Much like the promise of the Risorgimento itself, the images of the Macchiaioli proved transitory, and

Figure 8.4   Telemaco Signorini: *Sul greto d'Arno* (On the bed of the river Arno), *c.* 1863–5. Gallery of Modern Art, Florence
Source: Tonelli and Hart (1986, p. 139)

they were soon redefined as precursors of Impressionism or simply another school of provincial Italian painters. Only during Fascism (1922–43) were the Macchiaioli once again raised as proponents of an ideal Italy, this time, of course, as precursors of the chauvinistic and ultra-nationalist vision of an older rural Italy, so beloved of the most reactionary Fascists. With such unfortunate friends, rehabilitation has been a long time in coming.

Rome and the Roman landscape ideal

A better-known attempt than that of the Macchiaioli at creating a representative landscape for Italian national identity came to fruition after unification was achieved. This involved looking to the ancient past of Rome as the seat of empire to find inspiration for a new Rome around which the new Italy could be built. The selection of Rome as the capital certainly suggests that the Roman past was in the minds of Italy's unifiers, even before unification was finally achieved. As early as 1861, although not yet part of the new state, Rome was declared the capital. The annexation of the city and its surrounding region not only provided the last chunk of the national territory claimed by Italian patriots but also a 'neutral' centre, not associated, as were Turin, Milan and Florence, with the local élites who had taken hold of the process of Italian unification (Caracciolo 1956). In other words, as Birindelli (1978, p. 23) puts it, Rome 'became the capital not for the qualities that it had but for the ones it was missing'. This political advantage plus the obvious associations with a 'glorious' past gave Rome a lead over its competitors. Further, as

Italian unification was more the result of international diplomacy than of nationalist revolt, the city's international visibility and its critical role in attracting outside support were also important factors.

By way of contrast, German unification during the same period (1850–70) was much more internally oriented. The choice of Berlin reflected both the Prussian dominance of the new state and the Prussian state's prior commitment to economic and military growth as manifested in the growth of Berlin itself. Rome was so different. Rather than being a centre of national prestige or strength, Rome – an ecclesiastical capital without either manufacturing industry or modern bureaucracy – was widely viewed in the new state as a 'parasitic' city that consumed but did not produce (Scattareggia 1988, p. 43).

None the less, the choice of Rome was vital to the architects of a new Italian identity. The city itself was a unifying force across all of the movements for unification. If there was a single 'tradition' that the population of the peninsula and islands held in common, it was that of ancient Rome. The myth of a unified past, however different from the present, underwrote the unified future which the Savoyard monarchy and its aristocratic allies, who had taken control of the Risorgimento, saw for the new state. The city presented – at the least – a strong image for a group concerned that the new Italy might turn out to be too decentralized for their political and economic interests. It also portrayed a vision at odds with the more parochial ones emanating from local élites in Turin, Milan and Florence. Rome represented a central link in a country in which local and municipal attachments were strong; in Tobia's words (1996, p. 180), it was 'the meeting point through which it became possible for municipalism to be projected directly towards a *national* dimension'. Locating the capital in Rome also took on directly the claims of the Pope to be a temporal as well as a spiritual ruler. The Pope remained the one local ruler of pre-unification of Italy to reject the spirit and purpose of Italian independence and unification. As capital, therefore, Rome symbolically embodied various aspects of the jacobinism and centralism that were hallmarks of Italian unification. The myth of Rome was that of a strong centre that counteracted the centrifugal pressures emanating from the real political divisions of the country, represented by other places such as Florence and powerful institutions such as the papacy.

From the outset, the new rulers tried to make Rome a symbolic centre for their regime (Figure 8.5). Initially there was an attempt, under the patronage of the Piedmontese politician Quintano Sella, to establish a new centre of gravity for the city to the north-east, beyond its 1870 core. This largely failed, it being easier and more profitable to local interests to concentrate government offices in the historic core. In this they largely succeeded, expropriating convents, monasteries, palaces and other buildings from the previous papal regime. Another, and more important, symbolic method was by means of 'patriotic building'. This involved locating monuments and public buildings to celebrate the new regime, recall its historic connections, and challenge the singular association of the Roman Catholic Church with the most sacred sites

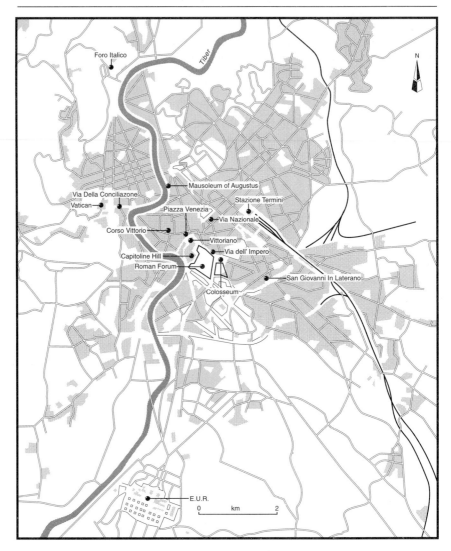

Figure 8.5   Rome

in the city. From one point of view, however, these efforts at securing a new monumental Rome in the years 1870–1922 came to nothing. As Bruno Tobia (1991) has argued in some detail, impressive ideological–rhetorical debate produced little physical change in the city's landscape. Within the historic centre, only the subversive placement of the monument to King Vittorio Emanuele II (the first king of the new Italy) on the edge of the Capitoline Hill (the historic core of the city, next to the seat of the commune and the Roman *fora*) and midway between the Pope's two seats (at the Vatican and, as Bishop of Rome, at San Giovanni in Laterano) provided a powerful symbol of the new national identity in the new Rome. Even then the monument was to a person rather than to some abstract ideal of the nation, although, during the Fascist era, attempts were made to turn the monument into something more representative. The identification with the person of the monarch was particularly problematic, since many of the proponents of unification had been republicans or opponents of the Savoyard monarchy. While the Vittoriano (the monument to Vittorio Emanuele II) may well have been the 'only true national monument' that 'aroused a common national feeling' (Tobia 1996, p. 182), its symbolic power drew attention to the *lack* of commemoration of the real heroes of the Risorgimento: Giuseppe Garibaldi and Camillo Cavour.

From another point of view, however, less focused on individual monuments, the changes in the fabric of the city representing the arrival of a new nation can be seen as more considerable. The reorientation of the axis of the city and the placing of monuments did create a new secular image for the city at odds with the ecclesiastical one that had hitherto predominated. In particular, the placement of the Vittoriano, and its construction in white Brescian marble at odds with the brown tones of surrounding buildings, provided a new visual anchor for the city. Via Nazionale and its western extension, Corso Vittorio, ploughed a new east–west axis through the historic centre, making Piazza Venezia, in front of the Vittoriano, the central hub for traffic as well as the new symbolic centre of the city. Other changes, such as the embankment of the river Tiber, the straightening of streets and the 'regularization' of piazzas into Euclidean shapes, and the transformation of archaeological sites into monumental settings (Figure 8.6), also represented successful attempts at both remaking the city and associating the changes with the glories of the angular and rational city built by the ancient Romans before the 'decadence' of later times (Agnew 1995a).

In the 1920s and 1930s, Fascism continued what had begun under the Liberal regime (Figure 8.7). Two new anchors to the city as a whole emerged over time: the Foro Mussolini to the north-west of the historic core (where the Olympic Stadium now stands) and the Esposizione Universale di Roma (EUR) complex to the south-east; the latter was started in 1937 for an exposition that was never held, and was finally completed in the 1950s. Possibly Mussolini's most important act in terms of the manipulation of urban space for political purposes was the transfer in 1929 of his office from Palazzo Chigi to the Palazzo Venezia in Piazza Venezia. Thereafter, Piazza Venezia became the key space in Rome for performing the ceremonies and the ritual speech-making that were the hallmark of Italian Fascism. Broadcast to central piazzas in towns and cities throughout Italy, Mussolini's speeches from the balcony of Palazzo Venezia created a sense of national 'togetherness' that Italy had never had previously, nor has enjoyed since (apart from the occasions when the national football team is playing).

Mussolini, more and more the personification of Fascism as the years wore on, increasingly turned to ancient imperial Rome to provide a pedigree for his otherwise modernist movement. Reconstructing Rome according to an

Figure 8.6   Excavations of the Roman Forum at the turn of the twentieth century
Source: Sanfilippo (1993, p. 181)

imperial image became a vital part of the agenda of Fascism. Plans were often compromises between different factions and architectural viewpoints. As a result, the outcome in terms of real changes was not always coherent. Segments of roads were built but often never went all the way to their intended destinations. For example, a road was punched through the Roman *fora* between Piazza Venezia and the Colosseum (today called the Via dei Fori Imperiali), but the extension of this road, intended to lead inland and to the Adriatic, was never completed. Drawing the city towards the sea, to celebrate a renewal of an outward, imperial orientation and claim the Mediterranean for Italy as *mare nostrum*, was perhaps the most successfully realized goal, once it was defined. The *autostrada* linking the outskirts of Rome to Ostia, opened in 1928, was the earliest manifestation of this strategy. This was followed by the Via del Mare, linking Piazza Venezia to the southern outskirts, and the EUR project to pull the growth of the city seawards.

However successful as architectural projects, the impact of both Liberal and Fascist attempts at rendering Rome as a representative landscape for the new Italy was severely limited. For one thing, Rome was naturally polycentric. In 1870, the city had a complex structure, a legacy of its past eras of

Figure 8.7 The Via dell'Impero (now the Via dei Fori Imperiali) shortly after its opening in 1933. Mussolini could now see the Colosseum from 'his' balcony in the Palazzo Venezia, just to the right of the Vittoriano Monument (back right). This new road best symbolized Mussolini's choice of associating Fascism with the Roman past
Source: Sanfilippo (1993, p. 389)

expansion and contraction. One consequence was that it lacked a single pre-existing monumental centre that could be captured for the new national identity. Again, the city was still the seat of the Pope, who, until 1929, refused to recognize the new state. As the headquarters of the Roman Catholic Church, Rome was still symbolically connected to the world 'in between' ancient Rome and modern Italy, which the architects of the new Italy had wanted to erase from popular memory in order to celebrate the creation of the infant state. Another problem with Rome as the setting for a representative landscape for Italy as a whole was that the city's morphology evoked too many pasts to offer singular interpretations of what was there. Effectively layers of ruins built up over the centuries, Rome lends itself to the image of Eternal City. But this representation was at odds with that of a new national identity. The eclectic mixture of epochs and influences in the physical fabric of the city leads towards universalistic more than nationalistic interpretations. Rome is a city for the ages, and for all (at least, all Christian) peoples.

The difficulty of realizing a singular landscape ideal in Italy

Italy's very Europeanness worked against achieving a long-lasting association between a particular landscape ideal and an Italian national identity. It remained forever associated with the glories of ancient Rome and the Renaissance, phenomena that the whole of Europe (or, even more expansively, the whole of Western civilization) claims as parts of its heritage. Italianizing these also suffered from four aspects of Italian geography and society that point up the difficulties of realizing singular landscape ideals.

The first is the obvious one that Italy does not have an 'integrative' physical geography. Its geographical identity as a singular unit is undermined by the sharp demarcation between the Po basin in the north and the mountainous spine/coastal plain pattern to the south. As a result, the physical landscapes 'available' for expropriation are remarkably varied, reflecting the terrain, climate and vegetation of a peninsula stretching from the heart of continental Europe almost to the shores of Africa. This geography, working against an effective integration of the modern state (as compared, say, to the France described by Braudel 1988), also produced a widely accepted continental–Mediterranean dichotomizing of Italian population and society that made it difficult to accept the proposition of a singular landscape image. By way of example, Schama (1995, p. 263) points out the importance of the ancient metaphor of rivers as the 'arterial bloodstream of a people' in bringing together facts of physical geography and national landscape images. Unlike in England (with the Thames) France (with the Seine and Rhône) and the Austro-Hungarian Empire (with the Danube), in Italy the major rivers such as the Po and Tiber divide the country rather than bring it together.

Second, certain features of Italy's historical geography also worked against the successful creation of a national landscape ideal. One of these is the absence of a dominant city, such as London in England or Paris in France, to subordinate the country to a singular vision. Rome was only the fifth city of the new state in 1871, exceeded in population by Naples, Milan, Genoa and Palermo. As the capital city it grew vigorously, but still remains politically and culturally predominant only in its immediate hinterland and in parts of the south. Indeed, it suffers from a very negative reputation in other parts of the country, particularly in northern Italy, where it is associated with corrupt politics and inefficient bureaucracy. Related to this is the continuing importance of local and regional identities in Italian culture and society. Dialect differences, local economic interests, and attachment to local customs and traditions remain very strong in Italy. Unlike the case in England, Scotland, France and Germany, but rather like in Spain, class and status distinctions in Italy are expressed in local as much as in national terms of reference. 'Folk' religious beliefs with strong localist connotations have remained strong, and resilient, even in the face of massive urbanization and social change (see, for example, Filippucci 1996; Pratt 1996; more generally, see Badone 1990). Rather than fading away in the face of pressures for nationalization, this powerful '*campanilismo*' or localism has persisted (Agnew 1988b; Levi 1996). It has also proved impervious to co-optation into reinforcing a larger national identity. Thus at the moment in northern Italy a political movement, the Northern League, is attempting to use local identities as the basis for a programme of radical federalism or, indeed, outright secession (Agnew 1995b).

The two words for 'country' in standard Italian, *paese* and *patria* – the first used easily and frequently, the second self-consciously and infrequently – give support to this polarizing of local and national identities. The first is essentially used to describe the local area. The second refers to Italy as a whole and is a formal term that rarely arises in everyday conversation. In neither usage does it translate as 'countryside', which is the alternative meaning to that of 'country'; as *patria* in English. Part of the problem has been that the national institutions created at the time of unification have been seen by significant minorities as 'foreign' impositions. Not only were the Savoyard monarchy and its affiliated institutions carried to Rome, but the state also brought novel practices to regions where the writ of any sovereign was historically weak (as in Sicily, for example). Further, many groups (such as serious Catholics and anarcho-syndicalists) regarded the state itself as illegitimate. The absence of a widely accepted civic nationalism or patriotism made inventing a singular vision of the state next to impossible. Fascism was, among other things, a way of trying to *force* unification; of bringing about the national uniformity and autarchy that the dominant royalist strand of the Risorgimento had promised. Its failure, then, is revealed by the ready reversion to localism and particularistic identities that followed its wartime defeat.

Third, and perhaps most significantly in accounting for the absence of a singular landscape ideal, Italy has lacked the dominant heroic event or experience upon which many singular landscape ideals are based. In England the shock of industrialization produced a romantic attachment to a rural/pastoral ideal that has outlasted the original historical context. For the United States, the myth of the frontier and the subjugation of 'wilderness' has likewise served to focus national identity around themes of survival, cornucopia and escape from the confines of city life (Agnew 1988a). In Italy only the recycling of the Risorgimento serves a similar purpose (Dalle Vacche 1992). The problem, however, is that this concept has multiple messages that have varied from its origin, depending which 'side' of the unification process one chooses to emphasize. The landscape legacy of the Risorgimento is likewise divided, effectively becoming Florence versus Rome. When allied to disputes over later mythic episodes such as the impact of Fascism, the resistance to Fascism (1943–45) and the unrest of the period 1968–85, the net effect is to produce multiple interpretations of Italianness and its essential landscape that persist but mutate over time in response to dramatic events, rather than a stable, singular interpretation that serves to knit all Italians together.

Finally, Italian unification was never able successfully to capture the religious beliefs and practices of the Italian population. From 1870 until 1929 the state remained alienated from the Church, denied access to its spiritual authority. Fascism tried to 'sacralize' the state by building an alternative civic religion. But this had to coexist with existing religious affiliations and the physical presence of the Pope. As a result, the ritualistic power of the Italian state remained compromised, never able to obtain that symbolic investment in its attempts at designating certain sites as sacred to the nation and landscapes as representative of the spirit of the people that seems to arise so much more effortlessly in, say, the English or German cases.

Conclusion

As a new, more politically integrated Europe seems set to arise, Italians are among its most fervent supporters. They have long been ready to operate at the geographical scale of Europe. Their most important myths – ancient Rome and the Renaissance – are broadly European more than they are narrowly Italian. The difficulty in achieving the singular landscape ideal, more readily acquired by some other European nation-states, suggests that such ideals are historically contingent and not without their own problems. When British 'Eurosceptics' lament the loss of sovereignty in the new Europe, they point almost intuitively to a material or *physical* England that is also under threat. In this view, therefore, the strong visual image of an unchanging rural England has become a major political resource *and* liability, encouraging intransigence over further moves towards European unification, yet undermining rational discussion about the pros and cons of centralized and decentralized governance at the level of Europe as a whole. Italy, whose small-firm manufacturing industry was also always seen (particularly by English commentators) as a symptom of 'backwardness' before it became the engine of the country's recent economic renaissance (see Chapter 6), has no such singular landscape (or other) ideal of 'how it is' and 'what it is like' to hold it back. This not only fits better the evolving world-economy as it is reorganized around localities and city hinterlands more than national economies (Strassoldo 1992), but also undermines the simple story about national identities and representative landscapes that many geographers (and others) tell themselves. History does not 'end' with a representative landscape for a secure national identity. This modernist story is one we have been telling ourselves for so long that we have forgotten that it is only one story (Hastrup 1992). Italy tells another one – one that has much more to say about the Europe of tomorrow.

References

Agnew, J. A. 1988a: La città nel contesto culturale e i valori ambientale. In Muscarà, C. and Pagani, L. (eds), *Natura e cultura nella città del futuro*. Bergamo: Consorzio del Parco Regionale dei Colli di Bergamo, 31–5.

Agnew, J. A. 1988b: 'Better thieves than Reds'? The nationalization thesis and the possibility of a geography of Italian politics. *Political Geography Quarterly* 7, 307–21.

Agnew, J. A. 1995a: *Rome*. Chichester and New York: John Wiley.

Agnew, J. A. 1995b: The rhetoric of regionalism: the Northern League in Italian politics, 1983–1994. *Transactions of the Institute of British Geographers* NS 20, 156–72.

Agnew, J. A. 1996: Liminal travellers: Hebrideans at home and away. *Scotlands* 3, 31–42.

Applegate, C. 1990: *A nation of provincials: the German idea of Heimat*. Berkeley and Los Angeles: University of California Press.

Badone, E. (ed.) 1990: *Religious orthodoxy and popular faith in European society*. Princeton, NJ: Princeton University Press.

Birindelli, M. 1978: *Roma italiana, come fare una capitale e disfare una città*. Rome: Savelli.

Boime, A. 1986: The Macchiaioli and the Risorgimento. In Tonelli, E. and Hart, K. (eds), *The Macchiaioli: painters of Italian life, 1850–1900*. Los Angeles: Wight Art Gallery, UCLA, 33–71.

Boime, A. 1993: *The art of the Macchia and the Risorgimento: representing culture and nationalism in nineteenth century Italy*. Chicago: University of Chicago Press.

Braudel, F. 1988: *The identity of France*, vol. I: *History and environment*. London: Collins.

Caraccioli, A. 1956: *Roma capitale: dal Risorgimento alla crisi dello stato liberale*. Rome: Rinascita.

Chittolini, G. 1994: Cities, 'city-states,' and regional states in north-central Italy. In Tilly, C. and Blockmans, W. P. (eds), *Cities and the rise of states in Europe, A.D. 1000 to 1800*. Boulder, CO: Westview Press, 28–43.

Churchill, K. 1980: *Italy and English literature, 1764–1930*. Totowa, NJ: Barnes & Noble.

Conversi, D. 1995: Reassessing current theories of nationalism: nationalism as boundary maintenance and creation. *Nationalism and Ethnic Politics* 1, 73–85.

Dalle Vacche, A. 1992: *The body in the mirror: shapes of history in Italian cinema*. Princeton, NJ: Princeton University Press.

Daniels, S. 1993: *Fields of vision: landscape imagery and national identity in England and the United States*. Princeton, NJ: Princeton University Press.

D'Azeglio, M. 1966: *Things I remember*, trans. E. R. Vincent. London: Oxford University Press.

Filippucci, P. 1996: Anthropological perspectives on culture in Italy. In Forgacs, D. and Lumley, R. (eds), *Italian cultural studies*. Oxford: Oxford University Press, 52–71.

Gillis, J. R. 1994: Memory and identity: the history of a relationship. In Gillis, J. R. (ed.), *Commemorations: the politics of national identity*. Princeton, NJ: Princeton University Press, 3–24.

Graham, B. J. 1994: No place of mind: contested Protestant representations of Ulster. *Ecumene* 1, 257–81.

Handler, R. 1994: Is 'identity' a useful cross-cultural concept? In Gillis, J. R. (ed.), *Commemorations: the politics of national identity*. Princeton, NJ: Princeton University Press, 27–40.

Hastrup, K. (ed.) 1992: *Other histories*. London: Routledge.

I macchiaioli nelle collezioni pistoiesi e le evidenze culturali dell'epoca 1983. Pistoia: Edizoni del Commune di Pistoia.

Johnston, W. M. 1987: *In search of Italy: foreign writers in northern Italy since 1800*. University Park, PA: Penn State University Press.

Lanaro, S. 1989: *L'Italia nuova: identità e sviluppo, 1861–1988*. Turin: Einaudi.

Levi, C. (ed.) 1996: *Italian regionalism: history, identity and politics*. Oxford: Berg.

Lowenthal, D. 1991: British national identity and the English landscape. *Rural History* 2, 205–30.

Penman, B. 1972: Introduction. In Manzoni, A., *The betrothed*. London: Penguin.

Pocock, J. G. A. 1997: What do we mean by Europe? *Wilson Quarterly* 31 (Winter), 12–29.

Pratt, J. 1996: Catholic culture. In Forgacs, D. and Lumley, R. (eds), *Italian cultural studies*. Oxford: Oxford University Press, 129–43.

Rose, G. 1995: Place and identity: a sense of place. In Massey, D. and Jess, P. (eds), *A place in the world? Place, cultures and globalization*. Oxford: Open University/ Oxford University Press, 87–132.

Sanfilippo, M. 1993: *Le tre città di Roma: lo sviluppo urbano dalle origini a oggi*. Rome: Laterza.

Scattareggia, M. 1988: Roma capitale: arretratezza e modernizzazione (1870–1914). *Storia Urbana* 42, 37–84.

Schama, S. 1995: *Landscape and memory*. London: HarperCollins.

Signorini, T. 1952: *Caricaturisti e caricaturati al Caffè Michelangiolo*. Florence: Le Monnier.

Strassoldo, R. 1992: Globalism and localism: theoretical reflections and some evidence. In Mlinar, Z. (ed.), *Globalization and territorial identities*. Aldershot: Avebury, 35–59.

Tobia, B. 1991: *Una patria per gli italiani: spazi, itinerari, monumenti nell' Italia unità (1870–1900)*. Bari: Laterza.

Tobia, B. 1996: Urban space and monuments in the 'nationalization of the masses': the Italian case. In Woolf, S. (ed.), *Nationalism in Europe, 1815 to the present: a reader*. London: Routledge, 171–91.

Tonelli, E. and Hart, K. (eds) 1986: *The Macchiaioli: painters of Italian life, 1850–1900*. Los Angeles: Frederick S. Wright Art Gallery, UCLA.

Wilson, K. and van der Dussen, J. (eds) 1995: *The history of the idea of Europe*. London: Routledge.

9

The question of heritage in European cultural conflict

JOHN E. TUNBRIDGE

Introduction: culture, nationalism and heritage in Europe

Within the book's broad concerns with identity and multiculturalism, this chapter pursues the issue of heritage and its role in the renegotiation of the meanings of European place, already outlined briefly in Chapter 1. It is argued that heritage, as a political instrument and economic resource, both reflects and perpetuates the divisions within European society and culture. The latter is interpreted as a set of human qualitative characteristics (creativity, communication, belief systems), which, although nebulous, is deep-rooted and reflective of social structures. The discussion focuses primarily on northern Central Europe, a region which epitomizes, in the very problems of its own volatile definition, the contested nature of European culture and heritage, past and present.

Heritage is that which we have inherited from the past and, as such, its basic meaning appears straightforward enough. Since the 1970s, however, the concept has become interwoven with other critical contemporary issues such as environmental amenity, lifestyle diversity and the promotion of place images, ultimately under the rubric of 'postmodernism' with its reassertion of values such as human scale and distinctive local identity. One important consequence of the escalating intellectual and economic profile of heritage lies in its expanding meaning, the term being applied not only to our historic environment, both natural and built, but to every dimension of material and intellectual inheritances and cultural identities (see, for example, Herbert 1995). It remains the case, however, that landscapes are among the most pervasive and, for many, emotive expressions of visible heritage (Lowenthal 1994). In part as a consequence of this much expanded role in contemporary societies, heritage is regarded increasingly as a contested concept; one, moreover, that is substantially implicated in human conflict (Tunbridge 1984). The straightforward question 'whose heritage?' engages with every facet of human differentiation, be it class, religion, ethnicity, gender or location.

The centrality of heritage within the contested nature of societies has recently been conceptualized through the notion of dissonance, which can be defined as 'a discordance or a lack of agreement and consistency' in the definition of heritage (Tunbridge and Ashworth 1996, p. 20). This intrinsic contestation ensures that heritage is implicated in the constructs of inclusion and exclusion – of inheritance and disinheritance – that are characteristic of Europe's contemporary multicultural societies. Many diverse reasons can be advanced to explain why particular individuals and social groups do not identify with officially designated heritages, often – but not necessarily – constructed at the national scale. The manner in which heritage is commodified to target an array of markets involves the creation of a variety of place products at different scales, carrying with them a cacophony of messages that reflect contrasting values, priorities and agendas. Frequently, conflict occurs between the uses of heritage as a record of cultural creativity, an instrument of political control and a source of economic enrichment. Again, human heterogeneity remains fundamental to the nature of heritage but virulent hatreds may be rooted in such diversity, the source of atrocities which in themselves become a singularly tormented form of heritage dissonance over who inflicted what upon whom. Cultural and political schisms are also central to the concept of heritage dissonance, which often reflects other forms of social, political and cultural contestation.

In contemporary terms, heritage is socially constructed, whether by intent or default, and it is beyond dispute that such reconstruction of the past also perpetuates its conflicts. These latter are expressed frequently in terms of competing national identities, or in conflicting regional identities at variance with these – a discordance, indeed, which is almost the defining cultural feature of Europe. Before we see in this diversity a recipe for despair, however, it follows that heritage can be manipulated to produce both positive and negative outcomes. This emphasizes that the importance of heritage rests less in the objects that comprise that nebulous concept than in the meanings that are attached to them. The same objects can carry an array of meanings that change through time as societies and their renditions of place are continuously renegotiated.

Europe has experienced successive invasions of culturally distinctive peoples over more time but less space than most other world regions. Its intricate topographic detail and interpenetration of land and sea have exacerbated the resulting overlays and intermixtures of cultural elements, most obviously those of language and religion. The longevity of recorded history has further sensitized local identities, not to speak of grievances over territories lost and other perceived affronts between neighbours. These rival identities are often language-based, particularly between a residual 'Celtic fringe' in the far north-west, Latin languages in the south, Teutonic in the north and centre, and Slavonic over most of Eastern Europe. Often combined with religious animosities, such differences provided the raw material for the construction of the various manifestations of Otherness, fundamental to the rise of the nation-state (see Chapter 7). So much of Europe's heritage is structured through the medium of language, and thereby compartmentalized into mutually unintelligible packages based upon widely divergent interpretations of history, that this factor alone underscores the potency of the concept of a continental heritage, still, however, largely comprising conflicting solitudes.

The emergence of the historic conservation movement, which has its origins in early-nineteenth-century Romanticism, was quintessentially a European-led phenomenon and one closely associated with the creation of nationalisms. Heritage institutionalization, even sacralization, and the creation of national identities have crystallized the internally centripetal, but externally centrifugal, forces among the states of Europe. Nowhere is this more obvious than in their capital cities, the core morphologies of which exude national glory, their monumental magnificence typically accentuated by imperial imagery. Inter-state rivalry found its ultimate expression in the global wars which wreaked such havoc upon that same built heritage in the twentieth century. The subsequent memorials of war – often, although not necessarily, cast in nationalist terms – can provide a fertile heritage stimulus to repeated cycles of conflict.

In the late twentieth century, Europe has striven to overcome the legacy of nationalism that so nearly destroyed it. Subcontinental defensive and economic orders, albeit dichotomized into two rival ideological camps for four decades, have sought to subordinate national to wider identities. Tunbridge and Ashworth (1996) interpret this as a phase in a longer-term oscillation between non-national and national identifications in Europe, which may or may not lead to a future subordination of nationalistic heritage interpretations. But in the 1990s, despite both continental and globalizing influences, European national identities are proving obstinately durable if not resurgent (Hooson 1994; Johnson 1995); indeed, some that were latent have vigorously reasserted themselves following the removal of the ideological veneers of the Cold War.

National perceptions of the Otherness of different states are as durable as those of one's own, a factor especially pertinent in North-Central Europe, where perceptions, definitions and heritage identities are notoriously divisive. As we have seen in Chapters 3 and 4, the end of the East–West dichotomy has given new life to the spirit of the Germanic expansionist concept of *Mitteleuropa* (Brechtefeld 1996) and, notwithstanding the pan-European commitment of the German government and its preoccupation with the internal difficulties of reintegrating its own east and west after forty years apart, many fear the accelerating influence of a newly resurgent Germany, reaching as far as the former Soviet republics. In examining the relationships between heritage and identity, we should not lose sight also of the economic allure of the past: Western Europe is the long-standing global leader in heritage tourism and the countries of Central and Eastern Europe are seeking to emulate its success. In political terms, however, heritage thus far has been negatively implicated in perpetuating the divisions and mutual suspicions within Europe. The central question therefore arises: how might this ambiguous concept be recast – or commodified – to promote the opposite effect of unifying Europe? One critical dimension of the issue is the existence of minorities in virtually all states, for whom the incumbent heritage projections have limited, if any, meaning.

European minorities old and new

The length and intensity of human interaction within the narrow confines of Europe have, not surprisingly, given rise to a profusion of minorities to whom national heritages are more or less dissonant in the sense that they reflect exclusion and disinheritance. While some social elements within the majority population might now be so perceived, we will concern ourselves here with the principal issue of cultural (linguistic and religious) minorities. In terms of nation-state cohesion, the most problematical minorities are those which are regionally concentrated, typically the remnants of antecedent populations displaced or assimilated elsewhere, or peoples arbitrarily federated by external power. Elements within such groups may actively, or potentially, nurture grievances over both minority status and historic injustice, and may take the search for a separate identity to the point of claiming independence. These processes have been central to the 1990s reordering of Eastern Europe, where Slovenia, Slovakia, Estonia and potentially Chechnya point to contrasting ways in which independence might actually be achieved.

Minority conflicts are a potent source of conflict, even in the oldest and apparently most stable states of Western Europe. The UK is beset by complex and enduring 'Celtic fringe' problems as the Irish, Scots and Welsh continue long-standing quests for degrees of autonomy or independence. France has a Breton and (recently violent) Corsican issue, while the linguistic divide between its Teutonic north and Latin south is manifested in the idea of an Occitan cultural region that stretches from the Pyrenees to the Alps and seeks to challenge the hegemony of Paris and the Île de France. Northern Scandinavia has its Lapp minority, while Spanish national identity is challenged by regional nationalisms in Euskadi (the Basque Country) and Catalonia. Among regionally concentrated minorities, distinctive heritage expressions in landscape and townscape typically perpetuate both a territorial sense of identity and also – and more dangerously – the importance of conflict in identity constructions. Ireland and Wales are classic examples where the built environment contains elements that can be interpreted, if one so wishes, as symbols of English conquest and occupation, even after Irish post-independence iconographic adjustments and Welsh linguistic reassertion (Carter 1989; Johnson 1994).

Ancient but spatially dispersed minorities present a different and distinct geographical problem. Vance (1977) documents the diffusion of trading minorities through Europe during and after the Roman Empire. In medieval Europe, these peoples – despite having alien cultures – were valued for their commercial contacts and were allotted 'quarters of tolerance' in mercantile cities. They included Italian Lombards in North-West Europe and German merchants in the east. The latter were responsible for the Hanseatic League, which in the fourteenth century comprised more than 200 mercantile cities, largely but not entirely located along the Baltic and North Sea littorals. It was the Jews, however, who formed the oldest, most numerous and widely dispersed minority. Despite religious antagonisms, their commercial skills had enabled them to settle as far as Germany even before the fall of Rome. Notwithstanding a history of cyclical persecution, they later diffused more widely, becoming concentrated in Eastern Europe in more recent centuries (Golding 1938). Their geographical distribution remained largely a matter of official control within the cities to which their livelihoods mostly restricted them, again by official edict. Jews became synonymous with the 'ghetto' (a term originally derived from their designated quarters in Italian cities). Their former presence is still recalled by street names such as 'Judengasse' or 'rue des Juifs' (Vance 1977). The greatest concentration of Jewry developed in Poland, where they accounted for one-sixth of the population in 1939 and even constituted a majority in some towns (Gruber 1992).

It is with respect to ancient dispersed minorities, most particularly the Jews but also smaller groups such as the Romanies (Gypsies), that culture and heritage have been most tragically an issue of conflict beginning within European states. In the nineteenth century, growing anti-Semitic resentment stimulated violent pogroms against Jews in Eastern Europe. Anti-Semitism reached its apogee under Nazism. The German occupation of Poland in 1939 and the western Soviet Union in 1941 provided the opportunity for reghettoization, which was followed by the 'Final Solution', the extermination of 6 million Jews in what is now known as the Holocaust. This nightmare and its aftermath remain inseparable from the contestations of heritage, identity and multiculturalism in Europe today. The disappearance of nearly all Jews has created a ghostly non-presence throughout Central and Eastern Europe (see Figure 3.7). The physical fabric of their cultural markers has often been abandoned or put – frequently insensitively – to some other use. The post-Cold War era has seen a surge of Western tourist interest to discover this lost heritage (Gruber 1992), which, in the case of Kraków's 'Schindler tourism' (created largely by the success of the film *Schindler's List*), is being appropriated economically by non-Jewish Poles who might previously have rejected such manifestations of Judaism in their past. A further example is provided by Prague's Jewish Museum, a multi-building repository of Jewish culture, initiated ironically by Hitler with looted artefacts to create a museum of a vanished race. Its contemporary display of children's art from Auschwitz encapsulates the reality that the Central and Eastern European city is but one aspect of Jewish heritage of Europe. The other is the sacralization of the places of genocide. There were four actual extermination camps at Chelmno, Belsec, Sobibor and, the largest of all, Treblinka, all located in Poland. These were destroyed before the Germans began their retreat west, the site of Treblinka, for example, being marked only by a memorial. These places and the concentration camps that do survive, including Majdanek and, most infamously, Auschwitz/Birkenau (the latter being the extermination section), have become part of a European heritage of atrocity (Tunbridge and Ashworth 1996). But even the death and concentration camps are contested heritage, Auschwitz – now perhaps the pre-eminent symbol of the Holocaust – also being the subject of a 'steady Catholicizing process' in which it significantly stands as a 'symbol for Poland's role in Catholicizing Europe, in the past, now, and in the future' (Charlesworth 1994, p. 591). Furthermore, the continent's heritage of atrocity has been extra-Europeanized, most notably, for example, in the US Holocaust Memorial Museum in Washington, DC, and above all in the state of Israel, where its legacy is one vital factor in exacerbating the Israeli–Palestinian conflict.

Seen from a broad continental perspective, it is a trenchant irony that Europe's unwanted and nearly eliminated dispersed minority has been replaced by others in the half-century since the Holocaust. The need for '*Gastarbeiter*' to power German industrial recovery was met chiefly by the importation of Turks, who now constitute over 2 per cent of the national population and form (other cultural differences aside) a larger non-Christian presence than did the Jews in 1939. Following the unification of the two Germanies in 1990, other external cultures have also been imported. Furthermore, by virtue of its post-1945 constitutional commitment to admit refugees, Germany has borne the brunt of post-Cold War immigration from the East and much of the Third World during the 1990s. This has fuelled violence from right-wing extremists, disturbingly reminiscent of the 1930s, as well as competitive tensions between the minorities themselves. Other western European countries have echoed these developments, particularly the immigration from former colonies to the UK, France and the Netherlands. Source countries are as diverse as the former empires, generating similarities with Germany in the proportions of non-Christian populations (Asian Muslims, Hindus and Sikhs in Britain, North African Muslims in France) but marked differences in the regions of the Third World with which cultural links are thus perpetuated. While right-wing extremism has been widely provoked, the sheer diversity of cultures and directions of origin has impeded a co-ordinated European response, particularly as Germany looks east while France and its Mediterranean neighbours look south.

What role does heritage play in this new arena of European minority cultures and conflict? How far have newcomers adapted existing heritage expressions and asserted their own (both widely encountered in North America), and found acceptance from mainstream Europeans in the process? Unlike many of the former (and remaining) Jews, these newcomers arrived as low-wage workers and, decades later for some, disproportionately remain so. Their imprint on the identity of the poorer urban districts, where most live, tends to be expressed more in streetscape and lifestyle details than in structural creativity. Signage and graffiti, modes of dress, and – most commonly – 'ethnic' restaurants and other businesses provide a distinctive cultural ambience to such areas. But often this is no more than a superficial addition to a built townscape which – some places of worship aside – continues to assert the indigenous identity. For example, Kreuzberg, now a mainly Turkish inner-city district of Berlin, portrays such a veneer on traditional low-rise German apartment blocks. The 'insurgent' tone of posters and graffiti, however, expressing solidarity with other oppressed minorities against civil disabilities and right-wing extremism, signals dissonance rather than an orderly blending of cultures. Kreuzberg is mirrored to varying extents in minority urban districts throughout Western Europe and may be increasingly so in Eastern Europe, as either the economic attractiveness of its cities or denial of access to those further west intensifies. Such immigrant districts, which can be inner-city or (as in Paris) peripheral, may not comfortably express the heritage of newcomers and are also effectively written off by the mainstream population whose national symbolism lies elsewhere. In such contexts of dissonance, archetypal English rural landscapes can assume a more sinister significance. Kinsman (1995) discusses the reactions of a black Englishwoman to the English countryside, which she sees as dangerous because its residents regard her as an alien from the inner city. The reluctance of many indigenous Europeans to expand their deeply ingrained sense of cultural identity and heritage to include newcomers is unsurprising in light of their past failure, more or less, to embrace minorities of many centuries' standing.

As discussed in Chapter 1, multiculturalism provides the only viable basis on which a European heritage can be constructed. The continent's cultural and heritage conflicts cannot be resolved without, first, a mutually acceptable renegotiation of diverse regional and national identities; second, reconciliation with the survivors of old dispersed minorities and their now extra-European descendants; and finally, the recognition of new citizens within the European fabric. The first makes progress (as in Britain and Spain); the second is well advanced (necessarily led by penitent Germany); but the third – involving perceptions of race as well as culture – is still at an early stage. While these issues complicate the task of achieving a form of European identity, paradoxically this further infusion of multiculturalism could provide one means of finally reconciling such dissonances at the national level.

Misplaced peoples, misplaced heritages: the case of 'Lost Germany'

Human ebbs and flows, both within and beyond the continent's shores, have characterized Europe's population for millennia. Although most are now a matter of emotional amnesia, albeit occasionally revived by the nostalgia of heritage tourism, some more recent involuntary migrations remain central to European cultural memories. The Irish, for example, have adopted the Jewish term, diaspora, to refer to the scattering of the island's peoples in the wake of famine and poverty during the eighteenth and nineteenth centuries. In Eastern and Central Europe, it is the more recent demographic upheaval associated with the final throes of World War II that is still pervasive. Almost at a stroke, millions of people were displaced from their cultural roots during the closing months of the war, above all from east to west. As the victorious Red Army pushed Soviet national boundaries and ultimately its sphere of influence far to the west, perhaps as many as 12 million Germans were forced out of East Prussia, Silesia and Pomerania and the Czech Sudetenland. Many Poles, ejected from the eastern territories lost to the Soviet Union, moved into those areas so vacated, together with many other Eastern Europeans who were driven out, dislocated by the chaos of war, or opted to leave (Figure 9.1). In this involuntary stepwise human detachment from its heritage, Germany's lost Breslau (Wrocław) was filled with Poles mourning a lost Lwów (once Lemberg, later L'vov, and now Ukrainian L'viv). The resulting dissonance has been aggravated by more recent westward moves by surviving minorities, most notably ethnic Germans from Romanian Transylvania.

Many factors complicated the attitudes of these incoming populations towards the retention or even reconstruction of the heritage of others thus acquired. Townscape features are sometimes common to different language groups as a result of shared regional physical environment or shared cultural overlays left by now defunct empires. For example, there are distinct similarities in the architectural heritages of the former Hanseatic trading cities, irrespective of the countries in which they are now located. In any case, linguistic Otherness does not necessarily coincide with alienation in other values, most notably those of religion. Such factors were compounded by the superimposition of a new ideology, as Eastern European cities were modified or reconstructed in a more or less visibly socialist mould after 1945. Two examples of lost German cities serve to demonstrate the different outcomes for heritage thus misplaced.

Gdańsk in Poland had been Danzig in the German Empire, and remained German-speaking after being made a Free City in 1919. Its forcible repossession by Germany was in fact the immediate cause of World War II, with the invasion of adjacent formerly German territory, then in Poland. Germany's defeat in 1945 resulted, however, in the city, largely in ruins after a long siege by Soviet forces, being taken over by Poland. Contrary to superficial logic,

Figure 9.1 The changing political geography of North-Central and North-Eastern Europe during the twentieth century

the Poles did not raze the ruins, but gradually reconstructed the Gothic/ Baroque city centre essentially as it had been in 1939 (except for the absence of German monuments and a pragmatic reduction in population density behind the visible main streets) (Figures 9.2 and 9.3). They did likewise in some other formerly German cities in Poland, including Breslau/Wrocław (Tunbridge 1994). The fundamental reason lay in Polish identification with the heritage of such cities, Gdańsk originally having been a Polish port before its take-over in the Middle Ages by the Hanseatic League. Subsequently reverting to Poland again, it was acquired by Prussia only in the eighteenth century to become part of the unified German state in 1871. In its morphological identity, Gdańsk, like other Hansa cities, has a Baltic regional quality, the abundance of red brick reflecting the realities of the local building material as much as broadly Germanic motifs. Poles perceive the city as relatively Germanic but, none the less, identify it as rightfully Polish.

Figure 9.2 Gdańsk: St Mary's Church (Gothic red brick) and the statue of Neptune in the central square
Source: Author

Figure 9.3 Gdańsk: the old city and recently reconstructed harbourside warehouses located in the area still ruined from World War II
Source: Author

Germans, however – close, affluent and typically wishing to see the Danzig of their birth or ancestry – form the principal contemporary tourism market for Gdańsk. Its Polish reappropriation has to be reconciled with this economic commodification through a shared identification with the restored fabric, an abundance of German literature compensating for the general absence of German streetscape marking, and the comparative neutrality of the tourism goods on sale (e.g. Baltic amber). However, heritage is not frozen in time. While Gdańsk is a fine example of Renaissance Europe's urban culture, during the 1980s the city also became the symbol of Poland, its shipyard being the home of the Solidarity trade union resistance to Communism, which contributed so much to the reordering of Eastern Europe and thus to German reunification.

Until April 1945, Kaliningrad – across the Gulf of Danzig from Gdańsk – had been German Königsberg almost continuously for some seven hundred years, prior to its surrender, following a bitter siege, to the Red Army. It was the seat of the medieval Teutonic Order of German knights, and the base for their colonization of East Prussia. Its great castle became the place of coronation for the later Prussian kings, and the most visible symbol of German power thrusting eastwards into the Slavic lands. In unified Germany after 1871, Königsberg retained its military importance as the eastern bastion of German naval power in the Baltic. It too, like Danzig, had been a member of the Hanseatic League and remained the mercantile focus of East Prussia. Königsberg, moreover, was an important academic centre, notably associated with Immanuel Kant (1724–1804), perhaps the greatest European philosopher since classical times. During World War II, Königsberg's role as a symbol of German nationalism reached its apogee as Hitler (whose Eastern Front headquarters, the *Wolfsschanze* (Wolf's Lair), was located close by at Rastenburg) used the city as a repository for looted Russian cultural treasures. The most famous was the legendary Amber Room panelling, originally made, ironically, in Königsberg for Frederick I of Prussia and sent to St Petersburg in 1716 as a gift for that city's founder, Peter the Great. Its disappearance from Königsberg in 1945 remains one of the war's great heritage enigmas and a source of continuing tension between Germany and Russia (although one fragment was recovered in Bremen in 1997).

This brief summary explains why the Soviet capture of Königsberg in 1945 involved no ordinary misplacement of population and heritage: even by the brutal standards of the Eastern Front, Königsberg was a special case, and has differed markedly from Danzig in its subsequent fate (Tunbridge 1996). The city was essentially empty, most of its civilian population having fled, the remainder being deported, if not killed or shipped to Siberia. Already badly damaged by Allied bombers the previous year, Königsberg was virtually destroyed during the final desperate German defence in 1945. Subsequently, the ruins – including those of the symbolically charged royal castle – were largely bulldozed, a virtual eradication of Germanness, which must be seen in the context of the bitter relationships of post-war Europe and the absence (unlike Danzig/Gdańsk) of any Soviet cultural identification with the city. It must also be seen, however, in ideological terms. Capitalism was being displaced by Communism, which had no use for a reconstructed commercial inner city. This was replaced with some cultural and political facilities and especially open space, recast around socialist iconography, including a statue of Kalinin, the nominal Soviet President after whom the city was renamed. This fate contrasts with that of Gdańsk, in which socialist iconography was marginalized by national repossession and also the reconstruction of churches to maintain a religious continuity (although now Catholic). Furthermore, the eradication of Königsberg in favour of Kaliningrad must be seen in terms of the military need to re-establish the city to serve – in a total reversal of its previous role – as the westernmost Soviet naval/military bastion in what became the Cold War. Because of this, Kaliningrad was shrouded in secrecy between *c.* 1950 and 1990. The last embers of the heritage of Königsberg were apparently extinguished, its legacy enduring only in the mementoes of its displaced surviving population in West Germany and beyond.

The end of the Cold War, however, has produced a hitherto unimaginable resurrection of that heritage. Not only was Kaliningrad's military role now obsolete, but the city and its region (the northern half of former East Prussia) were cut off from the rest of Russia by the independence of Lithuania and Belarus'. The city was quickly confronted with ideological, economic and national uncertainties, which have persisted through the 1990s. While it experiments with new roles as an ice-free internationally oriented free port, and an economic meeting point between Western Europe and Russia, Kaliningrad is struggling to re-establish both its economy and its identity on a durable basis. In this process a newly united Germany – a source of aid, trade and, most significantly, tourism motivated by personal nostalgia – has become more central to Kaliningrad's vision than a disorganized Russia. Unlike Danzig, however, only remnants of Königsberg remain, and some among the city's present population have little empathy with them. Although a willingness to sell misplaced former residents their own lost heritage can be understood in economic terms – and indeed occurs widely (Tunbridge and Ashworth 1996) – more curious is the surprising appropriation by many, mainly younger, residents of the city of what hitherto was an alien misplaced heritage (for a recent discussion see Vesilind 1997) (Figures 9.4 and 9.5).

This trend began as early as the 1960s, focusing on the universal figure of Kant, a self-declared *weltbürger* (world citizen) whose entire life was none the less rooted in Königsberg (Malter and Staffa 1983). Building upon his legacy, the University of Kaliningrad has sought to establish an intellectual continuity with the world destroyed in 1945. By the 1990s, Kant's tomb by the cathedral ruins had become a shrine upon which flowers were regularly laid, while quotations from him and other Königsberg literary figures appeared on roadside plaques, in both German and Russian. Most significant, a replica of a lost statue of Kant was installed by the university, with

Figure 9.4 Kaliningrad: reconstructed cathedral on former city-centre 'island' in the Pregolya river. This is now in gardens and is marked by Soviet iconography
Source: Author

Figure 9.5 Kaliningrad: eastern gate of the old city (Königs Tor) with trees growing from ruins
Source: Author

German assistance. Encouraged by the collapse of the Soviet Empire, this process of re-evaluation has evolved into a more general appreciation of relict German features in the townscape of Kaliningrad and the landscape of surrounding East Prussia. In the dilemma of confused identity since 1990, these remain as the clearest markers of a sense of place, one very different from Russia proper. Accordingly, decades of decay of Germanic relicts have been arrested. In the city these include fragments of Baroque architecture, the Gothic medieval city gates, inner defensive walls and forts, the central underground German command bunker (housing the Museum of Capitulation) and twentieth-century villas in the north-western suburbs. Most remnants are of red brick, entirely distinct from and far more imageable than the repetitive Soviet modernist apartment blocks that generally separate them. Most significantly, the cathedral has actually been reconstructed from its surviving shell, a project that has involved co-operation between Russian and German governments and finance from private sources. (As in Moscow itself, contemporary Russian reconstruction of churches also reflects the resurgence of earlier ideological heritage.) Furthermore, the Kaliningrad Museum occupies a restored German civic building and, again with German contributions, now addresses the German past and (in conciliatory terms) the time of transition. As tourism increases, the trappings of capitalism are also being restored in the recently near-vacant city centre, initially involving an improvised shopping centre using portable structures and additional hotel accommodation in ships at riverside moorings.

Whether the re-creation of Königsberg's heritage could extend to wider reconstruction, as in Gdańsk, and – more fundamentally – Warsaw or Dresden (Soane 1994), remains an open question. Many seek the restoration of the name, in place of its ideologically obsolete successor, but such changes have so far been resisted by Moscow (Vesilind 1997). However, the reincorporation of the city within a German economic sphere seems inexorable and a small 'German' presence has reappeared in the form of ethnic Germans recently misplaced *to* the region from Central Asia by post-Cold War political upheaval there (Vesilind 1997). The progressive re-creation of the lost heritage could take many forms from simple marking through archaeological presentations to selective reconstruction, and a combination of economic and identity interests are likely to guarantee that some of these will occur. The now ideologically misplaced heritage of Kaliningrad will have to be renegotiated if the people of Kaliningrad/Königsberg and its region are to resolve their uncertain identity within the multicultural diversity of the new Europe.

German heritage: misplaced ideology as cultural conflict

Culture, as learned collective behaviour, ultimately includes political ideology. Central and Eastern Europe, in particular, have spent the twentieth century navigating shifting ideological sands which have interacted with the deeper strata of culture so as to leave complex and confusing messages in landscape and townscape. Nowhere is this more apparent than in Germany. As a result, the two-thirds of the Germany of 1900 which still remains cannot escape internal conflict and contradiction over culture and heritage. Regional identities have been confused as a result of the inward migration of Germans from the lost eastern provinces. More fundamental, however, are the internalized identity conflicts resulting from a succession of ideological displacements, particularly in the former GDR. A 90-year-old Leipziger in 1997 will remember the authoritarian Second Empire of Kaiser Wilhelm II; the first democratic Weimar Republic; the totalitarian Nazism of Hitler's Third Reich; the state socialism of the GDR; and the contemporary democratic Bundesrepublik. For such an individual the wrenching ideological shifts have typically been exacerbated by personal trauma. Thus a population ostensibly described by cultural homogeneity is fraught with identity conflicts, which inevitably resonate throughout Europe, given the renewal of Germany's pivotal role in the continent.

The sharpest contradictions occur in the former GDR, where the catastrophic destruction of the Third Reich was succeeded by a Communist state ideologically opposed both to the Nazi legacy and to democratic West Germany. The GDR demonized both, and also the Western Allies, while suppressing memories of the rapacity of the Red Army and lionizing the Soviet Union. The Prussian royal palaces at Potsdam, for example, were allegedly spared by Soviet magnanimity, while the Bach House at Eisenach and the Zwinger Palace at Dresden were reconstructed with generous Soviet assistance, underscoring both the criminal acts of the Third Reich and the barbarity of Allied bombing. Paradoxically, however, the very poverty and backwardness of the East preserved the built environment of the Third Reich far more effectively than was so in reconstructed West Germany, even while the GDR's leadership alleged that the latter harboured Nazism's legacy.

The Thuringian town of Weimar (Figure 9.6) is an outstanding illustration of the impact of these confusing ideological shifts. Two centuries ago it was the cultural centre of the German world, in the sense of high culture patronized by the local princely rulers and personified by the commanding figure in the German cultural pantheon, Johann Wolfgang von Goethe (1749–1832). He lived in Weimar for fifty years and, together with Friedrich von Schiller, made it a centre of European culture as post-Enlightenment Romanticism was again challenged by a revival of Classicism. Goethe's legacy in Weimar was perpetuated after his death by figures such as the musician Franz Liszt, and the town was later inherited by a united Germany as a treasured icon of national identity. It acquired a particular association with modern democratic values when, as a safe refuge from the revolutionary Berlin of 1919, its National Theatre witnessed the birth of the Weimar Republic. In this period, its cultural traditions were perpetuated in a new form as the town became the first home of the Bauhaus school of architecture. In the Third Reich,

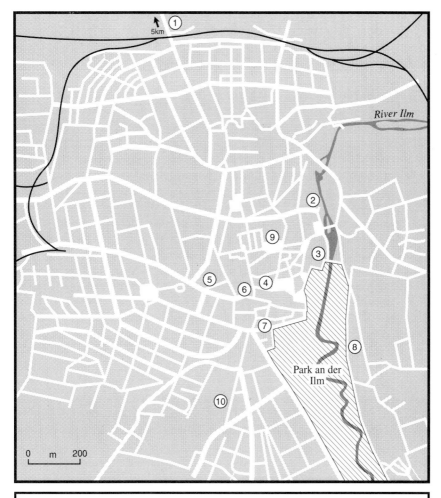

1	Buchenwald (national memorial)
2	Goethe and Schiller archives
3	Castle, art collection & National Research Centre (now Weimar Classics Foundation)
4	Town Hall
5	German National Theatre
6	Schiller house/museum
7	Goethe National Museum
8	Goethe's Garden House (*Gartenhaus*)
9	Herder Church
10	Goethe and Schiller graves

Figure 9.6 Weimar

however, its identity was appropriated as an icon of nationalistic aggrandise-ment, Weimar being one of the German towns and cities designated by Hitler to be reconstructed as a centre of Nazism. Buchenwald concentration camp was built on the Ettersberg, which overlooks the town, where it could serve as a source of slave labour.

The GDR can be credited with recognizing the subsequent significance of Weimar as expressing the essential paradox of German heritage. It sought, however, to appropriate the town's virtues within its own ideology and to dis-appropriate the unsavoury as the product of a demonized Other, conveniently transferred to the persona of West Germany. A national research centre of clas-sical German literature was created in Weimar Castle, Goethe's war-damaged house was carefully reconstructed and his early *Gartenhaus* in its bucolic setting in the nearby Park an der Ilm preserved. The East German government could not afford, however, to complete the reconstruction of the city's central square or maintain most of its fabric (occasioning an appeal to UNESCO in the GDR's final days). In any case, war damage was witness to the Western Allies' bar-barism in response to the Third Reich's crimes, the latter of course epitomized by Buchenwald. Since this had been built largely to imprison Communists, the GDR could claim legitimacy by endowing the camp's victims with the mantle of resistance to Nazism, linked with Weimar's humanist traditions.

Democratic united Germany is faced with a demanding and delicate chal-lenge in recasting the heritage of Weimar. The country's capacity to fund such reconstruction and refurbishment in the 1990s (Weimar being the 1999 European City of Culture) is perhaps the least of this challenge, although it has been accompanied by all the stresses of tourism, which increasingly impacts on the fabric of post-Cold War Eastern European cities. The reappropriation of the positive aspects of Weimar's heritage for a united nation is an obvious advantage. Buchenwald, however, is another matter. Like other former con-centration camps in the east (notably Sachsenhausen near Berlin), it has been the subject of a protracted heritage 'reconceptualization' during the 1990s (Koonz 1994; Tunbridge and Ashworth 1996). While the capitalist collusion with Nazism in criminal medical experiments and slave labour cannot be denied, some of the camps – including Buchenwald – continued in use in the early days of the GDR as 'denazification' centres, their inmates often subjected to arbitrary punishment and death. The reforging of a balanced and evolution-ary interpretation of the concentration camps from the 1930s to the present day involves the most sensitive of perspectives both internal and external to Germany. Oddly, in reconciling these crimes, Buchenwald could now become a heritage monument for the new Europe, since its victims were international and the crimes perpetrated upon them cannot entirely be attributed to a single ideology. Its memorial tower stands on the Ettersberg slope above one of the principal European east–west *autobahns* and, in physically overlooking the town of Weimar, which encapsulates so much of the virtues of European civi-lization, stands as a powerful warning which can scarcely be eradicated either from the German heritage paradox or that of humanity as a whole.

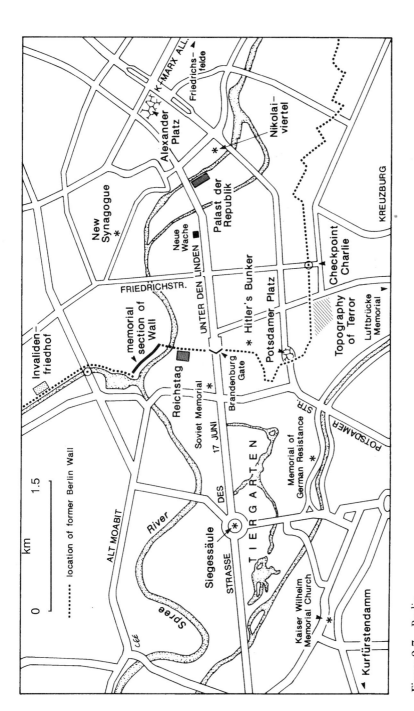

Figure 9.7 Berlin

It is, however, in Berlin that ideological shifts exert the greatest impact upon heritage, not only in Germany but perhaps in Europe (Figure 9.7). The city's recent history, and the frenetic pace of its contemporary redevelopment as the soon-to-be capital of united Germany, are permeated with issues of contested heritage, an inevitable outcome of its status as national capital within which successive ideologies have had to root themselves and then visibly suppress their predecessors (Tunbridge and Ashworth 1996). The heritage of Prussia and Imperial Germany provides the backdrop. The central axis of Unter den Linden was developed to symbolize the military and cultural rise of the Prussian state after 1650. It led to the triumphalist Brandenburg Gate, built at the end of the eighteenth century. Today the focus of heritage contestation is chiefly between the surviving remnants of the Third Reich and the GDR and the forging of an identity for democratic united Germany, with its unforgettable legacy of past atrocity. Although the impending return of the German government to Berlin and the Reichstag parliament building is all too redolent for some of unsavoury symbolic associations with the imperial past, it is fortunate that Hitler's grandiose plans to rebuild Berlin as the monumental capital of the thousand-year Third Reich did not advance far beyond Albert Speer's 1938 plan.

Those Nazi reminders that do survive can be used as a means of reconciliation. Thus the ruins of the Gestapo and SS headquarters now constitute the museum Topographie des Terrors. Again, the GDR-initiated reconstruction of the New Synagogue destroyed in Kristallnacht (1938) conveys a message concerning respect for minorities, while the wartime ruin of Kaiser Wilhelm Memorial Church projects its constant reminder of the dangers of the past along the length of the Kurfürstendamm (the heart of the former West Berlin). However, to avert the possibility of neo-Nazi enshrinement of symbolic sites, some places in the city which had very specific associations with leading Nazis have been deliberately destroyed. Thus, after its last surviving inmate, Hitler's deputy Rudolph Hess, died in 1987, Spandau Prison, which had housed those leading Nazis not sentenced to death at the Nuremberg Trials in 1946, was demolished. Places having direct associations with Hitler himself are particularly suppressed. Most important is the Chancellery bunker where he died in 1945. Formerly in the inaccessible zone behind the post-war Berlin Wall, its unmarked but still visible ruins have become a vexed redevelopment issue since 1990. Despite the desire of some to transform the bunker into a museum, it has now been sealed off to prevent neo-Nazi enshrinement.

The heritage of the GDR is scarcely less vexed, because it centres on the line and remnants of the Berlin Wall, built in 1961 as Cold War tensions escalated. Between then and 1989, East German border guards killed 263 people who were trying to escape over the Wall to West Berlin (and a further 653 elsewhere along the Iron Curtain border between the two Germanies), killings remembered after 1989 at spontaneous popular memorials and now by a formalized, permanent memorial on a section of the Wall facing the

Reichstag. The same atrocities have long been documented at the Haus am Checkpoint Charlie museum. More generally, but still contentiously, the memorialization in street names of leading figures in the Communist pantheon has been reversed, as is also the case in Weimar and elsewhere. However, much of the GDR's socialist–modernist reconstruction of Berlin, as at Alexanderplatz, has been appropriated into the democratic pluralism which now prevails.

The forging of a new identity for Berlin and its role as the capital of a unified Germany, one moreover that is consistent both with democracy and with penitence for the past, has been the central preoccupation in the city since unification (Morris 1994; Soane 1994). Much controversy has surrounded the ultimate decision to redevelop generally in a 'New Vernacular' style, which maintains broad cultural continuity with the past but, to some, is too resonant of 'a certain past' and insufficiently open to the new pan-European democratic role. The choice of foreign architects, however, has deflected the charge of unhealthy nationalism in such specific cases as the reconstruction of the Reichstag, while the re-creation of the destroyed Potsdamerplatz, near the former Wall, is the most visible expression of the city's integration into a wider European cultural realm. But the redesign of the Neue Wache, the National Memorial on Unter den Linden, has been the most sensitive challenge, following its ideological appropriation by Nazism and Communism. Since 1993 it avowedly identifies with the lost Weimar Republic, but with a sculpture and inscription recognizing – in the most inclusive terms – the internal and external victims of the nation's intervening crimes (Tunbridge and Ashworth 1996).

Cultural and heritage conflict resolution: a postmodern Europe?

The cases considered so far have isolated two cognate themes. The first concerns adjustment to culturally dissonant heritages. While the economic exploitation of the alien is widespread in the tourism industry, its political appropriation is more subtle. The resolution of dissonance may be eased at scales both smaller and larger than contemporary nation-states, but the case of Kaliningrad also suggests that the unequivocally alien might be appropriated by degrees, beginning with the most broadly serviceable and least offensive artefacts. This is similarly suggested in the post-colonial Third World by the survival or resurrection of its European heritage. The same mechanism could build a European heritage out of the present conflicting national solitudes, even if the mutual appropriation of the more hostile elements would require reinterpretations not yet formulated. A second theme in the examples cited here concerns the interrelationships between ideological power and heritage.

The case-studies above all speak eloquently of the need for significant re-evaluations in terms other than, or in addition to, national heritages. As discussed in Chapter 1, some commentators anticipate the disintegration of the nation-state in favour of more ambiguous forms of sovereignty and identity at a variety of scales. The examples here variously represent regional and European identities and, to the extent that they also signify national and ideological misplacement, a sustainable reconciliation of their identities might best be achieved by focusing on these non-national scales, above all the European. To predict, however, the advent of cultural and heritage harmony in Europe in the foreseeable future would be considerably more hazardous than predicting a similar resolution of monetary union. Let us briefly consider the litany of more serious obstacles to its attainment. The disintegration of the former Yugoslavia has demonstrated the persistence of culturally based hatreds and the impotence of Europe as a whole to stop them. Worse, it compounded human with heritage atrocity, the destruction of the Bosnian National Library in Sarajevo being but one of hundreds of attempts to extirpate the cultural memory which the archival and built heritage represents. Again, right-wing rejection of new minorities, particularly rampant in France, appears to bode ill for the acceptance of cultural diversity; new minorities may also import ancient exotic cultural quarrels to compound this problem.

In contrast to these unpromising realities, academic and official thought in the 1990s remains deeply influenced by the postmodern revaluation of human scale and variety, which includes the promotion – indeed celebration – of human diversity in the built environment. The concept of multicultural societies vested in heritages with multiple meanings is not obviously restricted to Europe. A number of Western societies, particularly the diverse 'settler' societies of Canada, the United States and Australia, have established effectively multicultural social policies with legal constitutional backing, and it is appropriate to consider what relevance they might have for contemporary Europe.

Canada adopted a 'multicultural' policy in 1972, formally recognizing the diversity of its society in the wake of a liberalized immigration policy and rapid changes in the source areas of its immigrants. This has evolved from the promotion and financial assistance given to 'ethnic' festivities and general positive awareness, to the entrenchment of minority rights in the national Constitution of 1982. These developments gave belated recognition to aboriginal 'First Nations' and progressively to groups in mainstream society who could be defined as different, notably in terms of gender, sexual and ability characteristics. There can be little question that many Canadians have benefited from these developments and that a widespread positive appreciation of diversity has evolved, considerably aided by the infusion of colour and life into cities which were previously insufferably bland.

Serious problems do exist, however, with Canadian multiculturalism, sufficiently intractable indeed to jeopardize national survival. In the first place, the policy was devised for pragmatic electoral advantage without sufficient thought being given to its logical contradictions (see Tunbridge and Ashworth 1996), particularly that between equal and different treatment. Tolerance may permit minority pursuit of intolerance and equality perpetuate minority practice of inequality. Furthermore, a competitive, even victim, group mentality has been fostered, which is implicated in socially divisive pressures for affirmative action to redress alleged imbalances. Financial stringency during the 1990s has underlined the impossibility of reconciling such problems at a cost acceptable to mainstream society, and the consequent risk of once more exacerbating cultural and racial tensions. In addition, multiculturalism is now criticized for hypersensitizing minorities, causing them to reject majority values and ultimately obstruct the creation of any credible collective identity (Mallet 1997). The most serious difficulty with Canadian multiculturalism, however, lies in the exacerbation of the problem of national unity created by the alienation of the francophone majority in Quebec, for whom the *survivance* of their Frenchness remains the overriding priority. None the less, official heritage management is deeply committed to multilateral reinterpretation of the Canadian historical inheritance, as expressed in the built environment, museums and all other media.

The same is essentially true of Australia and the United States. In the former, however, the reality of a core British/Irish heritage is demographically unassailable (despite its fashionable denial), and thus Australia's multicultural heritage experience is of less relevance to Europe than the Canadian 'mosaic' of minorities. On the other hand, the United States has sought to blend its unparalleled diversity into a perceived 'melting-pot' contained within an all-pervasive national mythology and iconography, which commands a degree of allegiance that Europeans would not foreseeably accept on a continental scale. But in all three countries, democratic rejection of minority favouritism has now been encountered. To what extent the 'revenge of the ballot box' over manipulative extensions of multiculturalism will percolate through to the recasting of heritage remains to be seen; heritage disinheritance is not a matter of immediate life opportunities, but its relationship to them is not hard to trace.

Despite the sobering observations above, the case of recently democratic South Africa possibly presents a more optimistic argument for some kind of institutionalized multiculturalism in the contemporary effort to construct unified political identities. The urgent necessity to entrench reconciliation between the races, and also different African ethnicities, demands equity both of opportunity and of heritage representation. The importance of heritage was recognized by the African National Congress before it came to power and South Africa is now engaged in a gradual process of diversifying and reinterpreting its essentially white heritage imagery, without eliminating this resource, which is economically invaluable as well as politically sensitive. The aspirations involved have been discussed for Pietermaritzburg by Haswell (1990), and most recently for Cape Town, the European 'Mother City', by

Worden (1997), who documents the practical difficulties of appropriately recasting the city's heritage in the presence of disproportionately powerful tourism-oriented international and local capital.

Conclusion

In short, the experiences of European settler societies overseas (some, like Canada, the product of exported cultural rivalries) offer no conclusive multicultural precedents for European adoption, even though they may provide lessons to Europeans on avoiding the pitfalls in what must necessarily be their own distinctive adaptation to very different future demographic, historical and geographical circumstances. At best, most diverse societies are in a trial-and-error stage of multicultural adjustment. Ultimately, all will have to make their own locally appropriate adjustments to the difficulties of multiculturalism on a continuum somewhere between naively uncritical acceptance and rejection. As Tunbridge and Ashworth (1996) argue, any multicultural heritage response needs to consider the merits of the typical 'inclusivism' of something-for-everyone as against a more pessimistic 'minimalism' which de-emphasizes (or conceivably eliminates) sectional heritages in favour of a common denominator to which all, hopefully, can relate.

Nothing in these equivocations is intended to suggest that the Europeanization of culture and heritage is a lost cause, although admittedly it is a very difficult one (Ashworth and Graham 1997). There are positive movements in this direction and, as in so much else, the task is to accentuate the positive. Partially common cultural heritages – of Greece and Rome, Hansa, Habsburg and Ottoman Empires among others – are ready instruments of division but could equally be harnessed to the sense of continental identity. The harmonization of national histories in educational curricula, as President Mitterrand vainly recommended to Prime Minister Thatcher in the 1980s, will be very hard to achieve but is at least a clear target. The existence of specific bilateral heritage reconciliations in the wake of enormous national sacrifice, as between France and Germany at Verdun, may in time suggest one way out of the ex-Yugoslav impasse. As discussed in Chapters 1 and 3, the continent's shaping by war has also left the most enigmatic of heritages. The remnants of World War II, which motivated European union in the first place, still litter the continent from the Atlantic Wall into Russia: much is dissonant to local populations but, particularly with the passage of raw living memory, could provide the most compelling shared heritage upon which to focus a European identity. This would be consolidated by a continental reinterpretation of those 'corners of a foreign field' which most war cemeteries of 1914–18 represent. In extreme cases of atrocity, as in the Nazi death and concentration camps, the consequences of intolerance so painfully memorialized may likewise provide a necessary warning of the pressing need to achieve a European, if not a global, resolution of cultural and heritage conflicts. The development of European institutions through the growth of the EU also naturally favours centripetal as against centrifugal tendencies. The next century may see a time when it does not matter that Beethoven's manuscripts are (since 1945) in Polish Kraków rather than German Berlin, for they will be universally recognized as European heritage; if indeed not world heritage.

However, when these broad considerations are focused down to the renegotiation of the urban and rural places at which most heritage messages are ultimately mediated (Ashworth and Tunbridge 1990), Europe-wide harmonization becomes more complex, if not intractable. As Ashworth argues in Chapter 10 with respect to the townscape as text, the dissemination and receipt of messages of identity and heritage is a pluriform affair; we should not draw simplistic conclusions regarding the nature of those messages or their role in generating and perpetuating cultural conflict. In concluding here, it is perhaps appropriate to refer to the fundamental distinction between the political and economic uses of heritage: the political harmonization of European heritage, to underwrite a common identity, would not guarantee its compatible exploitation by a profit-seeking tourism industry. A critical part of the Europeanizing task ahead is the definition of the freedoms and responsibilities of this mainly private sector in interpreting heritage (of war and atrocity most particularly) within bounds that are compatible with a free and democratic society, yet do not run counter to the building of a meaningfully European future.

Acknowledgement

I am grateful to B. Myslinski (Department of Geography, Carleton University, Ottawa, Canada) for insights on Polish heritage perceptions.

References

Ashworth, G. J. and Graham, B. 1997: Heritage, identity and Europe. *Tijdschrift voor Economische en Sociale Geografie* 88, 381–8.

Ashworth, G. J. and Tunbridge, J. E. 1990: *The tourist-historic city*. London: Belhaven.

Brechtefeld, J. 1996: *Mitteleuropa in German politics, 1848 to the present*. London: Macmillan.

Carter, H. 1989: Whose city? A view from the periphery. *Transactions of the Institute of British Geographers* NS 14, 4–23.

Charlesworth, A. 1994: Contesting places of memory: the case of Auschwitz. *Environment and Planning D: Society and Space* 12, 579–93.

Golding, L. 1938: *The Jewish problem*. Harmondsworth: Penguin.

Gruber, R. E. 1992: *Jewish heritage travel: a guide to Central and Eastern Europe*. New York: John Wiley.

Haswell, R. F. 1990: The making and remaking of Pietermaritzburg: the past, present and future morphology of a South African city. In Slater, T. R. (ed.), *The built form of western cities*. Leicester: Leicester University Press, 171–85.

Herbert, D. T. (ed.) 1995: *Heritage, tourism and society*. London: Mansell.

Hooson, D. (ed.) 1994: *Geography and national identity*. Oxford: Blackwell.

Johnson, N. C. 1994: Sculpting heroic histories: celebrating the centenary of the 1798 rebellion in Ireland. *Transactions of the Institute of British Geographers* NS 19, 78–93.

Johnson, N. C. 1995: The renaissance of nationalism. In Johnston, R. J., Taylor, P. J. and Watts, M. J. (eds), *Geographies of global change*. Oxford: Blackwell, 97–110.

Kinsman, P. 1995: Landscape, race and national identity: the photography of Ingrid Pollard. *Area* 27, 300–10.

Koonz, C. 1994: Between memory and oblivion: concentration camps in German memory. In Gillis, J. R. (ed.), *Commemorations: the politics of national identity*. Princeton, NJ: Princeton University Press, 258–80.

Lowenthal, D. 1994: European and English landscapes as national symbols. In Hooson, D. (ed.), *Geography and national identity*. Oxford: Blackwell, 15–38.

Mallet, G. 1997: Has diversity gone too far? *Globe and Mail* (Toronto) 15 March, D1.

Malter, R. and Staffa, E. 1983: *Kant in Königsberg seit 1945: eine Dokumentation*. Wiesbaden: Franz Steiner.

Morris, E. 1994: Heritage and culture: a capital for the new Europe. In Ashworth, G. J. and Larkham, P. J. (eds), *Building a new heritage: tourism, culture and identity in the new Europe*. London: Routledge, 229–59.

Soane, J. 1994: The renaissance of cultural vernacularism in Germany. In Ashworth, G. J. and Larkham, P. J. (eds), *Building a new heritage: tourism, culture and identity in the new Europe*. London: Routledge, 159–77.

Tunbridge, J. E. 1984: Whose heritage to conserve? Cross-cultural reflections upon political dominance and urban heritage conservation. *Canadian Geographer* 28, 171–80.

Tunbridge, J. E. 1994: Whose heritage? Global problem, European nightmare. In Ashworth, G. J. and Larkham, P. J. (eds), *Building a new heritage: tourism, culture and identity in the new Europe*. London: Routledge, 123–34.

Tunbridge, J. E. 1996: Spatial identity: a fundamental human need? The case of Königsberg. In Vanneste, D. (ed.), *Space and place: mirrors of social and cultural identities? Acta Geographica Lovaniensia* vol. 35. Leuven: Katholieke Universiteit Leuven, 119–30.

Tunbridge, J. E. and Ashworth, G. J. 1996: *Dissonant heritage: the management of the past as a resource in conflict*. Chichester: John Wiley.

Vance, J. E. 1977: *This scene of man: the role and structure of the city in the geography of western civilization*. New York: Harper & Row.

Vesilind, P. J. 1997: Kaliningrad. *National Geographic* 191 (March), 111–23.

Worden, N. 1997: Contesting heritage in a South African city: Cape Town. In Shaw, B. J. and Jones, R. (eds), *Contested urban heritage: voices from the periphery*. Avebury: Aldershot, 31–61.

|10|

The conserved European city as cultural symbol: the meaning of the text

G. J. ASHWORTH

Introduction: the city as text

This chapter pursues the argument that the European present is explicable only from the perspective of a European past and that this past (or, more accurately, these pasts) are used by various contesting political and social entities and ideologies to infuse meaning into structures which thus become symbols. Focusing this idea upon the renegotiation of European urban space, it could be argued that the urban landscape is self-evidently a cultural symbol, however culture is understood. In so far as it has been created over periods of past times by the deliberate actions of people, the urban landscape is in itself therefore part of their culture and, by its presence, expresses the needs, values and norms that shaped it in the past and maintain it in the present. The morphology of the city is thus a medium through which these attributes are transmitted, an artistic production expressing the past and present aesthetic values of the societies that deliberately created it and, incidentally, supporting important cultural industries. Given that Europe's societies are largely urban, the city as a physical structure can thus be regarded as the most prevalent, engaging and pervading cultural symbol of modern Europe.

The problem with these claims relates to the definition of each of the terms used so blithely above. What is meant by European cultures and the European city, and which Europeans, for whatever reasons, have created the latter as a symbol of the former? What exactly do these cities symbolize and for whom? In an effort to respond to these questions, the present discussion is confined to the deliberately preserved historic built environment of the city. This is, of course, not the only instrument for the creation and transmission of symbolic meanings. The form or morphology of the city – its shapes, buildings and spaces – conveys meaning, as Kostof (1991) and many others have argued at length. The justification for the restriction of focus, however, rests not just upon the practical observation that 'if anything can be a sign, then the study of signs becomes so broad as to be trivial' (Rapoport 1982, p. 37), nor upon the ubiquity, visibility and dominance in the urban scene of the preserved environment, but especially upon the deliberate nature of its creation. It is axiomatic here that the historic cities of Europe (and in practice that means all cities that make conscious reference to their historicity) did not emerge into the modern world through some natural process of Darwinian evolution. Rather, they were created by someone for some purpose, again raising questions as to who did this, why, and with what effect upon the present.

These are not merely esoteric academic debates, conducted largely by urban or architectural historians, because – as shown throughout this book – the contested meanings placed upon the past have a direct and urgent practical bearing upon conflict and identity in modern Europe. In practical terms, however, the past extends to an infinity of events and objects, and thus it is the questions concerning selection which become critical. What should be preserved? How many structures, or how much of our cities, do we need to preserve? In what way, and by whom, should this be done? For what uses and users is the urban landscape preserved? What is the end-state of such preservation and how do we know when it is reached?

The significance of such questions to the European city at the turn of the millennium has physical, financial and psychic connotations. In terms of physical extent, the number of buildings protected in some form or another, and, even more, the areas under some sort of protective designation, now include a significant proportion of the total urban space, not just in a few selected historic gems but in almost all European cities. For example, with only a very few exceptions, every Dutch city with over 20 000 inhabitants now has its central area under some form of blanket protective designation. On average, protected buildings comprise between 5 and 10 per cent of the building stock in a European city. A major world metropolis such as London has 35 000 nationally listed and a further 15 000 locally listed monuments, together with some 800 conservation areas that account for 25 per cent of the total metropolitan land area (English Heritage 1995), a figure that rises to almost 40 per cent of the City of London (Larkham and Jones 1993). Such protection is never absolute, and continuous trade-offs occur between the protected past and the projected future. The point here is not whether any particular quantity of conserved urban cityscape is desirable but only that almost any urban planning, in almost any European city, cannot now occur other than within the constraining context of the conserved urban landscape.

The implications of physical scale are paralleled by those of financial cost. This is not the place to argue the general economics of urban conservation, nor assess the balance of costs and benefits to cities or societies (Lichfield 1986). It is necessary only to observe that the conserved urban landscape

imposes high and rising permanent costs on governments, municipal authorities and individuals (whether directly through maintenance or indirectly
through other opportunities forgone). This in turn has many organizational
and budgetary implications far beyond those of historical architecture and
urban design.

Finally, there is a well rehearsed polemic that the heritage-dominated
European city has been produced by, and in turn reinforces, heritage-minded
societies (see, for example, Hewison 1987). In this argument, conserved
cities are seen as the products of a nostalgic retreat from an unwelcome present and uncertain future into a more comforting and reassuring real or imagined past. Thus, at its worst, the conserved city is symbolic of an inability to
manage presents or contemplate futures, the rich past of Europe producing a
'Eurosclerosis' that paralyses any attempt to adapt the city to meet the challenges of a changing world. A thousand years of European urban evolution
has been successfully stopped. Less dramatically, 'living in an old continent'
(to adopt Wright's phrase 'On living in an old country' (1985)) is a deliberate
choice that inevitably shapes the way the contemporary world is viewed and
constrains planning for the future.

Thus using the preserved urban landscape as a symbol in the transmission
of cultural values is not without various types of cost, not least in foreclosed
options. We have created conserved cities and now have to live in them and
also reconcile the contested statements which they make concerning the
nature of identity in modern Europe. In pursuing these issues, a linguistic
analogy seems to offer a convenient starting-point. Signs are a form of communication which may convey symbolic meaning (Tuan 1978), and thus
places, as Eco (1972) has argued, possess attributes of both functionality and
communicativeness; they are both denotative and connotative. For Barthes
(1986, p. 92), 'the city is a discourse and this discourse is truly a language'.
Such statements about places are part of a widely accepted conventional wisdom which holds that places not only are full of signs conveying symbolic
meanings but are, more fundamentally, symbols in themselves. Signs are what
makes one place distinct from another, localities rather than points in an
abstract spatial geometry. If the city is such a language, as the semiologists
constantly claim (Barthes 1970; Choay 1970; Eco 1986; Gottdiener and
Lagoulopolis 1986; Gottdiener 1995), then the next questions concern the
nature of the syntactics (structure of the systems of signs), the semantics (the
relation to the signified) and the pragmatics (the relations of the signs to the
behavioural responses of the reader). What story is being told through them,
for whom is the narrative intended and is this intention bring fulfilled?

In dealing with these questions, we encounter the immediate problem that
cities express and convey meanings through codes. Both encoding and decod-
ing are required, the difficulty being that such coding systems are neither universal in space nor stable over time. Different encodements are in use at
different places and times. In addition, the physical signs used to convey such
coded messages are generally quite deliberately designed to be robust enough

to survive over long time-periods. It is thus very easy to portray the city not as a Barthian intelligible discourse, but as a cacophonous city of Babel, where numerous 'languages' are loudly shouted or dimly muttered. Most are understood only by some citizens; many through their very familiarity are not heard any more; others are so obsolescent as to be only incompletely understood; while some are so dead as to be comprehensible only to scholars of defunct languages. This 'diversity, variety and richness of popular and local discourses' (Featherstone 1990, p. 2) is unlikely to 'play back systemicity and order' so much as confusion and disorder. The argument here, however, depends on the assumption that the cityscape can be renegotiated to produce such order.

The pioneers of semiotics have generally been either sociologists or linguists and – some architects excepted (such as Jencks 1980; Broadbent *et al.* 1980) – little interested in places as media of expression. They are even less concerned as to how such places have been planned and are being managed with this objective in mind. Conversely, those concerned with managing places, and with selecting, preserving and rehabilitating relict structures, have had little interest in the contemporary semiological implications of their actions. With one very important exception, the 'why' question has usually seemed an irrelevant distraction. In marketing, however, the cultural symbolism of the preserved urban landscape is approached through the identification and analysis of the place-producer, the place-product and the place-market, an approach that resolves itself into the answers to three direct questions that structure the remainder of the chapter: who has written the text? what does it say? and, most important, who reads it? (Ashworth and Voogd 1990).

Who is the producer of the conserved European urban landscape as cultural symbol?

Common sense might suggest that interrogating the text of our urban landscapes for 'authorship' would produce such an enormously long list of varied 'writers' as to be useless for our purpose. However, a brief consideration of the history of urban conservation reveals its specificity to a particular time-period, to Europe as a continent, and even to a particular group of European citizens. In terms of timing, it is salutary to remember that the desire to preserve large parts of the existing built environment is both recent and historically aberrant. While it is possible to cite odd cases from urban history of the deliberate preservation of particular buildings – usually for symbolic reasons – these remain exceptions to the general trend that what we now possess has survived through chance, neglect and lack of motive to redevelop, rather than the deliberate act of preservation, an idea that dates only to the late eighteenth and nineteenth centuries.

The key figures who were closely associated with Romanticism, the dominant intellectual trend in many European countries during the first part of the nineteenth century, included the famous British trio John Ruskin, A. W. N. Pugin and William Morris, together with the Frenchmen Eugène Viollett-le-Duc and Prosper Mérimée. Such individuals advanced the then novel idea that some buildings and even cityscapes should not be replaced when physical or functional obsolescence dictated. Rather, because they contained transferable values, whether architectural/aesthetic, social or moral, particular buildings and townscapes should be preserved, or even 'restored' back to some previous condition. (The argument, begun then, over what restoration implies still continues.) By the last quarter of the nineteenth century, many European governments had responded, albeit reluctantly, to such arguments by establishing small teams of officials whose task was to create inventories of valued buildings and thereby establish criteria that conferred the new attribute of officially recognized 'monument'. Such lists may have been incomplete (most are still in progress) and lacked the force of law, especially when confronted with the liberal principle of private property rights, but the logic was clear. Legislation, which at first protected individual buildings, soon followed – generally by the first quarter of the twentieth century – ultimately leading in the 1960s and 1970s to the conservation of urban districts as a part of local planning.

To describe the history of the urban conservation movement across a continent, in one short paragraph, when it has filled many books (Dobby 1978; Ashworth and Tunbridge 1990; Larkham 1996), is an absurd generalization, but it does graphically illustrate three key points. First, urban conservation is a recent phenomenon, with all the crusading characteristics of a 'movement', including righteous zeal in an unquestionably just cause, a lack of introspection and an impatience with opposition. The results in terms of the urban landscape can only now be assessed.

Second, the origins of urban conservation were European in two important senses. In the first instance, the movement occurred across almost all the nation-states of Europe in much the same way and at much the same time, resulting in broadly similar organizational structures, legislation, practice and thus conserved historic cities. The countries around the North Sea may have led, at least in legislation, and those around the Mediterranean followed, but the gap was always short. We can thus argue that if the European city exists as an idea, then it is composed of conserved urban forms and the idealized urban lifestyle these contain. Such an archetype was created not by a few thousand years of an essentially unique urban historic experience, but by a century of continentally standardized conservation planning. The process was European in another sense, in that although other continents of course possess distinguished urban histories and architectural creativities, the European idea of urban conservation diffused later elsewhere, most usually first to cities of European settlement overseas as in the United States, Australia and South Africa (Ashworth and Tunbridge 1990). This has numerous implications. Some are commercial, European cities being the product leaders in the cultural and heritage tourism industries, while others are political, 'global heritage' and its international institutions being little more than European conservation extended on to a world stage (Ashworth 1997).

Finally, the will to conserve was the obsession of a passionate, educated and generally influential minority. Moreover, despite the growth in museum and heritage site visits (see, for example, de Haan 1997 on the Netherlands), the growing popularity of heritage-motivated tourism (Boniface and Fowler 1993; Costa and van der Borg 1993) and even the rare but revealing studies of mass popular attitudes towards historic buildings and artefacts (see, for example, *Belgica Nostra* 1990), the producers of the historic city remain a small élite. Their social, educational and political characteristics have changed little since the nineteenth century, despite the assumption by the state of the enormous financial and legislative responsibilities occasioned by urban conservation.

These points suggest that answers to the question of who created our now conserved cities might be conceptualized through three broad ideas concerned with the exercise of power in European societies. These are, first, the concept of political 'legitimation' (Habermas 1973, 1996), where present governments as well as individuals feel a need to justify their exercise of power, or just their very existence, through an appeal to a past which appears to confer that right; second, the 'dominant ideology thesis' (Abercrombie *et al.* 1982), which argues that a governing dominant group imposes its values upon a governed subordinate group; and finally, the 'cultural capital thesis' (Bourdieu 1977), which extends these ideas by postulating the existence alongside the economic realm of a cultural capital composed not only of the artworks and buildings of a society but, more fundamentally, of the standards of taste that selects and interprets them. These three ideas come together when the 'capture' of this capital allows it to be used to legitimate a dominant ideology.

At the level of urban planning practice the important implication is simpler and very visible in our cities, namely that a small group of producers has created the now dominant historic city for a relatively small group of users. The functional reality of our conserved European cities is that only a minority of Europeans wish to live in them and a minority of commercial functions to occupy premises in them. This leaves empty buildings and underused and often deprived districts enlivened only by some selective islands of inner-city gentrification, speciality retailing, restaurants, bars and heritage tourism. While we return later to the question of for whom the European conserved city was created, it is also worth noting at this point that the producers of the historic city have rarely pursued this issue, regarding the answers as either self-evident or irrelevant to their urgent practical tasks.

What culture is symbolized by the conserved European city?

The conserved urban landscape that is now so dominant a feature of European imaginations was thus endowed with symbolic values and meanings at a specific time by a particular set of producers. If we now know who wrote it, the ensuing question must be: what did they write? It would be easy to conclude from the above brief discussion of the creation of the European urban landscape that the cultural messages it contains are relatively simple, unambiguous, immutable and easily and universally intelligible. At its crudest, the 'dominant ideology thesis' assumes the presence of all of these attributes. A self-consciously ascendant governing group legitimates its dominance to an equally identifiable and passively receptive subordinate group through the messages it encodes in the symbolic cityscape. The content of the language is thus simply described and accountable; the rest is just the identification of the message and its expression at various spatial scales in an array of societies. In European terms, by far the most potent scale is that of the imperial nation-state. Such a perspective is implicit in much literature on heritage and cultural studies. For example, Wright (1985) demonstrates how 'national history' was invariably written in support of national legitimation, while Horne (1984) discusses how the content of European national museums expresses the various European nationalisms, present and extinct. Again, the colonial city is often depicted as a clear, unambiguous text, the precise form and orientation of its buildings, the street patterns, markets and the location of the key military, religious and administrative buildings all being conscious or unconscious reflections of the colonial order and acting as the visible symbols of an alien power (Duncan 1990). However, for two sets of reasons, such straightforward correlations do not adequately describe the symbolic meanings encapsulated in the conserved environment of the European city.

The nature of the message

As we have seen in Chapter 9, the messages being conveyed by conserved cities are frequently unintelligible, irrelevant, ambiguous in meaning, volatile and even distasteful. Most European cityscapes will simultaneously exhibit messages possessing various combinations of all these attributes to an array of social groups. Given that the deliberate encoding of symbolic meanings in the urban landscape has occurred over a period of time, much of what exists is just no longer comprehensible or, even if fathomable, is irrelevant to the readers and thus ignored. Most unintelligibility occurs through a long process of obsolescence, but there is at least one case on the periphery of modern Europe where the creation of unintelligibility was a deliberate policy of a government intent on breaking with the past and thus eradicating messages now considered undesirable. In its dedication to the creation of a modern secular society, the nationalist government of Mustafa Kemal (Atatürk), who ruled Turkey from 1922 to 1938, secularized the use and thus the meaning of historic religious buildings, most notably the church/mosque of Hagia Sophia/Aya Sofya. In 1927, in a deliberate act of disinheritance, the official Turkish alphabet was altered from Ottoman/Arabic to the Roman. 'We must free ourselves from these incomprehensible signs that for centuries have held our minds in an iron vice . . . our nation will show with its script and with its mind, that its place is with the civilised world' (Mustafa Kemal 1927, cited in Mansel 1995, p. 418). At a stroke, existing linguistic messages were rendered unintelligible to future generations, while – at a more trivial level – a public signage was created that appeared European rather than Oriental. Nor is this an isolated case. More recently, there have been shifts between the use of Roman and Cyrillic alphabets in public signage in Croat- and Serbian-occupied areas of Bosnia, Kraijina and Slavonia, regardless of the usages of the existing populations. Few parts of Europe are without similar examples, although, in some circumstances, a deliberate official unintelligibility in public signage may be quite acceptable to local populations: Irish Gaelic and Welsh are as incomprehensible to many residents as to visitors but make a statement of local distinctiveness.

An alternative to eradication is 'museumification', whereby the contemporary meaning of symbols is neutralized by their interpretation as objects possessing only historic artistic value, the nature of the message being changed to one that has less contemporary social or political relevance. Before the collapse of the USSR, this policy was used in a number of potentially dissident Soviet republics, especially in the interpretation of religious buildings and monuments, whose potential for expressing opposition to Soviet ideology had to be neutralized but which, in turn, were too impressive as artworks to be destroyed or ignored (Misiunas and Taagepara 1983). Thus in an atheist state, churches, mosques and religious art, including folk art expressing unacceptable ideologies and allegiances, were made accessible and even restored as artistic objects and museums.

As demonstrated by the example of Rome considered in Chapter 8, the cities of Europe are littered with a now-ignored iconography inscribed in the conserved built environment, whose message is not so much unintelligible as irrelevant, because the society that created it and the one that now reads it are quite different. For example, much Belgian statuary, content of museums and monumental building – particularly that located in Brussels – was intended to legitimate the largely liberal, French-speaking Belgian state created by the European powers as a result of the 1830 revolt against the union with the Netherlands. This state, and the nationalism that underpinned it, is now largely defunct and has been superseded by new state ideologies at either the new 'national' scales of Flanders and Wallonia (for which Antwerp and Liège respectively rather than Brussels tend to be the central 'show-case' cities) or, more weakly, at the European continental scale. Few *Bruxellois* would be even aware that the landmark 'Congress Pillar' in Rue Royale celebrates the first Belgian Parliament; 'Martyrs' Square' was named for the patriots of the barricades of 1830 but is now ironically better known as the address of the Flemish Regional government; Jubilee Park, laid out in 1880 to commemorate fifty years of the Belgian state, and its grandiloquent triumphal arch, built to house the new Belgian national museums, now incongruously adjoin the

'Eurocrat' city of the Berlaymont and Place Schumann, whose ideology they flatly contradict.

Moreover, the messages themselves are rarely unambiguous statements. In most European cities they were plurally encoded by socially pluralist societies and are now also decoded pluralistically (Figures 10.1 and 10.2). Wren's London, Haussmann's Paris, Cerda's Barcelona or even Speer's Nuremberg were never simple statements, intended only to legitimate a single ruler or ruling idea, however strongly expressed. Equally, they now convey different messages to different people. Thus a medieval Gothic cathedral in Europe conveys a divergent message to a Catholic, a Protestant, a Muslim, an atheist, and frequently even to a regional separatist, nationalist or European internationalist.

It might be thought that conferring protective district or monument status automatically validated a particular symbolic message, which then remains permanently inscribed upon the landscape, whether it remains intelligible or not. This is of course to misunderstand the role of interpretation in the creation of such messages and also to ignore the instability of the symbolism of the urban landscape through time. Consider, for example, the not untypical case of the conserved Dutch city, where, even in a relatively centralized and culturally uniform country, the symbolism of the urban landscape has been deliberately changed (Figure 10.3). The national model for urban conservation, pursued by national agencies and applied by local practice, favoured very strongly the national historical founding mythology of the so-called 'Golden Age', an idea coined in the nineteenth century to legitimate the Orangist, Protestant, liberal-capitalist, bourgeois nation-state formed around the dominant core of the western provinces (especially Holland). The result has been the duplication of the seventeenth-century 'Holland' city, as typified by conserved Delft or Leiden, throughout the Netherlands, almost regardless of local cultural, historical or even physical circumstances. More recently the cities of the southern Catholic provinces, including Maastricht and Roermond, conquered by armed force during the Eighty Years War against Spanish rule (1566–1648), and thus 'disinherited' by the messages contained in the 'Golden Age' style, have been stressing what has been termed a 'Burgundian' style of conservation, purporting to reflect the distinctive historical experience, religion and more generally relaxed and hedonistic way of life of the provinces 'below the Rhine'. The result is both a 'remarking' (MacCannell 1976) of existing conserved structures and a 'reconservation': the removal of previously restored or reconstructed seventeenth-century plasterwork to reveal or reconstruct a fifteenth- and sixteenth-century brickwork. Similarly the north of the country beyond the IJsselmeer, equally spatially and culturally distant from the 'Holland' prototype, has seen the recent 'rediscovery' of the late medieval 'Hansa' context, often expressed through warehouse conservation. This can be interpreted as an act of assertion of regional distinctiveness in relation to the western Netherlands national core and an avowal of historic and social links with a newly rediscovered Baltic and North Sea region.

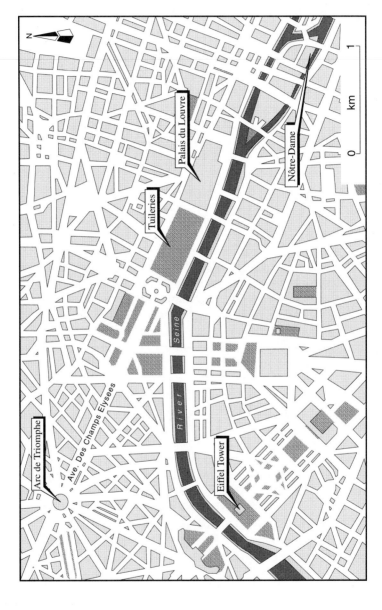

Figure 10.1 Paris: the urban text is dominated by the monumental layout of the area between the Arc de Triomphe and the Louvre. Begun during the seventeenth century, this was not fully developed until the nineteenth, when the city's morphology was further transformed by Baron Haussmann's *Grands Boulevards*

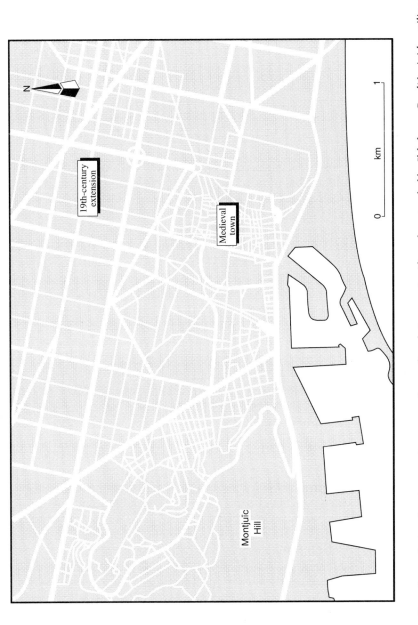

Figure 10.2 Barcelona: during the nineteenth century, the medieval city was significantly extended by Ildefonso Cerdà's rigid rectilinear plan

Monumental features do not have to be changed physically; they may require only reinterpretation. For some fifty years, the dominating statue on Budapest's Gellert Hill represented the 'liberation' of the city by the Red Army in 1945 and thus the triumph of Soviet ideology. It was erected originally, however, as a personal memorial to the son of the regent Mikós Horthy, who came to power in 1919, and was indirectly a statement of his regime's nationalist right-wing ideology. Today, in post-Communist Hungary, the monument remains as a symbol of urban pride and renaissance.

It must also be remembered that many messages fall into that wide and varied collection of heritage that has been labelled dissonant (Tunbridge and Ashworth 1996; see Chapter 9), in that the heritage is displaced, disinheriting or just distasteful to some people in some circumstances. Thus many of the messages being transmitted by the conserved landscape are ones that the present producers would rather not convey, or that the present recipients would rather not read. Such messages are powerful evidence in one of the strongest arguments against a too simplistic acceptance of the legitimation role of conservation. In almost every European city, many messages are being transmitted by conserved buildings and districts that do not legitimate present governments; rather, the reverse holds in that they are reminders of past errors, crimes and stupidities in which contemporary governments or societies are implicated as perpetrators or accessories, and which therefore tend to undermine their present legitimacy.

As we have seen in Chapter 9, the problem of managing atavistic messages is most graphically illustrated by the enshrinement in the urban fabric of that most European of historical crimes, the Jewish Holocaust of 1933–45. Major controversies surround the preservation and interpretation of the ghettos, as typified by Kraków-Kazimierz in Poland (Kazimierz was Kraków's Jewish quarter before World War II) (Ashworth 1996a). This pattern is repeated at more modest scales throughout Europe. Like most commercial cities of continental Europe, the medium-sized city of Groningen in the northern Netherlands (Figure 10.4) had, until 1941, a sizeable Jewish population of some 3000 people (around 8 per cent of the city's population). Today, the interpretation of their preserved synagogue (a national monument) and its surrounding district presents two problems, one technical and one emotional. Both stem from the contemporary reality that there is no longer a Jewish community in Groningen to read the symbolic language, while the city's present population either cannot or would rather not interpret it. For both the Dutch population of Groningen and the German tourists who are the city's principal visitors, the uncongenial message is that of complicity in genocide. Ironically enough, the former 'ghetto' ('Folkingestraat') is now partly inhabited by new ethnic minorities, notably from the Islamic Middle East, to whom the only remaining intelligible message – the synagogue – is irrelevant, or even provocative.

Although the Holocaust provides only the most dramatic evidence, the wider point is that much of the meaning of the text conveyed through the

Figure 10.3   Historic cities in the Netherlands

symbolism of the urban landscape is inevitably distasteful, to some degree, to some people. As heritage – by definition – is selective of pasts, then both that which is selected and, equally, that which is not may be dissonant to readers whose experiences and identities are written out of the symbolic text of history. This is a universal condition and quite implicit in the process of creation of the historic city. However, a range of management options exists for those cases in which dissonance is regarded as particularly serious.

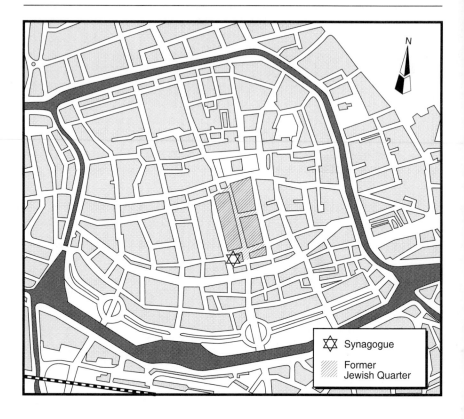

Figure 10.4   Groningen: late nineteenth-century boulevards, built upon dismantled fortifications, interrupt the late medieval plan

Policies for the marginalization, concealment or even destruction of the symbols in the hope of inducing a collective amnesia about the messages they contain are relatively common. Deliberate disinheritance through the destruction of symbols, and even whole cities, leading to what is now termed 'cultural' or 'ethnic cleansing', is not a new technique for neutralizing threatening ideas or people by demolition of their physical structures. The Old Testament contains numerous examples, and Cato's 'Delenda est Carthago' was literally accomplished. More recently, this 'traditional' European heritage management policy was widely applied during World War II and also the recent Bosnian war (with, for example, the deliberate shelling of Dubrovnik and the destruction of Mostar bridge). As we have seen in Chapter 9, much the same policy may be applied by democratic governments to architecture strongly associated with – and thus actually or potentially expressing – the ideological messages of the National Socialist regime in Germany or the Communist governments of Central and Eastern Europe (Tunbridge and Ashworth 1996; Ashworth 1996b).

Alternatively (and, while more difficult, probably more sustainably in the long run), the original messages can be reinterpreted in the existing symbols, thereby rendering them less painful or dissonant. The success of such policies is evidenced by countless castles, city walls, weapons and warships, prisons and even concentration camps that could potentially convey dissonant messages of suffering, oppression and general nastiness of Europeans to each other but which are now actually interpreted as simply historically, artistically or technically interesting old structures to be exploited as entertainments in the European tourism industry. The quantity of potentially dissonant heritage renders reinterpretation the only practical policy in most cases, although this presents, of course, a continuing series of hostages to changing fortune. There is no guarantee that future generations will read such reinterpretations in the same way. For example, cities such as Coventry, Dresden or Caen are currently debating the reinterpretation of their destruction during World War II in the changing context of national relationships within Europe, while the multiracial composition of cities such as Bristol or Amsterdam is compelling a reassessment of their role in the African slave trade and the built environments created from its profits. The only certainty is that the management of dissonant messages will be a continual process.

Place identities and scale

Discussions on the topic of place identities and scale rarely proceed very far without using the noun 'identity' and its verbal derivatives, 'identify' and 'identification'. The proposition is that the conserved urban forms exist simultaneously as places, which also actively help shape the distinctiveness of places and thus place identities. The dominant ideology thesis is spatial in so far as ideologies and governments are identified with specific areas, and their legitimation in those areas can be established through the identification of individuals with the place and thus, by transference, the governing idea associated with that place. Small wonder, then, that the strengthening of this identification was the principal motive, whether expressed in these terms or not, for the creation of the conserved city and that the conserved built environment is seen as one of the most effective means of conferring distinctive identities upon places.

Urban conservation has been, and still is, so strongly linked to this idea of the legitimation and validation of identity that it is worth developing a counter-argument to this conventional and largely unchallenged wisdom. It is entirely possible that the practice of the conservation of the forms of the built environment contains powerful intrinsic trends in exactly the reverse direction; that is, towards the standardization of forms and thus homogenization of places. At the micro-scale, it is easy to recognize and lampoon in what could be termed 'catalogue heritagization', in which the same street furniture (lamp-standards, traffic bollards, litter-bins), paving materials and even signage are selected as being recognizably 'historic', while new building is treated to a 'neo-vernacular historicist' embellishment. What was distinctive and admired when it first appeared during the 1960s and 1970s in, for example, Norwich's Elm Hill, Colmar's 'Petite Venice', Deventer's 'Bergkwartier' or Bremen's Bottcherstrasse was subsequently reproduced in numerous projects during the 1980s and 1990s, regardless of the local characteristics or historic periods preserved. Consequently, the historic districts of conserved cities have often become indistinguishable from each other, although each is quite clearly recognized as 'historic' through the use of such standardized heritage markers. This paradox, which 'involves at once the devaluation and the valorization of tradition and heritage' (Robins 1991, p. 40), is exemplified in globally operating financial and real estate companies working on locally commissioned urban development projects. Much urban waterfront regeneration, for example, has been labelled 'Rousification' after the pioneering developer, Rouse, as variants of San Francisco's Fisherman's Wharf and Ghiradelli Square were reproduced across Europe during the past twenty years (Tiesdell and Heath 1996).

It may be possible also to recognize national styles of 'heritagization'. Conservation architects and planners are as fashion conscious as any other such practitioners and shared professional training and subsequent networking help establish and transmit the acceptable style of the period. Consequently, building restoration work or townscape schemes can often be recognized by their period of execution, and presumably convey messages about how pasts were viewed by practitioners at that time, which may have little to do with the antiquity or symbolic meanings of the structures themselves. This pattern also gives rise to distinctive national schools. For instance, the 'French School' of conservation (Kain 1981) may be characterized by its relatively lavish public expenditure on set-piece national projects, each cloning Parisian archetypes such as the Place des Vosges in many a provincial town centre. The 'Polish School' (Milobedzki 1995) painstakingly and self-consciously reconstructs an idealized 'Polish' urbanism, usually reflecting the seventeenth-century 'Jagellonian' period. According to Soane (1994), the 'German School' has recently created and reproduced, in numerous urban conservation projects, a rediscovered German 'vernacular' style which is both an artistic reaction to 'modernism' as an international architectural style and an assertion of a new national confidence after a generation's submission to international political dominance. There may even be a 'Dutch School', recognizable through its almost obsessive regard for domestic detail and small-scale structures and spaces (and drawing heavily upon urban images from painters like Vermeer). This creates what has been termed the 'Anton Piek' effect, after the illustrator of romantic medieval tales for children. If these and many other such possible 'schools' actually exist – and are nationally dominant – then they will clearly produce a national urban heritage identity at the expense of locally distinctive identities.

It is perhaps somewhat premature to go further and argue that the same variables operate with increasing effect at the international scale. The transnational propagation of 'best practice', not least through global governmental organizations such as UNESCO or the Council of Europe (with its naming of European 'Years' and its annual designation of cities of culture), may lead to increasing standardization around global, or at least European, heritage archetypes. We can observe, for instance, the recent proliferation of various European associations of cities (whether under semi-official European patronage such as 'The League of Car-free Cities' or '*Quartiers en Crise*', or 'unofficial', including 'The Walled Towns Friendship Circle' or 'The European League of Historic Towns') which, by the process of transfer of technical and artistic practice, will surely tend to reduce rather than increase the local distinctiveness of the conserved urban form. We may be some way away from expressing the idea of 'world heritage' (Ashworth 1997) or even 'European heritage' (Ashworth and Graham 1997) through the conserved city, but enough is now evident at least for us to state the paradox at the centre of this argument. Urban conservation was – and remains – motivated by the desire to enhance distinctive identity, especially at the local scale. The more it is practised, however, the less locally distinctive that identity is likely to become. Thus the difficulty lies not in accepting that the main content of the message being communicated by conserved urban landscape concerns place identity, but in determining which identities are being shaped by which communications.

The stress on place leads directly to consideration of the dimensions of spatial scale and spatial hierarchy, which was introduced above in the context of the more general discussion over standardization or distinctiveness. Here the question is: what is the relationship between messages about place identities expressed at different spatial scales or at different points in a spatial hierarchy? Such expression can result in either different objects and localities being selected in support of different scales, or the same landscape elements being multi-interpreted in support of identities operating at different scales.

The 'nationalization' of the past, implicit in much of the above account of the deliberate creation of the conserved urban landscape in Europe by, and largely for, the nation-state, has determined that this is almost everywhere the dominant scale of interpretation. The dominance of the national scale, and thus nation-legitimating messages, has not gone unchallenged, however, by two other scales; namely the subnational, whether regional or local, and, at the other extreme, the supranational. The first is easier to recognize and relate back to the arguments about local identity raised above. The 'localization' of the past stems from the perceived needs of localities to express their distinctiveness to themselves and to others. At a mundane level, the conserved past is so widely used in the shaping of local place images within the 'Europe of the cities' (see the many cases discussed by Gold and Ward 1994) that further elaboration is unnecessary. The degree to which this occurs may often relate as much to matters that include the spatial hierarchical organization of decision-making in this field, and the division of responsibilities for the selection, financing and interpretation of heritage, as to existing perceived local differences in cultural identity. It is possible to construct a spectrum of the dominant level of operation of urban conservation policies that ranges from the nationally centralized at one extreme, through the regional to the locally decentralized at the other. France and, to a lesser extent, the Netherlands and Denmark would appear close to the 'national' pole; countries with federal or regional governments such as Belgium, Germany and – increasingly – Spain and Italy would appear in the centre; while the UK and Ireland would define the more local pole. In the past decade or so, however, there has been a recognizable trend towards more local accountability, if only for national budgetary reasons, which should be reflected in more local distinctiveness in selection and possibly interpretation.

At the other extreme, messages legitimating supranational identities are less easy to recognize in the cities of modern Europe, simply because of the weakness of any such scale of identity. Indeed, the most obvious examples in the built environment are relics of now defunct supranational entities that pre-date the nation-state, especially the dynastic regimes of Habsburgs, Romanovs and Ottomans in Central and Eastern Europe. The messages which these convey have little extant political relevance. Popular identification with pan-European organizations is minimal (Ashworth and Larkham 1994; Ashworth and Graham 1997), while the official 'producers' of world or European heritage, in policy-making bodies such as UNESCO, the Council of Europe or the EU, are in practice co-ordinating, recognizing and designating national or local heritages rather than shaping a competing alternative at a new spatial scale.

It is not sufficient, moreover, simply to recognize that the conserved urban landscape conveys messages supporting identities at different spatial scales, because the relationship between these messages is not always clear. They may, for example, be complementary and mutually reinforcing or alternatively competing and mutually incompatible. The logic of Bourdieu's argument (1977) about the need to control 'cultural capital' (including in this respect the 'culture' of the conserved cityscape) is that a competing or alternative government will need to challenge the existing dominant scale by asserting its own and competing legitimacy, which – in the case of the EU – may be at a different spatial scale. In this case messages shaping regional identity may lead to a justification for political separatism.

Equally, the same places, buildings and historical narratives can be used to convey spatially contrasting messages. Consider, for example, the case of the *Camino de Santiago* (Graham and Murray 1997; Murray and Graham 1997; see also Chapter 1), where the traditional pilgrimage route leading across Europe to the Galician city of Santiago de Compostela is being simultaneously promoted and interpreted at an array of different spatial scales. The older Spanish national message of the *Reconquista* and the forging of a Catholic nation-state through war against the Moors is now being supplemented or even replaced, on the one hand by Galician regionalism and, on the other, by the idea of a trans-European tourism pilgrimage which is both ecumenical and international.

For whom is this European culture symbolized?

Now that we have tried to decipher some of the messages encoded in the conserved urban cityscape, the last important question must be: who reads them? This points to several supplementary questions. Who was intended to read them, and with what understanding? Who actually does read and understand them, and in what way? (It would be especially interesting, of course, if there were to be a discrepancy in the answers to these questions.) In some ways, these are easier questions to answer than those posed earlier. The supply of conserved buildings and cities is one important part of a wider commercialized past in which buildings, artefacts, associations, memories and the like are converted by the process of interpretation into contemporary tradable commodities. The consumption of these commodities can be described and measured by simple market analysis.

The uses and users of the conserved city

The current land uses of the conserved city can be described and mapped with little conceptual difficulty, and thus the actual modern users of the symbolic buildings and districts can be related to the conserved fabric that they occupy or use. Inventories of public services, commercial activities and residential functions can be drawn up and the reasons for their specific location investigated (Ashworth 1992; Tiesdell and Heath 1996). However, occupation of preserved premises in conserved areas is not necessarily linked directly to their historicity and thus the symbolic values that they represent. Many shops and counter services, for example, often appreciate the centrality of particular locations as distinct from their projected messages, while valuing the presence of potential customers who may themselves have been attracted to such areas by their historicity.

The residential function in particular, the most important occupier of preserved buildings in the European city, is frequently attracted to buildings and areas because of their centrality, availability or cost, rather than any potential benefits obtained by the occupiers from the messages contained in the built environment. While some residents do value the perceived personal advantages, such as social status, acquired from the symbolisms of the buildings or historic districts they occupy, and thereby provide much of the financing for the renovation of such buildings and districts through rising property values, many do not. In the Dutch city of Groningen, not untypically, 20 per cent of residents of listed state monuments were unaware of the existence of such a designation, a clear majority did not regard historicity as an important variable influencing their choice of residence, and some even considered it to be detrimental as it restricted development options and imposed extra maintenance costs (Seip and Ashworth 1997). The idea that monument status simultaneously imposes a personal burden upon occupiers, as well as a collective benefit to areas, lies at the heart of the argument for public subsidy for renovation and maintenance (Stabler 1996).

Even if the historicity is an attractive locational variable, the link between the modern activity and the symbolic messages may be very generalized and vague. For example, personal services dependent upon client confidence, such as legal or medical services, may well value the occupation of conserved premises as conveying messages of continuity, probity and reliability, as well as conferring some psychic benefits of artistic patronage, which by inference may be transferred to the modern activity. Similarly, many representative functions, whether public or private, choose to locate in buildings or areas conveying what are regarded as appropriate symbolic meanings deriving from their conserved status. Town halls, court-houses and even private front offices may receive some legitimacy in the exercise of their functions from the buildings they occupy. But in all this, the messages will not be directly experienced; it is enough to know that they exist and will be transmitted to an imagined future.

What is being read by whom?

Thus the above brief inventory of modern users of the conserved city is an incomplete answer to the question: why do they use it? To pursue the importance of the conserved 'historicity' and thus the messages conveyed by it to users, we must return to the idea of the legitimation of dominant ideologies by dominant groups, which at least provides a convenient structure upon which to hang a counter-argument of three main points.

First, the direct consumption of museums, galleries, monuments and even historic cities is a minority pastime. This may seem to be contradicted by the quite dramatic growth in visitor numbers to most cultural and historic attractions experienced throughout Europe. None the less, however impressive this growth may be, such visitors still remain a minority and, for many traditional cultural attractions, a small, albeit influential, minority. The importance of this point is that if the messages of the built environment are intended to validate the legitimacy of a dominant group, then, in the absence of the subordinate group, this group is carrying out a form of self-validation.

Second, there is very little evidence that those who intentionally visit the conserved city and its artefacts actually read the projected messages as intended by either the original encoders or modern interpreters. Simply, one of the hallmarks of human-environment research 'is the realization that designers and users are very different in their reactions to environments and their preferences, partly because their schemata vary' (Rapoport 1982, pp. 15–16). A prosaic but widespread and ultimately disastrous example of this difference is the high-rise residential architecture which was constructed in many European cities during the 1960s and 1970s (Jencks 1980). To its designers and the politicians who commissioned such buildings, high rise was seen as representing an exciting and commendable modernity, but to its users, who were generally neither the designers nor indeed had chosen to occupy the design, the buildings were often functionally inferior to the structures they replaced and ultimately conveyed a stigma of social inferiority. It can be added that some recent attempts to revitalize such 'problem estates' have included the alteration of their architectural meaning through a 'postmodernization' that includes appending functionless decorative details and embellishments. Previously rejected under the 'less is more' slogan, these are intended to convey a historicist association with a past, now idealized, residential environment.

A number of researchers into the visitor experience of the exhibited past (e.g. Merriman 1991) suggest that the messages may be either ignored completely or incorporated into an unintended interpretation. In the terms of this argument, the subordinate group is either not receiving the messages of the dominant group, or is even receiving a different message from the exhibited object that accords better with the existing constructs of the visitor. The broad conclusion from Prentice's investigations (1993, 1996) into the experiences of visitors to historic sites is that although visitors expected to be instructed in what was regarded by them, at least partially, as an educational experience, the factual results of this instruction often bore little resemblance to the intent of the instructors. In addition, visits to historic sites tended to be associated with visits to non-historic attractions, rather than to the pursuit of historical topics elsewhere. This suggests that the consumption of the past at a historic site is viewed not as an aspect of historical scholarship, but as a recreational excursion in which historicity is not a major motive.

Third, and to advance the argument about what message is being projected and received, a specific group of users so far neglected in this discussion must be considered. One of the most obvious, visible and economically important groups of consumers of commodified history are heritage tourists. These users, and the tourist-historic cities they have been largely instrumental in creating, are a well researched group (Ashworth and Tunbridge 1990; Boniface and Fowler 1993), providing, for the purpose of this argument, a good example of what in marketing terms is called the multi-selling problem. This situation, in which the same physical product simultaneously serves different consumer markets, ensures that different messages are sought and received for the same conserved buildings, districts and cities.

Heritage tourists are important to this argument because they are numerically and economically influential and also have made conscious, and often expensive, decisions to consume the conserved cityscape. They are generally receptive to information, frequently arrive well briefed and are usually continuously instructed by various 'markers' on the meanings of the iconography they experience (MacCannell 1976). Their relationship with the messages contained in the conserved cityscape is thus neither marginal nor accidental, and the difference between such tourists and local residents, in this respect, is important.

There are two principal and contradictory positions comparing the consumption of the past by tourists and by residents. The first, and more widely held, argues that the tourist is, in this respect, clearly and obviously different from the resident by virtue of being a tourist. The tourist on a more restricted time–space budget is collecting quickly recognizable pre-marked sights which can be incorporated into existing constructs of understanding, inevitably different from those of residents. Motive is also assumed to be different. Visitors consume the past to satisfy a variety of perceived needs, which have engendered considerable discussion among the sociologists and psychologists of tourism (Cohen 1979; Dann 1981). Regardless of the outcome of these debates, it is tourist needs that predominate and not those of the residents for local social, cultural or ethnic identity (Urry 1990). Thus not only are different messages read, but they are read differently in what MacCannell (1992) refers to as 'empty meeting grounds'. In terms of heritage promotion, therefore, the axiom is that you can never sell *your* heritage to visitors, only *their* heritage back to them in *your* locality. The tourist–resident dialectic becomes just one more battle in the wider conflict, discussed throughout this book, for a contested past, colonized by different colonizers, in which therefore the questions: whose heritage? whose conserved city? whose symbols? and whose messages? receive different and contradictory answers.

However, a second and contrary position argues that heritage tourism and cultural tourism (of which heritage tourism is a part) are but one type of 'special-interest tourism', or, more accurately, a whole range of varied 'special interests' that can be bundled together. If tourism 'special interests' are merely the pursuit in a new locality of existing leisure activities and preoccupations, then the tourist and the resident may well be all but indistinguishable in their actual pattern of consumption behaviour. Further, this category of tourism is 'place specific', consuming the heterogeneity rather than the homogeneity of places. Consequently, it is attracted by the quality of uniqueness in the same way that residents pursue local identity. It may not of course be the same *genius loci* that is sought by the two groups of consumers, but each has the same interest in distinctiveness. Thus a tourist is only a resident in a different place. This idea underlies the so-called 'turnstile' model, so comforting to managers of historic resources, in which only the actual consumption of the historic resource matters and the origin of the customer is unimportant. Similarly, the 'windfall gain' model, strongly favoured in tourism development, argues that very few urban historic resources were actually preserved, and are now maintained, as heritage for the sake of tourism. This market does not appear in the justifications of the original creators of the conserved city, discussed earlier, nor as anything more than an afterthought in most local management policies for conservation areas (e.g. the exhaustive discussion of the evolution of UK conservation area policy provided by Larkham 1996).

The only resolution to these conflicting ideas is that there are different markets, between tourists and residents as well as within each of those groups. The characteristics of both product and consumer will determine the answers to the questions: how different? and is such a difference important? All this reiterates the warning against an over-simplified view of the users of the past and their reception of the meanings conveyed by its conserved structures.

Conclusion: the conserved European city as the expression of power, culture and identity

This chapter began with the unexceptional idea that every act of preservation of the fabric of the European city makes a statement, whether this was intended or not. The rest of the chapter has qualified, complicated and even contradicted this simple idea. There remains, however, an aura of unreality in the attempt to bridge the gap between the local practice of urban planning and management, and the broader sweeping issues of power, identity and conflict that are the focus of most chapters in this book. The attempt to relate the detail of the conserved built environment and its consumption, on the one hand to the wider exercise and legitimation of power, the creation, control and dissemination of culture, and, on the other, to the shaping of place identities at various scales, still leaves an enormous credibility gap. Are the local area planners solving the local problems of the historic city, architects and urban designers applying professional norms of aesthetic practice to building and area restoration and enhancement, museum curators selecting, displaying and interpreting historic artefacts, and even theme park managers profitably entertaining visitors, all somehow unwitting agents of much broader processes that some have seen as a Machiavellian conspiracy in the exercise of dominance?

Fortunately, not least for those who would otherwise bear such a heavy burden of guilt, variety and complexity counsel against such simplistic conclusions. Simply, there are many actors, agencies and motivations in the renegotiation of European urban place. Nor can the issues be reduced to a straightforward national versus regional dichotomy. The conserved European city is equally serving local, regional and perhaps even global demands and needs. The encoders, the original messages, the subsequent interpreters and the readers of the conserved cityscape are pluriform. Time itself has either changed meanings beyond recognition or rendered most statements illegible, irrelevant or just trivial. The European city, itself an unreal abstraction, does not lend itself to simple deterministic arguments. This alone renders it a text worth reading and thus an environment worth living in.

References

Abercrombie, N., Hill, S. and Turner, B. S. 1982: *The dominant ideology thesis.* London: Allen & Unwin.

Ashworth, G. J. 1992: *Heritage planning*. Groningen: Geopers.

Ashworth, G. J. 1996a: Holocaust tourism and Jewish culture: the lessons of Kraków-Kazimierz. In Robinson, M., Evans, N. and Callaghan, P. (eds), *Tourism and culture towards the 21st century*. Newcastle upon Tyne: University of Northumbria, 1–13.

Ashworth, G. J. 1996b: Realisable potential but hidden problems: a heritage tale from five central European cities. In Purchla, J. (ed.), *The historical metropolis: a hidden potential*. Kraków: International Cultural Centre, 39–64.

Ashworth, G. J. 1997: Is there a world heritage? *Urban Age* 4, 12.

Ashworth, G. J. and Graham, B. 1997: Heritage, identity and Europe. *Tijdschrift voor Economische en Sociale Geografie* 88, 381–8.

Ashworth, G. J. and Larkham, P. J. (eds) 1994: *Building a new heritage: tourism, culture and identity in the new Europe*. London: Routledge.

Ashworth, G. J. and Tunbridge, J. E. 1990: *The tourist-historic city*. London: Belhaven.

Ashworth, G. J. and Voogd, H. 1990: *Selling the city: marketing approaches in public sector urban planning*. London: Belhaven.

Barthes, R. 1970: Semiologie et urbanisme. *Architecture d'Aujourdhui* 42, 11–13.

Barthes, R. 1986: Semiology and the urban. In Gottdiener, M. and Lagopoulis, A. P. (eds), *The city and the sign: an introduction to urban semiotics*. New York: Columbia University Press, 87–98.

Belgica Nostra 1990: *Meepraten over monumenten: enquete onder de belgisch volk*. Brussels: Belgica Nostra.

Boniface, P. and Fowler, P. J. 1993: *Heritage and tourism in the global village*. London: Routledge.

Bourdieu, P. 1977: *Outline of a theory of practice*. Cambridge: Cambridge University Press.

Broadbent, G., Bunt, R. and Jencks, C. (eds) 1980: *Signs, symbols and architecture*. Chichester: John Wiley.

Choay, F. 1970: Remarques a-propos de semiologie urbaine. *Architecture d'Aujourd'hui* 42, 9–10.

Cohen, F. 1979: A phenomenology of tourist experience. *Sociology* 13, 179–201.

Costa, P. and van der Borg, J. 1993: *Management of tourism in the cities of art*. Venice: CISET 2, University of Venice.

Dann, G. 1981: Tourism motivation: an appraisal. *Annals of Tourism Research* 8, 187–219.

de Haan, J. 1997: *Het gedeelde erfgoed*. The Hague: Sociaal en Cultuureel Planbureau.

Dobby, A. 1978: *Conservation and planning*. London: Hutchinson.

Duncan, J. S. 1990: *The city as text: the politics of landscape interpretation in the Kandyan Kingdom*. Cambridge: Cambridge University Press.

Eco, U. 1972: A componential analysis of the architectural sign. *Semiotica* 24, 97–117.

Eco, U. 1986: Function and sign: semiotics of architecture. In Gottdiener, M. and Lagopoulis, A. P. (eds), *The city and the sign: an introduction to urban semiotics*. New York: Columbia University Press, 99–112.

English Heritage 1995: *Conservation in London: a study of strategic planning policy in London*. London: English Heritage.

Featherstone, M. 1990: *Global culture: nationalism, globalisation, identity*. London: Sage.

Gold, A. and Ward, S. (eds) 1994: *Promoting places*. London: Belhaven.

Gottdiener, M. 1995: *Postmodern semiotics: material culture and the forms of post-modern life*. London: Routledge.

Gottdiener, M. and Lagopoulis, A. P. (eds) 1986: *The city and the sign: an introduction to urban semiotics*. New York: Columbia University Press.

Graham, B. and Murray, M. 1997: The spiritual and the profane: the pilgrimage to Santiago de Compostela. *Ecumene* 4, 389–409.

Habermas, J. 1973: *Legitimationensprobleme in Spätkapitalismus*. Frankfurt am Main: Suhrkamp.

Habermas, J. 1996: The European nation-state: its achievements and its limits. In Balakrishnan, G. and Anderson, B. (eds), *Mapping the nation*. London: Verso, 281–94.

Hewison, R. 1987: *The heritage industry: Britain in a climate of decline*. London: Methuen.

Horne, D. 1984: *The great museum: the re-presentation of history*. London: Pluto.

Jencks, C. 1980: The architectural sign. In Broadbent, G., Bunt, R. and Jencks, C. (eds), *Signs, symbols and architecture*. Chichester: John Wiley, 71–118.

Kain, R. 1981: *Planning for conservation: an international perspective*. London: Mansell.

Kostof, S. 1991: *The city shaped: urban patterns and meanings through history*. London: Thames & Hudson.

Larkham, P. J. 1996: *Conservation and the city*. London: Routledge.

Larkham, P. J. and Jones, A. N. 1993: The character of conservation areas in Great Britain. *Town Planning Review* 64, 395–413.

Lichfield, N. 1986: *Economics in urban conservation*. Cambridge: Cambridge University Press.

MacCannell, D. 1976: *The tourist: a new theory of the leisure class*. New York: Schocken Books.

MacCannell, D. 1992: *Empty meeting grounds*. London: Routledge.

Mansel, P. 1995: *Constantinople: city of the world's desire, 1453–1924*. Edinburgh: Murray.

Merriman, N. 1991: *Beyond the glass case: the past, the public and the heritage in Britain*. Leicester: Leicester University Press.

Milobedzki, A. 1995: *The Polish school of conservation*. Kraków: International Cultural Centre.

Misiunas, R. J. and Taagepara, R. 1983: *The Baltic States: years of dependence*. London: C. Hurst.

Murray, M. and Graham, B. 1997: Exploring the dialectics of route-based tourism: the *Camino de Santiago*. *Tourism Management* 18, 513–24.

Prentice, R. 1993: *Tourism and heritage attractions*. London: Routledge.

Prentice, R. 1996: Tourism as experience, tourists as consumers: insight and enlightenment. Inaugural Lecture, Edinburgh: Queen Margaret College.

Rapoport, A. 1982: *The meaning of the built environment: a non-verbal communication approach*. Beverly Hills: Sage.

Robins, K. 1991: Tradition and translation: national culture in its global context. In Corner, J. and Harvey, S. (eds), *Enterprise and heritage: the cross-currents of national culture*. London: Routledge, 38–51.

Seip, M. and Ashworth, G. J. (eds) 1997: *Binnenstadsbeheer*. Groningen: Geopers.

Soane, J. 1994: The renaissance of cultural vernacularism in Germany. In Ashworth, G. J. and Larkham, P. J. (eds), *Building a new heritage: tourism, culture and identity in the new Europe*. London: Routledge, 159–78.

Stabler, M. 1996: Are heritage and tourism compatible: an economic evaluation of their role in urban regeneration. In Robinson, M., Evans, N. and Callaghan, P. (eds), *Tourism and culture towards the 21st century*. Newcastle upon Tyne: University of Northumbria, 417–46.

Tiesdell, S. and Heath, T. 1996: *Revitalising historic urban quarters*. Oxford: Architectural Press.

Tuan, Y. F. 1978: Sign and metaphor. *Annals of the Association of American Geographers* 68, 363–72.

Tunbridge, J. E. and Ashworth, G. J. 1996: *Dissonant heritage: the management of the past as a resource in conflict*. Chichester: John Wiley.

Urry, J. 1990: *The tourist gaze: leisure and travel in contemporary societies*. London: Sage.

Wright, P. 1985: *On living in an old country: the national past in contemporary Britain*. London: Verso.

|11|

The European countryside: contested space

Hugh D. Clout

Introduction: the perspective of time and space

As the twentieth century draws to a close, it may be argued that Europe's countrysides have entered a new phase in their long and complex evolution. Throughout history the imperative has been to produce more food in order to satisfy a growing number of mouths, or – in phases of demographic stability – to feed the population more effectively. The traditional means of attaining this objective was by increasing the cultivated surface at the expense of woodlands, marshes, moors, rough grazing and other marginal environments. The continent's rural landscapes bear ample testimony to centuries of effort as, for example, entrepreneurs transformed great stretches of watery land into polders, groups of peasants converted steep hillslopes into cultivation terraces, and humble family farmers reclaimed patches of commonland for cropping (Smith 1978; Butlin 1993; Schama 1995). A second process, which was well established in late medieval Flanders and Lombardy but gained in importance from the seventeenth century onwards, was to intensify food output from the existing cultivated surface. This was accomplished by devising complex rotations, increasing livestock densities to generate more manure, introducing new strains of crop, using chemical fertilizers and machines, and, in some instances, implementing land reform, which affected property ownership, tenancy rights, farm size, layout, and of course landscape (Overton 1996).

For many centuries, most systems of commercial exchange operated at a local scale, the only exceptions being the hinterlands of great cities, such as Paris and London, which covered ever-widening areas, and the riverine and coastal trade routes used to transport rare commodities, including salt and wine (de Planhol 1994). As towns and cities continued to expand, ever greater quantities of food were needed, with carriage of goods overland or by river or canal being joined in the nineteenth century by the new opportunities offered by railways and by steel-hulled steamships (Price 1983). These transport innovations, coupled with the colonial and commercial policies of Britain and some other nations, introduced a completely new dimension to the international food economy. The advent of cheap supplies of grain and meat from the New World shook the foundations of traditional rural economies in Britain, Denmark and other European states, whose farming activities were not protected – as in France – behind tariff barriers (Lowe and Bodiguel 1990).

Through the centuries, generations of farmers exploited Europe's highly diverse resource base, extending from the Arctic to the Mediterranean and from the western Atlantic fringes to the plains of Russia, in order to set in place its traditional mosaics of farms, fields, villages and market towns (Lebeau 1969). Changes in technology, social organization and political power modified the way Europe's farmers perceived their highly diverse resources through time, as they sought to bring 'nature' under control (Thomas 1983). Large reclamation and improvement schemes required abundant resources of labour and finance, which could be mustered only by the rich and powerful, including the Crown, the Church, members of the nobility, and groups of urban entrepreneurs. Their great accomplishments in making the European landscape stand alongside the cumulative achievements of countless numbers of humble farming families who reclaimed scraps of land and fertilized patches of soil (Nitz 1978). The story of human mastery of Europe's resources was, of course, not one of continuous 'progress'. Many of the impressive achievements of the Roman era and of the great medieval period (*c.* 1100–1350) of population growth, land clearance and settlement creation fell into ruin and disrepair in subsequent phases of economic and demographic decline. These latter were characterized by the abandonment of outlying farms and isolated villages and the shrinkage of cultivated space (Duby 1970; Abel 1980).

Traditional patterns of rural settlement, and the agrarian landscapes of which they formed part, were recorded by military surveyors, cartographers and tax inspectors during the seventeenth, eighteenth and nineteenth centuries (Kain and Baigent 1992; Sporrong 1994). Large nucleated villages on the North European Plain contrasted with scatters of hamlets in Ireland; ancient twisting lanes were at variance with recent roads laid out ruler-straight by government engineers; and centuries-old fields in Celtic areas contrasted with the regular enclosures of lowland England and southern Scandinavia (Figure 11.1) (Kjaergaard 1994). The vast arable fields in market-oriented parts of the Paris Basin contrasted with tiny strips in regions of peasant farming and testified to the power of great landowners, ecclesiastical and lay, in acquiring and amassing property at the expense of impoverished family farmers (Moriceau 1994). The components of the rural scene have always been a reflection of the continent's social inequalities. These diverse rural patterns fired the enthusiasm of early geographers, who sought both to

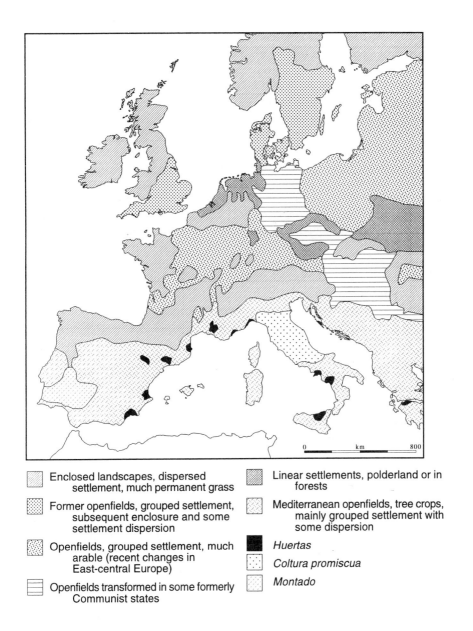

Enclosed landscapes, dispersed settlement, much permanent grass

Former openfields, grouped settlement, subsequent enclosure and some settlement dispersion

Openfields, grouped settlement, much arable (recent changes in East-central Europe)

Openfields transformed in some formerly Communist states

Linear settlements, polderland or in forests

Mediterranean openfields, tree crops, mainly grouped settlement with some dispersion

Huertas

Coltura promiscua

Montado

Figure 11.1   Rural landscapes of Europe
Source: After Lebeau (1969)

record their complexity and to analyse the changing relationship between human groups and the natural world which led to their creation and modification (see, for example, Vidal de la Blache 1903). Rural Europe in the nineteenth century was indeed an assemblage of thousands of *pays*, each with its distinctive cultural landscape, local economy and social structure.

Inevitably, the maps produced by early surveyors were snapshots in time and presented a deceptive picture of rural stability. Europe's countrysides had always experienced change and were continuing to do so as rates of demographic growth fluctuated, new opportunities were offered to supply food to growing towns and cities, and harsh interventions of war ravaged many districts. None the less, two fundamental points remained relatively constant. Despite the multiplicity of local trades and crafts which flourished across the European countryside, the essential function of rural areas was to generate greater quantities of food both through the age-old process of land reclamation, and through the increasing suite of innovations which contributed to agricultural intensification. In addition, Europe's countrysides acted as human nurseries to supply workers to the continent's towns and cities. Migration flows were overwhelmingly one way, as teenagers and young adults quit their parents' farms and villages for the fabled opportunities of provincial cities and national capitals. The implications of this age-selective out-migration were so severe that local birth rates fell and some areas experienced net depopulation during the second half of the nineteenth century. At the same time, small groups of scholars, artists, poets and town dwellers lamented the march of urbanization and mechanized farming at the expense of rural life and landscape. However, nature conservation and protection of the countryside were to remain minority concerns until the present century.

World War I brought appalling material destruction of a magnitude previously unknown in human history across some of Europe's most fertile farmlands (Clout 1996). It arguably had an even more devastating effect on its rural population since farmers and the sons of farmers comprised the greater share of those who were killed, injured or maimed. Most academic studies of the European countryside continued, however, to dwell on such well established themes as agricultural systems, the structure of rural settlements and landscapes, rural-to-urban migration, and the presence of communities of farmers and other country folk who were portrayed as living and working in harmony (Short 1992). With the exception of official reports expressing disquiet at the growing quantities of farmland being consumed by suburbs and factories in lowland England during the 1930s, it seemed as if the eternal order of the fields continued to reign supreme up until World War II (Cherry and Rogers 1996). Indeed, the virtues of family farming and of close identity with one's home locality were central to the political ideologies of Nazi Germany. They were echoed, too, in the rhetoric of the French Popular Front and again in that of the Vichy regime in the early 1940s, which identified the 'true France' as a composite of numerous *pays*, each with its distinctive landscapes, popular arts and traditions, vernacular architecture and farming systems (Faure 1989; Lebovics 1992). Long-established myths of national identity being rooted in the land and in farm husbandry were encountered in varying forms throughout the continent – even in England, where the level of urbanization was most advanced. Not surprisingly, the rural, family-farming inheritance figured prominently in the process of nation-building in Ireland.

The dynamics of rural change

It was not until the early 1960s that these ways of thinking were challenged effectively in arguments which demonstrated that rural areas in many parts of Europe were undergoing rapid, unparalleled and profound change. Farmers and farm workers were leaving the land in droves as industrializing and urbanizing economies required abundant supplies of labour. New technologies were enabling the rapidly declining number of farmers to produce increasing quantities of food year by year and to transform inherited rural landscapes with unparalleled rapidity. Schemes for farm enlargement and plot consolidation changed both the appearance of vast areas of rural Europe and their fundamental ecology (Clout 1984). Certain stretches of countryside were being repopulated on a permanent or seasonal basis by commuters (sometimes travelling long distances by car each day), retired townspeople (perhaps with family roots in local farming) and owners of second homes (Warnes 1982; Kayser 1989; Buller and Hoggart 1994). In addition, greater quantities of farmland than ever before were being converted to urban uses, including housing, industrial estates, airports and motorways.

At the same time, European attitudes toward the environment were changing, with national parks and nature reserves being established by national governments for a combination of scientific and aesthetic reasons in response to a wide range of interest groups. In broad terms, nature was to be cherished rather than mastered as in past generations. Wetlands and moorlands were no longer to be 'improved' but rather to be protected. Conservationists and ecologists heaped their wrath on farmers who uprooted hedgerows, reclaimed marshes, felled copses or transformed the traditional rural scene in other ways (Shoard 1980). In Eastern Europe, the creation of state farms and collectives in the late 1940s and 1950s, and subsequent widespread industrialization, had also given rise to massive changes in landscape, ecology and social structure, about which most West Europeans remained singularly uninformed.

Research by some social geographers and sociologists heralded a new strand of academic thinking about the countrysides of Western Europe and especially about the people who inhabited them (George 1964; Rambaud 1969; Pahl 1975). While approaching the theme in distinctive ways, each of these scholars demonstrated that rural phenomena were not isolated from urban life. In spite of enduring differences in land use, which helped distinguish towns from rural areas, the countryside certainly could not be considered as synonymous with agricultural production. Some rural areas were experiencing rapid transformation in terms of social composition and economic activity as well as landscape. Notions of community and social harmony were shown to be largely inappropriate since people who lived in the countryside had many different reasons for residing there, belonged to differing classes, possessed varying amounts of wealth, had different needs and aspirations, and held sharply distinctive opinions about defending the traditional rural scene and conserving the so-called natural environment (Newby 1978).

Countryside features, including farm buildings, habitats and whole landscapes, which had been taken for granted by generations of farmers, were being reappraised by in-comers as having considerable potential for exploitation and enjoyment. Vernacular buildings and other historic features, long objects of academic study, were now perceived as heritage components to be cherished and acquired. In a striking way, they had become part of the real-estate market. It was patently obvious that Europe's countrysides were not unpopulated landscapes to be dissected for scholarly analysis but were, in fact, arenas of harsh competition and markedly unequal social and economic opportunity. Many of the continent's rural areas were most certainly neither stable nor being subjected to only gentle modification. Fundamental changes in farming, coupled with the impact of accelerating urbanization and all that implied for land-use change, were sweeping away traditional landscape features at an alarming rate, threatening to destroy the intricate and distinctive environments which geographers had long investigated.

During the twenty years following 1965, the fundamentalist belief that farming was pivotal to the rural economy, and even to national identity, undoubtedly lost ground. Over the same period food production in Western Europe continued to increase massively, with agricultural training, enlargement of fields and farms, mechanization, fertilization, biological innovation, land drainage, irrigation and a host of other processes being encouraged by the CAP and associated funds of the then European Community (EC), as well as by schemes operated by individual states. At the same time the rise of biotechnology and agribusiness made further contributions to the transformation of the farming economy. Growth of rural tourism and recreation associated with rapidly rising rates of car ownership, retirement to the countryside, industrial development in rural areas, and the changed perception held by growing numbers of West Europeans that the countryside was a desirable place in which to live wrought nothing less than a social revolution (Mathieu and Duboscq 1985; Hoggart *et al.* 1995; Béteille 1996). This affected not just peri-urban sections of Western Europe's countrysides but also remote localities in the 'deep countryside'. The rather ambiguous term 'counter-urbanization' was soon to encompass the cumulative expression of these processes (Champion 1989). In sections of the Mediterranean sunbelt, traditional forms of social structure and economy were displaced by highly commercialized tourism, stimulated by national and regional governments as a way of escaping from underdevelopment and financed by multinational corporations, including travel firms, airlines and property developers. Such changes were mostly concentrated in stretches of countryside fringing the Mediterranean and Atlantic littorals of southern Europe. At the same time, new irrigation schemes refocused commercial agricultural activity in lowland areas, with former agricultural workers migrating away from less privileged rural environments where stretches of dry farmland were abandoned.

In general terms, material conditions were different in Eastern Europe, but second homes proliferated in the countryside around Prague and other major cities, and rural tourism was well established in the Tatra Mountains and other attractive environments (Coppock 1977). In some regions, large numbers of factory workers continued to live on their farms and travelled to urban work each day. National parks and nature reserves were established in Eastern Europe, standing in sharp contrast to the widespread pollution that spewed from factories and mines (Figure 11.2). Social links between town and country remained strong in Poland since the nation's family farming

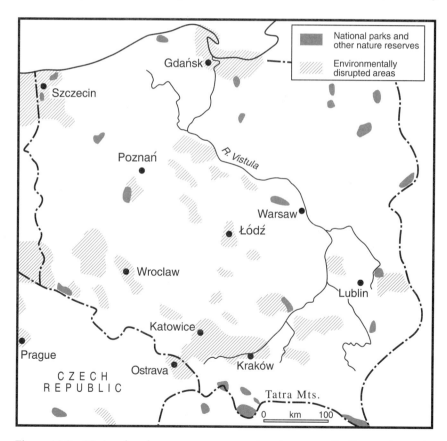

Figure 11.2 National parks, nature reserves and environmentally disrupted areas in Poland

system had survived the introduction of state and collective farms, unlike the experience of other parts of Eastern Europe (Huigen *et al.* 1992).

Food production and forestry accounted for sharply decreasing proportions of Europe's economies, but agricultural output and levels of productivity continued to increase as the logic of agribusiness and intensification was pursued with a vigour arguably surpassing that of commercial producers in medieval Flanders or the gentlemen-farmers of eighteenth-century England. The past had always been characterized by a growing demand for food; indeed, Western Europe's precocious urbanization and industrialization would have been impossible without continuing improvements in agricultural yields and total production. In contrast, by the early 1980s the relationship between demand and supply had changed, in Western Europe at least. 'Mountains' of surplus grain, dairy products and fruit, and 'lakes' of milk and wine were accumulating at many points in the EC, incurring huge expense as they had to be put into store, destroyed or sold for minimal return to Eastern Europe (Winter 1996). Donation of non-perishable surpluses to the hungry of the developing world was unacceptable politically, with trade being judged preferable to food aid, which, it was feared, might stifle indigenous farming. A new approach to agriculture in Western Europe had to be found if the crisis of costly over-production was to be contained.

This kind of dilemma had, of course, been predicted in the late 1960s when the then EC Agriculture Commissioner, Sicco Mansholt, proposed that stretches of marginal farmland should be withdrawn from food production and converted to nature reserves, national parks, wilderness areas or forests. He recommended that moderately intensive farming should be encouraged in specified districts, with intensive farming being focused in other zones. In addition, Mansholt advocated an innovative, wider view of countryside management, which recognized the existence of numerous non-farming activities in the countryside and their potential to help diversify the rural economy in conjunction with innovations in manufacturing, tourism and service activities. In short, the erroneous belief that 'agriculture' was an adequate synonym for 'countryside', which seemed to have settled in the minds of policy-makers in Brussels and in the national capitals of Western Europe, needed to be dislodged. Mansholt's proposals were rejected amid howls of fury from many sections of the farming community. Nevertheless, an alternative vision of the countryside and of the desirable scope of rural management had been delivered and, despite the opposition, something very similar to it was to reappear in a surprisingly short time.

As well as growing concerns about the dilemma of over-production, other trends in European society were converging to reinforce the needs to revise agricultural policy. The first was the rise of environmentalism and green politics, complex movements that emerged in varying ways and with differing emphasis in the component states of Western Europe. Increasing numbers of citizens expressed their concerns about the harmful effects of modern farming on wildlife, landscape, and the quality of food and water. In Britain the emphasis tended to be on nature and landscape (calling for the protection of wild birds and flowers endangered by the removal of hedgerows), whereas in Denmark, the Netherlands and the then West Germany, threats to public health through indiscriminate use of toxic chemicals in food production, polluting soil, water, air and food itself, were the primary causes for complaint (Clout 1993). In general terms, environmental arguments were articulated more emphatically in the urban, long-industrialized northern parts of Europe than in the rural, still industrializing south. None the less, the environmental case was made with increasing force, and in some respects coalesced with the scientific and aesthetic cases for nature conservation, which had been expressed as a minority opinion since the nineteenth century and had led to the designation of national parks and other protected areas in remote locations in many parts of the continent.

A second social trend was the growing perception of the countryside less as a place for food production and more as a recreational arena for members of the urban élite and also increasing numbers of car-owning city people who might enjoy country weekends, vacations, sporting and leisure activities. For some, the countryside was also seen as a place of permanent residence, either during their working lives or in retirement. Vacant farm cottages, redundant barns, land commanding enjoyable views, accessible lake shores, fishing rivers and many more elements of the countryside were being viewed as commodities which could be purchased by urban dwellers for sums that rural residents would be unable to afford. Rural places tended no longer to be recognized as interlocking components of a functional agricultural world and were instead perceived in completely different ways as discrete purchasable goods. The process of commodification had indeed taken hold. The competition was, of course, an unequal one and it was ordinary rural workers, in search of affordable homes, who were the losers.

New policies to meet new imperatives

Over-production, environmentalism, commodification and a suite of other issues promoted the revision of the CAP in the mid-1980s and early 1990s and stimulated a set of new statements about the future roles of the European countryside (Winter 1996). The European Commission's Green Paper, *Perspectives for the Common Agricultural Policy* (Commission of the European Communities 1985), reviewed the environmental implications of modern farming, the financial dilemmas associated with food surpluses, and the continuing problems of deprivation and depopulation that were encountered in 'deep rural' parts of Europe (especially in some sections of the Mediterranean South), where counter-urbanization or retirement in-migration were not being experienced. It emphasized that, in future, farming would not be viewed simply as supplying food and jobs and thus helping to support rural settlements, but would also be appraised in terms of its environmental role as a maker, sustainer and potential destroyer of rural landscapes and environments. The revised CAP which eventually emerged comprised a complex and evolving package of measures that cannot be itemized here. In summary, the long-established productivist emphasis, associated with the quest to promote intensification and to increase food output, was replaced by a controversial new approach. Outputs of certain commodities were to be curtailed by setting quotas, reducing guaranteed levels of price support, and encouraging farmers to produce only goods which remained in demand. In turn, these processes would facilitate the departure of some farmers from agricultural work altogether, and ensure that stretches of land would be withdrawn from conventional forms of production.

The Commission's document, *The future of rural society* (Commission of the European Communities 1988) formally recognized the complex interrelationship of economic, social and environmental elements in the countryside, and also stressed the growing importance in rural space of activities which might be considered non-agricultural or only marginal to farming. Policies in each Member State were required to embrace this new vision for the countryside, whose implications were not of course constrained within the then limits of the EC. For example, in 1990, Sweden introduced a thoroughgoing revision of agricultural and rural policies prior to entering the EU, which aimed at reducing arable land by one-fifth by 1995 through the planting of trees, extensive grazing, energy cropping and re-creation of wetlands (Lewan and Lewan 1993).

The states of Eastern and Central Europe, which aspire to join the EU in the not too distant future, will also have to adjust to the new rural order with its declining stress on conventional food production. Following the collapse of Communist regimes in 1989 and the subsequent closure of inefficient and polluting factories and mines, which could not withstand competition on the open market, some Eastern and Central European analysts, in emphasizing the traditional importance of farming in their national economies, have stressed the high quality of their food products in the hope that agriculture might play a rather more important economic role in the future, alongside high-tech and service-sector activities. Thus Polish commentators have recalled the vast extent of their agricultural territory and their role of supplying grain to industrializing countries in North-West Europe during the nineteenth century and of exporting meat, vegetables and fruit products in more recent years. From this reading of economic history, they argue that Polish farms might play a greater role in the twenty-first century. Such a scenario, however, plays down the relative inefficiency of Polish agriculture, composed largely of small family holdings whose labour resources have been inflated during the 1990s as unemployed former factory workers have returned to their rural roots to occupy parental farms. Eventual membership of the EU will subject Eastern and Central European states not only to the forms of economic policy that operate in the West, but also to forces of world-wide competition. In short, the role of the countryside will have to be reappraised right across the European continent in the next century.

The planned retreat

What may be summarized as the planned retreat from both intensification and the enlargement of the farmed surface builds on trends that were already in evidence, albeit on a relatively modest scale, across some stretches of European countryside (Bolsius *et al.* 1993). None the less, over the past ten to fifteen years, national and supranational institutions have had to reappraise both the advice which they give and the sources of funding allocated to farm-based enterprises. In turn, 'farming families' have been forced to devise new survival strategies, rather than relying on income from agriculture as the sole or predominant source of support.

This transformation in attitudes and action among farmers and their households is proving difficult both to conceptualize and to achieve. For ease of discussion, a rather artificial distinction may be drawn between activities to be abandoned or scaled down, and those to be adopted or accepted more widely. Thus the introduction of quotas on dairy herds and schemes to promote despecialization and extensification may be thought of as representing the former kind, whereas a multitude of alternative ways of trying to make a living reflects the latter. This challenging shift from conventional agriculture requires a complicated reappraisal of the rural resource base, embracing not only its conventional grounded components, such as soil, slope and climate, but also its positional characteristics (including relative proximity to a major city or transport route). To these can be added its heritage, which may be expressed in terms of the combination of settlement history and scenery that forms the 'cultural landscape', and the worth attributed to those features by society. Some of its constituent parts (such as old cottages) may already be perceived as appealing and attractive to outside investment. Together with relatively ignored elements, they also have the potential of being packaged, promoted and marketed as products for tourism consumption.

The implications of reductions in grain prices or quotas on livestock may seem obvious enough, but what is to be done with former arable or grazing land, or with surplus farm buildings? How may former grain producers or dairy farmers and members of their households readjust their lives to earn acceptable incomes? Each of these factors of production offers scope for reappraisal and innovative response in order to contribute to a set of survival strategies. This process will require a great deal of creative energy, training, and personal and familial adjustment, as well as the injection of necessary investment, which may have to be raised in the form of grants or loans. Despecialization has been encouraged to reorientate farmers away from generating surpluses in order to engage in alternative forms of production. For example, a scheme was introduced to compensate producers of table wine in Bas-Languedoc for rooting up vines and switching to growing fruit or vegetables on freshly irrigated land. It soon became clear, however, that viticulture was not just an economic activity but was, indeed, a way of life firmly anchored in local culture and politics. The scheme was greeted with bitter resentment and those vinegrowers who did change to producing fruit and vegetables soon had to contend with powerful competition from Italy and Mediterranean Spain.

Extensification is a multi-layered notion in the new vision of the European countryside, which ranges from the complete withdrawal of land from agricultural use (for example, for forestry, wilderness or even construction), through 'setting aside' cropland for a specified period, to using farmland in less intensive and more environmentally friendly ways. These modifications are not merely economic adjustments but involve the confrontation and reformulation of long-established beliefs about the 'appropriate' use of rural land. Put simply, the traditional association between land and people in the form of family farming throughout Europe specified that the proper use of farmland was to produce increasing quantities of food, which not only nourished the household but also helped to sustain the nation. To give land over to scrub or woodland involved breaking a sacred obligation to the homeland as well as demonstrating failure and identifying a farming household as one that could not cope. In Vichy France, a field which fell out of cultivation had been described as a part of France that died. Yet under the revised strategy for the countryside, to reallocate farmland to economic forestry, recreational woodland or simply as an aesthetic backdrop of trees has been promoted as an appropriate policy option both for areas of 'marginal farming', such as the Limousin region in central France, which is rapidly becoming tree-covered, and for peri-urban areas of high agricultural productivity, as in parts of midland England, where the demand for recreational forestry is great.

Implementation of periodic 'set-aside', which affects areas of efficient, intensive farming (the areas that generate the largest surpluses) as well as less productive zones, has been greeted with varying degrees of opposition by farmers throughout the EU. Some have argued that earlier schemes in the United States revealed that the logic was fundamentally flawed since farmers often intensified crop production from areas which were permitted to remain in cultivation, and hence set-aside failed to curb food surpluses. Arguably the most intense hostility has originated among French politicians and farmers, who argue that to operate policies which remove land from food production not only has economic, social and environmental ramifications but also threatens the unique identity of the French nation, which has been rooted in family farming. In addition, they maintain that, as a vastly important producer of a wide range of food products, France depends on agriculture to a much greater degree than its long-industrialized neighbours in North-Western Europe. To reduce financial support from the established agricultural base of the rural economy, it is argued, will diminish job opportunities and encourage out-migration, thereby reducing the viability of services and raising their cost for those who remain. As a result, communal facilities will have to be withdrawn and the vicious spiral of out-migration will twist inexorably downward to generate an extreme form of depopulation known in evocative terms as *désertification*. Lower intensities of land management will degrade cherished landscapes, which are overwhelmingly 'cultural', and not 'natural', as is popularly supposed. The argument runs that as scrub invades upland pastures and hillslopes become covered with trees, these abandoned landscapes will not only lose their appeal to visitors but will destroy the 'image' of particular stretches of countryside (Neboit-Guilhot and Davy 1996). In more pragmatic terms, the risk of avalanches, landslips and other environmental disasters is known to be greater in abandoned, untended mountain areas. This point has been acknowledged for many years in policies to defend Alpine farming and to ensure that mountain settlements and services survive in Austria and Switzerland.

The third aspect of extensification involves land continuing in farming but being used in a less intensive way. Of course, this has happened already in depopulating areas throughout Europe and on many holdings run by ageing farmers. What is both new and extremely controversial is for planners and politicians to advocate de-intensification as a major feature of policy. In fact, Dutch authorities had employed minor options in EC legislation of 1975 to enable management plans to be drawn up with farmers in ecologically sensitive areas in order to compensate them for not intensifying food production. They were required to retain copses, hedgerows, ditches and pools, to maintain a high water table, and to abstain from applying artificial fertilizers. A more extreme expression of this trend toward 'renaturalization' involved flooding some stretches of surplus polderland which had been reclaimed painstakingly in past centuries. During the early 1980s, an environmentally friendly compensation policy was introduced in Bavaria and northern Germany, while in Denmark 'renaturalization' has involved converting grain areas back to grazing land along valleys. National and local variations on this theme have been implemented elsewhere in the EU in recent years. For example, since 1987, a suite of environmentally sensitive areas (ESAs) has been designated in the UK, whereby farmers are offered compensation provided they adhere to voluntary conservation contracts which restrain livestock densities, ban the application of artificial fertilizers, and require hedges, ditches, field walls, barns, woodlands and wetlands to be maintained. By 1995, one-sixth of the agricultural area of the UK was covered by designated ESAs. Landowners with property adjacent to extensively operated farms soon learn that what they recognize as weeds and pests may be perceived by other people as 'nature'. In the past few years mounting concern over pollution of watercourses through excessive application of agricultural fertilizers has caused restrictions on the use of nitrates in numerous parts of rural Europe. Policies to promote ecologically sympathetic farming have led to many farmers complaining that they are being reduced to the status of park-keepers who are paid to manicure the landscape into a kind of stage set for urban visitors to gaze upon, rather than being assisted and encouraged to perform their centuries-old duty of acting as wise stewards of the earth's resources to produce food.

Standing back from the present

This fundamentalist view of the 'proper' role of farmers in society and their obligation to increase food output, by bringing more land into cultivation or by pursuing intensification, has very deep roots. It was reinforced in this century by the need for Europe to produce more food during both world wars, when fighting prevented farming over some of the most productive agricultural areas, normal supplies of farm labour were disrupted by conscription, and hostilities at sea hampered intercontinental trade. Food was in desperately short supply and rationing was widespread during both world wars and for a time after 1945. In Britain a plough-up campaign converted stretches of grass and uncultivated land into productive arable, at least for the duration of World War II. Food production and farming were afforded special status in the array of national legislation promoting economic recovery after peace had been restored in 1945. The privileged place of farming in national life was reinforced in each Member State of the newly formed EC when the CAP was introduced in the early 1960s.

A broader historical view of Europe's farming offers a rather different perspective. First, it is clear that the mosaic of land uses that characterized Europe's countrysides at any date in the past few decades was simply a manifestation of economic, social, demographic, technological and political circumstances at that particular time. Looking back across the centuries, it becomes clear that the extent of cultivated land in relation to grazing and woodland has varied enormously at local, regional, national and continental scales (Bertrand 1975). Similarly, the number, distribution and morphology of rural settlements have been subject to dramatic shifts. In predominantly rural societies, without opportunities to import foodstuffs or to intensify agricultural production, population growth could be achieved only by increasing the cultivated surface at the expense of other major forms of land use. Thus the medieval phase of population growth was indeed an age of enormous clearance of wasteland and drainage of marshland, and also the time when the great eastward migration of settlers to new villages and fields set in the midst of forest clearings led to the spatial definition of Germanic East-Central Europe and the establishment of some of the localities which ethnic Germans have chosen to leave in order to live in the Federal Republic following unification in 1990 (Mayhew 1973) (Figure 11.3). The subsequent period of population decline in the early fourteenth century, associated with war, disease, climatic deterioration, soil exhaustion and the failure of farming systems to intensify, saw cultivation retreat, settlements shrink, wetlands reappear, and woodland and scrub recolonize what had previously been productive fields. 'Nature' regained territory which had been tamed by centuries of human effort. Such processes were not unique to late medieval times and have operated on local and regional as well as continental scales throughout history.

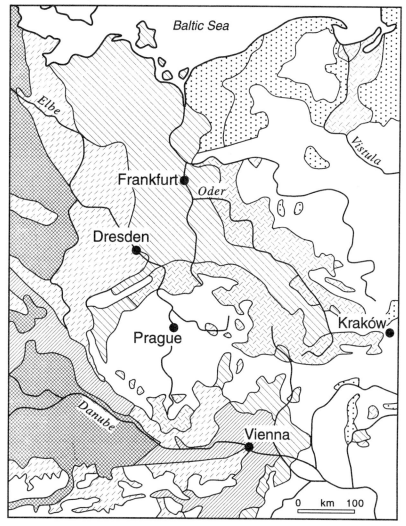

Early Germanic settlement

Expansion of German settlement

700 - 1100   1250 - 1300

1100 - 1200   after 1300

1200 - 1250

Figure 11.3   Medieval expansion of German settlement into East-Central Europe
Source: After Mayhew (1973)

Second, intensive production systems, involving intricate crop rotations, livestock rearing, repeated manuring and substantial inputs of labour, were well established in Flanders and parts of northern Italy in late medieval times. Not until a complex array of philosophical, technological, biological and tenurial changes (often replacing open fields with enclosures) was adopted was more widespread agricultural intensification to become a reality. This transformation spread through the Low Countries and England in the seventeenth and eighteenth centuries, then southern Scandinavia, and subsequently other parts of North-West Europe (Berglund 1991). After 1950, mechanization, biological innovations and the application of agrochemicals made intensive farming an even more widespread phenomenon in Eastern as well as Western Europe. What was perceived as 'marginal land' was farmed less intensively, just as it had been during the preceding century, with some cultivated areas being abandoned to scrub or woodland. Remote settlements became depopulated or even deserted as agricultural intensification continued to forge ahead close to major cities, in localities well served by efficient transport links, and on land that could be irrigated.

Third, agricultural activities in the regions that composed the various European states were subjected to varying degrees of competition and change during the nineteenth century, as different colonial and commercial policies were put in place. Changes were also linked to the technological opportunities afforded by sailing-ships, steamships and the railways. Industrializing Britain came to rely increasingly on grain and meat imported from imperial possessions in the New World. Danish agriculture was transformed to send dairy goods to Britain and the Netherlands, which both received additional supplies of grain from Poland and Russia (Tracy 1989). By contrast, France and some other West European states sheltered behind tariff barriers, and hence their agricultural systems were not exposed to the kinds of transformation that occurred in Britain or Denmark.

A retrospective view reveals that the historical geography of the European countryside is structured around temporal pulsations in food production and spatial fluctuations in land use. The mosaic of arable, grazing land and woodland was far from stable, but the imperative to strive to produce more food was omnipresent. What is striking, and arguably unique, about our own time is both the remarkable ability of late-twentieth-century farming to generate rapidly increasing quantities of food and the subsequent policy changes which require Member States of the EU to strive to produce less, rather than more.

The quest for diversification

The current reformulation of rural Europe involves the promotion and indeed creation of new ways of generating income for those who continue to occupy 'farm' properties and for others who live in other forms of rural housing. This broad objective is embraced by the notion of diversification, which, of course, is far from being new. One of the distinctive features of the human geography of pre-industrial Europe was the widespread distribution of dual or multiple job-holding, whereby part of the farming year (usually the winter), or some of a farmer's time, was devoted to craftwork or other forms of non-agricultural activity (Garrier and Hubscher 1988). Such by-employments were particularly numerous around woodlands, in areas with ore deposits, along seashores and major rivers, and on the fringes of urban areas. Detailed censuses from as late as the mid-nineteenth century reveal an array of rural crafts and by-employments, many of which have since disappeared as rural craft activities were replaced by urban-based factory work. None the less, forestry and mining continued to be combined with agricultural work in some areas and were joined by new opportunities for part-time work in tourism, notably in upland areas. In this way the Alps and other mountainous zones were appraised in terms of their environmental attractions, rather than being perceived as hostile areas to be shunned.

The provision of bus services and latterly the widespread ownership of private cars (in Western Europe at least) introduced new possibilities of dual employment, with some workers in factories or mines continuing to live on farm holdings and commuting to urban work each day. Farmwork was done in the evening or at weekends, or was left to other members of the household. The income of farm-based families might be augmented by some household members taking off-farm work on a permanent or a seasonal basis, or by new sources of income being developed on the farm itself (Gasson 1988). Despite this complexity, one point remained clear: part-time farmers were regarded as poor farmers by agricultural advisers and also by their neighbours who worked the land full-time. They were judged to be inefficient and their occupation of farmland – almost as a hobby – was condemned for hindering the enlargement of full-time farms and the desirable promotion of agribusiness. Some of them abandoned patches of farmland to 'social fallow', not because the soil was infertile but because part-time farmers lacked the time or inclination to work it 'properly'. During the 1980s, agricultural policies were revised and it was recognized that part-time farming could offer an important survival strategy for some rural families. This was simply acknowledging the obvious, since in the early 1980s one-third of all farm holdings in the then 10 Member States of the Community were obtaining income from non-agricultural sources, with the proportion exceeding two-fifths in West Germany.

Functional diversification might involve innovative forms of farming (such as growing 'organic' vegetables, which command a relatively high price and for which demand is expanding), or on-farm non-agricultural activity (such as operating some kind of light industry or service enterprise in redundant farm buildings or providing farmhouse accommodation for tourists). Before the revision of the CAP, farm-based tourism was already well established in many parts of Western and Eastern Europe, in the form of letting out rooms, converting surplus barns and other buildings for holiday accommodation, creating caravan or camp sites, and setting out countryside trails or farm walks. One-third of farms in the Austrian Tyrol offered accommodation to visitors and the proportion stood at one-sixth in the crofting counties of Scotland (Bramwell and Lane 1994). Farm-based tourism has the obvious advantage of making use of surplus space and can be relatively flexible in responding to changes in demand. However, it requires capital investment (for example, for additional bathrooms), necessitates a 'commercial' view of life centred on profit and loss among family members, and can place an additional workload on the women of the household (Whatmore *et al.* 1994). Farms close to, or easily accessible from, major cities, or located in more remote areas with particularly attractive scenery or cultural qualities, offer considerable potential for weekenders and long-vacation visitors respectively. By contrast, intermediate agricultural areas with unremarkable landscapes are more difficult to promote. Road-side signs offer the chance of capturing passing trade but a growing trend is for the farm holiday experience to be packaged for consumers. In the case of farmhouse holidays in France for British visitors, this includes ferry and intermediate hotel reservations as well as the accommodation on the farm. In common with all tourism, rural tourism is fragile and is subject to changes in fashion and financial circumstances. For example, there was a sharp decline during the 1996 summer season in the number of British bookings for French *gîtes*, as some long-established clients sought higher standards of accommodation and all complained about the poor exchange rate. Fortunately this decrease in demand was compensated by an increase from Dutch and Belgian clients.

Conclusion: the future of the rural past

Despite the dramatic changes that have affected Europe's countrysides during the past half-century as a result of changes in agricultural activity, new housing for commuters and retired migrants, conversion of farmland to alternative uses and many other trends, the continent's rural landscapes remain remarkably varied as the new millennium approaches (Champion and Watkins 1981). The distinctive *pays*, the diversity of which fascinated geographers earlier in this century, are still recognizable and provide immediate raw material for stimulating tourism and encouraging countryside conservation (Meeus 1995). Thus surviving remnants of field walls, hedgerows, neat enclosures, vast open fields, polders, terraces, vineyards and other human features in the rural landscape serve as critical promotional cues, along with mountains, moors, marshes, heaths, hillslopes and wide river valleys. The harsh lands and environments of difficulty, formerly perceived in 'negative' terms by countless generations of Europeans who spent their lives trying to conquer them for cultivation, are now appraised as valuable components in the growing drive toward rural diversification. Thus the Danish Heath Society, for example, which for over a century encouraged the reclamation of heaths and the drainage of wetlands in order to increase the area of land available for cultivation, has found a new mission in conserving heathland and restoring watercourses to a more natural state (Olwig 1984; Sporrong 1994).

Châteaux, landscaped parks, manors, abbeys and battle sites, all of which testify to the impact of social power and control in the landscape, as well as arguably more egalitarian market halls and 'unspoiled' villages, have long provided individual attractions for tourists. Increasingly, such sites are being assembled into distinctive themed itineraries, often associated with particular landscape features, habitats, regional products or even the works of fiction of local authors. Now it seems that every vineyard has its signposted circuit and every mountain range has a special route identified by traditional cheese-making farms, by shady forests or roaring waterfalls. Marketing of the countryside and or rural commodities has, indeed, become the *fin de siècle* growth industry in Europe (Hoggart *et al.* 1995; Murray and Graham 1997). For example, the Calvados area in Normandy boasts various themed circuits, as well as routes linking abbeys and the beaches of the Normandy Landings in World War II, not to mention scores of farm-based outlets retailing cheese, cider, Calvados brandy and many other goods (Figure 11.4). Farms may extend a rather quaint attraction to the visitor, but the local food-processing works can often put on more attractive displays. Videos and educational literature show visitors how cheeses, sausages, liqueurs and other wares were once made in local farmhouses (but are now produced in rows of steel vats in rural factories), and also ensure that desired brand names are recognized on the shelves of supermarkets and the counters of delicatessens in the weeks and years ahead. All this is a far cry from hand-crafted country goods, but if a rural food-processing works is to flourish there must be farms to supply it and a market (perhaps even located on the other side of the world in the case of the *foie gras* and Armagnac brandy of the Gers area in south-western France) to consume its products. Authenticity may have been sacrificed to ensure the vitality of the local economy.

To some extent, the relative success of rural diversification will depend on proximity to urban demand centres and other locational factors, but will also be conditioned by the synergy of innovation, investment and promotion, which operates at a finer spatial scale, namely that of the individual enterprise (Huigen *et al.* 1992; Mathieu 1995). As Europe enters the twenty-first century, its countrysides are undoubtedly contested spaces, subject to social, economic and environmental changes, which are to the benefit of some groups in society but to the disadvantage of others (Flynn and Lowe 1994; Murdoch and Marsden 1994). In varying degree, such developments may be encountered elsewhere in the urbanized, post-industrial world. Certainly the trends that have affected Western Europe for several decades are now being experienced further east. Despite the magnitude of these changes, Europe's countrysides continue to display an impressive measure of continuity from earlier times, as demonstrated by the survival of historic buildings, relatively unspoiled landscapes, and a range of specific places that are cherished for their cultural associations or their scientific interest. In this era of

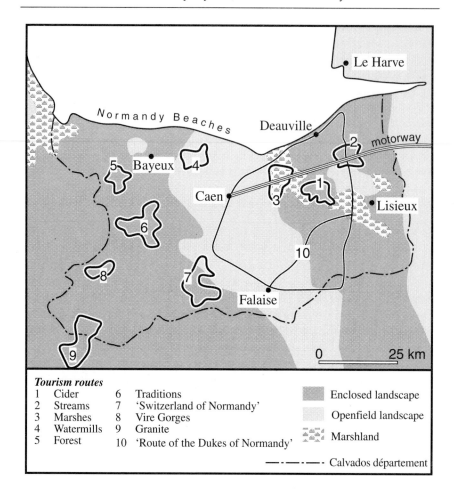

Figure 11.4   Tourism routes and rural landscapes in the Calvados area of Normandy

globalization, mass culture and international sameness in mass consumption (whether that be of food, clothes or other goods), authenticity is being sought – and sometimes fabricated – and distinctiveness is being purchased at a high premium (Marsden 1993).

The substance of history is being recognized not just for its academic interest but is being filtered, packaged and re-presented as 'heritage' for the enjoyment and education of all who can afford the price of an entrance ticket. The fabric of geography has been equated (albeit erroneously) in the public mind as 'nature', and that too is made available for us to experience and consume rather than to seek to tame as in times past. These changes in viewpoint are encapsulated in the debate on the future of Europe's rural landscapes and in exploring the essential tension between promoting change, in the name of short-term socio-economic development, and conserving countryside heritage for generations yet to come (Bethemont 1994).

References

Abel, W. 1980: *Agricultural fluctuations in Europe, from the thirteenth to the twentieth centuries.* London: Methuen.

Berglund, B. E. (ed.) 1991: *The cultural landscape during 6000 years in southern Sweden.* Copenhagen: Munksgaard.

Bertrand, G. 1975: Pour une histoire écologique de la France rurale. In Duby, G. and Wallon, A. (eds), *Histoire de la France rurale:* vol. 1: *Des origines à 1340.* Paris: Seuil, 34–113.

Béteille, R. 1996: *Le tourisme vert.* Paris: Presses Universitaires de France.

Bethemont, J. (ed.) 1994: *L'avenir des paysages ruraux européens: entre gestion des héritages et dynamique du changement.* Lyon: Presses Universitaires de Lyon.

Bolsius, E. C. A., Clark, G. and Gronendijk, J. G. (eds) 1993: *The retreat: rural land use and European agriculture.* Amsterdam: Royal Netherlands Geographical Society.

Bramwell, B. and Lane, B. (eds) 1994: *Rural tourism and sustainable development.* Clevedon: Channel New.

Buller, H. and Hoggart, K. 1994: *International counterurbanization: British migrants in rural France.* Aldershot: Avebury.

Butlin, R. A. 1993: *Historical geography: through the gates of space and time.* London: Edward Arnold.

Champion, A. (ed.) 1989: *Counterurbanization.* London: Edward Arnold.

Champion, A. and Watkins, C. (eds) 1981: *People in the countryside: studies of social change in rural Britain.* London: Paul Chapman.

Cherry, G. E. and Rogers, A. 1996: *Rural change and planning: England and Wales in the twentieth century.* London: Spon.

Clout, H. D. 1984: *A rural policy for the EEC?* London: Methuen.

Clout, H. D. 1993: The recomposition of rural Europe. *Annales de Géographie* 100, 714–29.

Clout, H. D. 1996: *After the ruins: restoring the countryside of northern France after the Great War.* Exeter: Exeter University Press.

Commission of the European Communities 1985: *Perspectives for the Common Agricultural Policy.* 8480/85, COM(85) 333 final. Brussels/Luxembourg: Commission of the European Communities.

Commission of the European Communities 1988: *The future of rural society.* Brussels/Luxembourg: Commission of the European Communities.

Coppock, J. T. (ed.) 1977: *Second homes: curse or blessing?* Oxford: Pergamon.

de Planhol, X. 1994: *An historical geography of France.* Cambridge: Cambridge University Press.

Duby, G. 1970: The French countryside at the end of the thirteenth century. In Cameron, R. E. (ed.), *Essays in French economic history.* Homewood, IL: Irwin, 33–41.

Faure, C. 1989: *Le projet culturel de Vichy.* Lyon: Presses Universitaires de Lyon.

Flynn, A. and Lowe, P. 1994: *The contested countryside.* London: UCL Press.

Garrier, G. and Hubscher, R. (eds) 1988: *Entre faucilles et marteaux: pluriactivités et stratégies paysannes.* Lyon: Presses Universitaires de Lyon.

Gasson, R. 1988: *The economics of part-time farming.* London: Longman.

George, P. 1964: Anciennes et nouvelles classes sociales dans la campagne française. *Cahiers Internationaux de Sociologie* 37, 13–21.

Hoggart, K., Buller, H. and Black, R. 1995: *Rural Europe: identity and change.* London: Edward Arnold.

Huigen, P., Paul, L. and Volkers, K. (eds) 1992: *The changing function and position of rural areas in Europe.* Utrecht: Royal Netherlands Geographical Society.

Kain, R. J. P. and Baigent, E. 1992: *The cadastral map in the service of the state: a history of property mapping.* Chicago: University of Chicago Press.

Kayser, B. 1989: *La renaissance rurale, sociologie des campagnes en mutation.* Paris: Armand Colin.

Kjaergaard, T. 1994: *The Danish revolution, 1500–1800: an ecological interpretation.* Cambridge: Cambridge University Press.

Lebeau, R. 1969: *Les grands types de structures agraires dans le monde.* Paris: Masson.

Lebovics, H. 1992: *True France: the wars over cultural identity 1900–1945.* Ithaca, NY: Cornell University Press.

Lewan, L. and Lewan, N. (eds) 1993: Landownership and landscape. *Rapporter och Notiser: Lunds Universitet* 121, 1–82.

Lowe, P. and Bodiguel, M. (eds) 1990: *Rural studies in Britain and France.* London: Pinter.

Marsden, T. (ed.) 1993: *Constructing the countryside: an approach to rural development.* London: UCL Press.

Mathieu, N. (ed.) 1995: *L'emploi rural: une vitalité cachée.* Paris: L'Harmattan.

Mathieu, N. and Duboscq, P. 1985: *Voyage en France pars les pays de faible densité.* Paris: CNRS.

Mayhew, A. 1973: *Rural settlement and farming in Germany.* London: Batsford.

Meeus, J. H. A. 1995: Pan-European landscapes. *Landscape and Urban Planning* 31, 57–79.

Moriceau, J.-M. 1994: *Les fermiers de l'Île-de-France, XVe–XVIIIe siècle.* Paris: Fayard.

Murdoch, J. and Marsden, T. 1994: *Reconstituting rurality.* London: UCL Press.

Murray, M. and Graham, B. 1997: Exploring the dialectics of route-based tourism: the *Camino de Santiago. Tourism Management* 18, 513–24.

Neboit-Guilhot, R. and Davy, L. (eds) 1996: *Les Français dans leur environnement.* Paris: Nathan.

Newby, H. 1978: *Property, paternalism and power: class and control in rural England.* London: Hutchinson.

Nitz, H. J. 1978: Small-holder colonization in the heathlands of north-west Germany during the eighteenth and nineteenth century. *Geographia Polonica* 38, 207–13.

Olwig, K. 1984: *Nature's ideological landscape: a literary and geographic perspective on its development and preservation on Denmark's Jutland heath.* London: Allen & Unwin.

Overton, M. 1996: *Agricultural revolution in England. The transformation of the agrarian economy, 1500–1850.* Cambridge: Cambridge University Press.

Pahl, R. E. 1975: *Whose city?* Harmondsworth: Penguin.

Price, R. 1983: *The modernization of rural France: communications networks and agricultural market structures in nineteenth-century France.* London: Hutchinson.

Rambaud, P. 1969: *Société rurale et urbanisation.* Paris: Seuil.

Schama, S. 1995: *Landscape and memory*. London: HarperCollins.

Shoard, M. 1980: *The theft of the countryside*. London: Temple Smith.

Short, B. (ed.) 1992: *The English rural community: image and analysis*. Cambridge: Cambridge University Press.

Smith, C. T. 1978: *An historical geography of Western Europe, 1500–1840*, revised ed. London: Longman.

Sporrong, U. (ed.) 1994: *The future of rural landscapes*. Uppsala: Swedish Science Press.

Thomas, K. 1983: *Man and the natural world: changing attitudes in England, 1500–1800*. London: Allen Lane.

Tracy, M. 1989: *Government and agriculture in Western Europe, 1880–1988*, 3rd ed. New York: Harvester Wheatsheaf.

Vidal de la Blache, P. 1903: *Tableau de la géographie de la France*. Paris: Hachette.

Warnes, A. W. (ed.) 1982: *Geographical perspectives on the elderly*. Chichester: John Wiley.

Whatmore, S., Marsden, T. and Lowe, P. (eds) 1994: *Gender and rurality*. London: David Fulton.

Winter, M. 1996: *Rural politics: policies for agriculture, forestry and the environment*. London: Routledge.

|*Epilogue*|

Europe's geographies: diversity and integration

Brian Graham

In addressing the meaning of place, culture and identity in contemporary Europe, this book has pursued three interlocking themes. It has shown how complex cultural and economic trajectories derived from the past have produced the geographical diversity of our modern era, while ensuring that the meanings of Europe and its places have been continuously renegotiated through time. It is clear that diversity or heterogeneity is the key to Europe's human geographies and that any efforts to evoke a European level of consciousness must be vested in notions of plurality. The divergent trajectories of modernization discussed here ensure that modern Europe is characterized, inevitably, by an array of contested agendas, particularly with respect to its further institutional integration and the geographical enlargement of the EU. It is important, too, to recognize that the policy of European union also involves an ideological commitment to ideals of social solidarity and a social economy that reflects the inclusive principles of convergence and cohesion. As *Agenda 2000* (Commission of the European Communities 1997) argues, the EU is not merely a political and economic entity but must also make a major contribution to peace, democracy and the defence of human rights and values in a wider and more diverse Europe. Thus the internal objectives of the Union must become more firmly focused on the attainment of a cohesive and inclusive society based on solidarity, as well as a high quality of life, a sound environment, freedom, security and justice.

While some of these sentiments might be dismissed as no more than high-minded – or even largely empty – rhetoric, it remains the case that the EU is a radical experiment in the reordering of political space. That its role in re-negotiating cultural space is less impressive is demonstrated by the argument that social cohesion and solidarity are transparently superficial, the EU being little more than a loose confederation of nation-states. In economic terms, national boundaries remain the principal constraint on the effectiveness of cohesion, there being no direct relationship between the prosperity of a region and its status as a contributor to a national budget or a beneficiary from it. The richer regions of poorer countries still have to pay for their poorer national regions (Yuill and Davezies 1997), a trend which remains a major contradiction in that EU regional policies are justified in part by the idea that economic disparities act as a barrier to further integration.

Diversity, however, is defined by more than economic criteria alone. Regional disparities also possess cultural connotations, again derived from the complex interplay of past and present. Thus the goals and objectives of European integration are often conflicting in themselves as the priorities of cohesion, convergence, competition and liberalization, and enlargement are not easily reconciled with cultural heterogeneity. While economic liberalization, for example, is a reductionist policy that ignores diversity and imposes a uniform strategy of policies almost irrespective of local conditions, the potential enlargement of the EU will make the Union even more culturally diverse and underline even more emphatically the dichotomy between it and those states left beyond its frontiers. Policies for enhanced European integration also conflict with national interests. Given that countries modernized in different ways, this mix of national agendas is quite inevitable. It is probably unrealistic, for example, to expect the UK with its offshore location and a – now somewhat debased – global perspective, derived from a history of empire-building and early modernization, to view Europe in the same way as France or Germany, desperately anxious to avoid any repetition of the twentieth-century wars that nearly destroyed them twice over. Whatever the perspectives of individual states on European union, however, national interests still predominate; the commitment to EU enlargement, for example, is combined with a desire not to increase the contributions of Member States to EU budgets, not least in the case of Germany – the largest contributor, but also carrying the enormous financial burden of reunification and the reconstruction of the former GDR.

As the discussions in this book have shown, the construction of nationalisms and the creation of nation-states constitute one of the most important – perhaps *the* most important – manifestations of the past in the present. The evolution of this interlocked mechanism of defining identity and governance in a sovereign, territorial state has taken centuries, and, clearly, present EU programmes and policies are in themselves insufficient to destroy the deep-rooted conceptualization of, and loyalty to, the nation-state. Indeed, as Bideleux (1996a) argues, the strategies of European integration were not created originally in opposition to the nation-state, but as part of the post-war rehabilitation of a number of those polities, while membership of the EU may actually have helped hold together fissiparous examples such as Spain or Belgium. The ethnic nationalism that has resurfaced in Central and Eastern Europe is also seriously at odds with the civic ideals of the EU and most of its current Member States – the 'citizens' Europe' at the heart of the 1997 Treaty of Amsterdam, which seeks to address the perceived distance between the Union and its citizens. Bideleux (1996b) believes that most states in Central and Eastern Europe are still in a state-building or nation-building phase, projects long since completed in most of Western Europe. He sees this as the fundamental qualitative difference between East and West, just as economic disparities are the basic quantitative difference.

If the nation-state as a reflection of ethnic identity and citizenship remains a potent force in manifestations and representations of European identity, it is also difficult to point to convincing evidence of the region as a post-nationalist alternative expression of identity and governance. It may exist – as in Catalonia – but not as a generally applicable model. As shown consistently in the discussions within this book, the diversity of trajectories of change in the past have produced an immensely complicated map of contemporary regional diversity. It is also apparent that people do identify with place at the regional scale, although the ways in which they do so are highly variable, as is the intensity of this scale of place belonging. Many regions defined by economic criteria possess little or no cultural meaning and – as, for example, the experience of northern Italy demonstrates – it is difficult to manufacture such representations of place (Agnew 1995). In other instances – as in Scotland – economic disparities may interact with factors including geographical peripheralization, the possession of clearly identifiable resources (in this case oil) and a distinct cultural identity to produce a clearly defined representation of place at this scale. In general terms, however, regional government is relatively weakly developed throughout the EU, decision-making remaining concentrated at the national level. Thus, beyond Spain and perhaps Germany, it is difficult to isolate more than a handful of convincing examples of region-states that stand as alternative expressions of belonging to the nation-state. Moreover, regional or local levels of governance may not equate spatially with regional levels of identity and consciousness. In France, for example, the Occitan cultural movement in the south of the country – the Midi – portrays a unifying regional identity that in actuality is fragmented by the concentration of local government at the smaller scale of the *département*. Moreover, as we have seen, cultural regions are themselves fragmented by even more diverse representations of identity at the local scale.

As the discussions here have demonstrated, the diversity of trajectories of change in the past have ensured that Europe has always been characterized by the tensions of diversity and integration. Fragmented feudal society, for example, was in part defined by omnipresent Latin Christianity. The remarkably complex geography of ethnicity in Europe has ensured that the integrative ideology of the nation-state could never cope entirely successfully with minorities left on the wrong side of boundaries. And so it is today. If diversity is the outcome of past processes, it follows that there will also be a heterogeneity of trajectories of change in the future. Diversity is the central condition of Europe's geographies. Thus the nation-state is in a process of change but not necessarily decline. Rather, it is the case perhaps that space is becoming more relative. The EU does not represent the nation-state writ large, nor is the region-state the nation-state in the minor key. Instead, we seem to be moving to a multi-level layering of governance – and of identity – in which the state may well remain the most important political and cultural level (Anderson 1995, 1996). Although the precise terms of its meaning remain contested, the notion of subsidiarity (see Chapter 6) means that the EU cannot regulate when policies can be made effectively at the local level. Although this concept was central to the 1992 Treaty of Maastricht, Bideleux (1996a) contends that the basic idea has been present since the inception of European union. Instead of presenting heterogeneity through the theme of disintegration versus integration, subsidiarity thus reflects the notion that Europe has long been characterized by 'continuous tension between forces of integration and disintegration, at national, subnational and supranational levels' (Bideleux 1996a, p. 15).

The arguments in this book clearly underline the contention that European integration, whether social, economic or cultural, cannot depend on policies of superficial homogenization. The essential diversity of Europe cannot, and should not, be subsumed within tropes of sameness. In precisely the same way that we seem to be moving towards multiple levels of governance and economic policy-making, this book has argued that we are also moving towards multiple levels of identity that contradict the simplifying synecdoches of ethnic nationalism that imposed a degree of uniformity upon the cultural diversity that has always existed in Europe – but only for a while. As we have seen, nationalism in Western Europe is now increasingly a civic affair. This model of the nation presupposes the existence of common institutions, a single code of rights, and a well demarcated and bounded territory with which the nation's members identify, and one to which they feel they belong (Smith 1991). Ethnic nationalism, by contrast, based on the idea of a single, ethnic identity, a community of birth and native culture, still prevails in Central and Eastern Europe, while its residues can be identified in contested places such as Northern Ireland and Euskadi (the Basque Country). At the civic level, the state must retain its importance because it embodies those elements of identity that equate to citizenship and democracy, of belonging to a community defined by shared values. But just as diversity and integration are not necessarily opposed sets of values, the maintenance of the centrality of the state does not negate the importance of European or various expressions of subnational levels of identity, not least because dissonance at the scale of the nation-state – and also within it – is one of the most potent barriers to the integration of Europe and the realization of a more competitive and efficient, but also more equitable, economy.

The founders of European unity such as Robert Schuman and Jean Monnet envisaged integration not only as a means of avoiding more wars, but also as a mechanism for creating a third force in the world that could compete with the United States and the USSR. This vision presupposed – but never consciously elaborated – a commonality of European interests that went beyond unity by systems and treaties. Charles de Gaulle, for example, 'believed that poets were greater unifiers than ideologues, artists and soldiers more creative of convergences than technocrats. He believed Chekhov and Bartók to be greater than Jean Monnet – and perhaps even than Karl Marx' (Lacoutre 1991, p. 398). But as we have seen here, that sort of common cultural veneer is not in itself a sufficient basis of European identity. The 'high cultural' heritage of architecture, music, literature and art is indeed part of a common European identity, but it may well be invisible to many citizens of Europe. Even more important, it ignores the hard questions and rampant instabilities that derive from the past, from the history 'that hurts', which is fundamental to the representations of Otherness that permeate national constructions of identity.

Hard questions have to be asked too about the democratic unaccountability of the EU and its mechanisms of government. Somewhat late in the day, the European Commission has recognized the distance that lies between the Union and its citizens and that there is a genuine call for an identifiable Europe (Commission of the European Communities 1997). Although such dimensions are often obscured by the petty bickering that too often dominates relationships between the Commission and Member States, the very idea of an identifiable Europe underscores the radical nature of the continent's grand project of integration. This is particularly so when we remember the alternative of war, national jealousies and immense human suffering that is Europe's past. But the problem remains that an identifiable Europe can be built only on representations of diversity. On the one hand, we have the increasingly powerful integrating mechanisms of the ECJ, the Commission and the European Parliament, soon to be joined by a central bank. But – as with high culture – it is facile to view these institutions as the integrative antithesis of diversity. Europe of the nation-states also failed to eradicate the heterogeneity that has characterized the continent past and present, and which simultaneously remains one of its greatest strengths as well as weaknesses.

The problem is that diversity is simultaneously a positive and negative force, a double-edged quality that defines the grounds both for inclusion and exclusion. Thus:

> Places create differences. Places . . . that have complementary mixes of elements, and possess permeable boundaries, can lead to mutual dependence, which can bring us together so that our differences enrich one another. But if . . . boundaries are virtually impermeable, these differences can be turned inward. They can narrow membership, isolate communities, create fear and hate of others, and push us in directions of inequality and justice.
>
> (Sack 1997, p. 254)

The impermeable boundaries of Ireland are one such case. In a deeply pessimistic analysis of the island during the late nineteenth and early twentieth centuries, the historian F. S. L. Lyons (1979, p. 177) argues for an 'essential diversity' that creates 'unbridgeable fissures' deeply embedded in the past and perpetuated by contemporary politics. He sees 'a collision within a small and intimate island of seemingly irreconcilable cultures, unable to live together or to live apart.

In addressing such dilemmas (and the enduring problem of ethnic conflict in Ireland is only Europe's essential diversity expressed as a – more extreme – microcosm) we need to accept Sack's point that differences can be positive and enriching, that dissonance in meaning can be seen as an inalienable quality of place (Tunbridge and Ashworth 1996). Indeed, it is a virtue in the same way that diversity is not the antithesis of integration. As the book has shown, European places and their meanings have been continuously renegotiated and remade through time. The result is to be found in the fragmentation of meaning and identity; nationalism may have subsumed this for a while but only in a transient way. Europeanness is not an alternative to the nation-state nor even an addition to that level of identity. As shown here, European dimensions to identity have always existed. What modern Europe requires, however, is a more conscious articulation of Europeanness, not as symphonic music or literature or art alone, but as one element in the many scales of meaning and identity that relate people to the places in which they live.

The enlargement of the EU and its further integration are part of the continuous remaking of place that defines the continent's history. The Union's commitment to cohesion, convergence and social solidarity also stands as a powerful alternative to the flexible, 'new serf' economy of what the French refer to as *le libéralisme sauvage*, currently trumpeted by the triumphalist economic moralism of the Anglo-Saxon world (Walden 1997). But Europe's history demonstrates only too well the transient and contingent nature of such trajectories of modernization and change. An ultra-free-market Europe can only force its peoples apart in the name of an ideology that is itself unsustainable in the longer term. Curious to think that the European Commission – unaccountable, anti-democratic, bureaucratic – is the principal institution that seeks to defend Europeans and their interests against the excesses of the free market and even their own governments. Such a dimension of Europeanness – values as well as heritage – must not subsume national, regional and local identities but reflect instead the complex dismantling of the synonymity of territoriality, sovereignty, nationalism and the state in the new Europe. Europe's geographies can be defined only through inclusivist notions of multiculturalism, plurality and diversity, presented not as unbridgeable fissures between peoples, but as manifestations of our mutual dependence in ensuring that the nightmares of Europe's past, if not forgotten, are never repeated.

References

Agnew, J. A. 1995: The rhetoric of regionalism: the Northern League in Italian politics, 1983–1994. *Transactions of the Institute of British Geographers* NS 20, 156–72.

Anderson, J. 1995: The exaggerated death of the nation-state. In Anderson, J., Brook, C. and Cochrane, A. (eds), *A global world? Reordering political space*. Oxford: Open University/Oxford University Press, 65–112.

Anderson, J. 1996: The shifting stage of politics: new medieval and postmodern territorialities. *Environment and Planning A: Society and Space* 14, 133–53.

Bideleux, R. 1996a: Introduction: European integration and disintegration. In Bideleux, R. and Taylor, R. (eds), *European integration and disintegration: East and West*. London: Routledge, 1–21.

Bideleux, R. 1996b: In lieux of a conclusion: East meets West? In Bideleux, R. and Taylor, R. (eds), *European integration and disintegration: East and West*. London: Routledge, 281–95.

Commission of the European Communities 1997: *Agenda 2000*. Com (97) 2000. Brussels/Luxembourg: Commission of the European Communities.

Lacoutre, J. 1991: *De Gaulle: the ruler: 1945–1970*. London: Collins Harvill.

Lyons, F. S. L. 1979: *Culture and anarchy in Ireland, 1890–1939*. Oxford: Clarendon Press.

Sack, R. D. 1997: *Homo geographicus*. Baltimore: Johns Hopkins University Press.

Smith, A. D. 1991: *National identity*. Harmondsworth: Penguin.

Tunbridge, J. E. and Ashworth, G. J. 1996: *Dissonant heritage: the management of the past as a resource in conflict*. Chichester: John Wiley.

Walden, G. 1997: France says no. *Observer*, 7 September.

Yuill, D. and Davezies, L. 1997: Economic and social cohesion in the European Union: the regional distribution of Member States' own policy. Paper read to Regional Studies Association Conference, Frankfurt an der Oder, September.

Index

Page references in *italic* refer to illustrations and tables

Absolutism 36–8
Agincourt, Battle of 37
Albania 123, 132
    population *10*, 138
Albigensian Crusade 37
Amber Room 246
Amsterdam 54, 71
Antwerp 71, 268
Asia
    Europe's relations with 22, 58, 124–5
    Russian expansion into 58–9, 127
Atlantic arc 55
Auschwitz/Birkenau 241
Australia 257
Austria 299, 304
    EU and 147
    GDP 170–1
    population *10*, 84
    unemployment 170
Austro-Hungarian Empire 83, 124, 129
Azerbaijan 34
    population *10*

Balance of power 92–5, 98
Barbarian invasions 22, 125
Barcelona 35, 172, 269, *271*
Basque Country, *see* Euskadi
Belarus' 135, 139
    population *10*
Belgium
    EU and 147, 158
    GDP 170–1
    linguistic divide 200, 268
    population *10*
Bellars, John 96–7
Belsec 241
Berlin 242, *253–5*, 259
    Berlin Wall 121–2, 254–5
    Kreuzberg 242
    recreation 255
Bonaparte, Napoleon 55
Bosnia 7, 128,
    population *10*, 138
Braudel, Fernand 4, 29, 34, 60–1
Bremen 246, 276
Brittany 43, 200
Brussels 148
    heritage 268–9
Buchenwald 252
Budapest 272
Bulgaria 83, 123, 133, *147*
    population *10*, 138
Byzantine Empire 55–6, 125–6

Canada 4, 256–7
CAP, *see* Common Agricultural Policy
Carbonari, *see* Italy
Carolingian Empire *56*, 66
Catalonia 35, 54–5, 83, 200
    Catalan language 204
Cathars 43

Cavour, Camillo 227
Charlemagne 126
Chechnya 7, 128
Chelmno 241
Christianity, *see also* Religion
    contemporary 189–90
    Council of Florence 126
    Council of Nicea 125
    in Central and Eastern Europe 125–8
    Crusades 36, 64
    Great Schism 7, 55–6, 59, 126, 192
    Latin 7, 36–7, 42–3, 55–6
    Orthodox 7, 55–6, 125–7
CIS, *see* Commonwealth of Independent States
City
    as text 261, 267–75
    conservation 262, 264–6, 275–8
Civil Rights 192–6
CMEA, *see* Council for Mutual Economic
    Assistance
Cohesion, *see* European Union
Cohesion Fund 165, 174, 181–2
Cold War 148, 247
Colmar 276
Common Agricultural Policy (CAP) 148, 167, 292,
    300, 303
    revision 295–6
    surpluses 294
Commonwealth of Independent States (CIS) 3, 124,
    135
Communism 59; *see also* Post-Communism
    collapse of 85, 121–4, 166
    command economy 132–3
    system of in Central and Eastern Europe 122–4
Competitiveness 168–9
Congress of Vienna 98
Conservation 264–6
    of nature 295
    'schools' 276
Constantine the Great 125
Constantinople 125–7
Convergence, *see* European Union
COR, *see* European Committee of the Regions
Córdoba 24, *56*
Coreper, *see* European Committee of Permanent
    Representatives
Corsica 178
Coudenhove-Kalergi, Richard 99–100
Council for Mutual Economic Assistance (CMEA
    or COMECON) 122–3, 132, 135
Council of Europe 278
Council of Ministers, *see* European Council of
    Ministers
Countryside, *see also* Rural tourism
    development of landscapes 287–91
    diversification 302–4
    dynamics of rural change 291–5
    extensification 298
    retreat from 297–9
Crécy, Battle of 37
Croatia 7, 57

population *10*, 138
Cumulative growth theories 175–6
Cyprus
 population *10*
Czech Republic 38, *147*
 population *10*
Czechoslovakia 123, 129, 132–3

Danzig, *see* Gdansk
D'Azeglio, Massimo 221
De Gaulle, Charles 155, 315
Delft 269
Denmark 55, 83, 299
 Danish Heath Society 305
 EU and 147, 158
 GDP 170–1
 population *10*
Deventer 276
Diversity 31–4, 41, 315–16
Dresden 249–50
Dublin 172

Eastern Front 246; *see also* Operation Barbarossa
ECJ, *see* European Court of Justice
ECSC, *see* European Coal and Steel Community
EEC, *see* Euopean Economic Community
EMU, *see* European Monetary Union
England 37, 84
 landscape images 214–15
English language 204
Enlightenment 22, 43, 191
ESF, *see* European Social Fund
ERDF, *see* European Regional Development Fund
ERM, *see* Exchange Rate Mechanism
Estonia 124, 135, *147*
 population *10*
Ethnic integration 195–6
EU, *see* European Union
Europe, *see also* City, Countryside, Modernization, Population, War
 as idea 3–4, 11, 91–5, 121–4, 139–40, 316
 countries *9*
 countryside 14
 ethnic groups 128–9, *187*
 heritage 41–5
 of the regions 34–5, 169
 physical geography 4, *5*, *6*, 230–1
 state formation 35–9
 vital axis 53–4, 172
 war in 11, 43–4
European Atomic Energy Agency 149
European Central Bank 148, 154
European Coal and Steel Community (ECSC) 145, 149
European Commission 145, 147–9, 151–2, 157, 316
 Commissioners 149
 competitiveness 168
 languages 203
 policy-making 153, 177
European Committee of Permanent Representatives (Coreper) 149, 157
European Committee of the Regions (COR) 158, 181
European Council 150
European Council of Ministers 145, 148–9, 151, 153, 157–8
 decision-making 154–5
 voting in 155–9

European Court of Justice (ECJ) 145, 148, 160–1
 functions of 150–1, 154–5
European Economic Community (EEC) 146, 148
European Free Trade Association 166
European Monetary Union (EMU) 148, 166, 177, 183
European Parliament 145, 149, 151, 157–9
 role 150, 153–4
European Regional Development Fund (ERDF) 167, 181
 Objectives 167
European Social Fund (ESF)167
European Union (EU) 12, 38–9, 130, 135; *see also* Cohesion Fund, Common Agricultural Policy, Structural Funds
 boundaries 7
 cohesion 165–70, 174, 178–81
 'Cohesion Four' 170
 convergence and divergence 174–8, 182
 creation 148
 'democratic deficit' 157–60, 312–13
 enlargements 146–7, 150, 166, 182, 203, 296
 fiscal transfers 179–81, 312
 GDP 170–*1*, 176–7
 ideology 42, 311–12
 institutional network 151–2
 institutions 148–51
 policy-making 153–5
 reform in 157–60
 regional disparities 170–4, 176–7
 Schengen Agreement 159
 Social Chapter 179
 sovereignty 152, 160–1
 subsidiarity 158, 179–80, 314
 unemployment 170, 172–3
Euskadi (Basque Country) 35, 178, 200
 language 205
Evans, Estyn E. 34
Exchange Rate Mechanism (ERM) 166

Fascism 107
Ferdinand, Archduke 57
Ferdinand (of Castile) and Isabella (of Aragon) 24, 37, 127
Feudalism 29–31, 36–7, 68
 in Eastern Europe 58–9, 68–9, 133–4
 transition to capitalism 65–6
Finland
 EU and 147, 158
 GDP 170–1
 language policy 200
 population *10*
 unemployment 170, 174
Flexibility 167–8
Florence 22, 219–24
Fordism 168
France 93–4
 départements 35
 diversity and identity of 31–4, 298–9
 EU and 147, 155, 157
 expansion of *32*, 37, *93*
 Napoleonic Wars 100–5
 population *10*
 pays 32–4, 290–1
 Revolution 83, 100, 104
 unemployment 170–1
 Vichy regime 290, 298
Franco, General 24, 107
Frisian language 204

Garibaldi, Giuseppe 221, 227
Gastarbeiter 241
GATT, *see* General Agreement on Tariffs and Trade
Gdańsk (Danzig) 243–7
  reconstruction as Polish city 244–6
GDP, *see* Gross Domestic Product
General Agreement on Tariffs and Trade (GATT) 152
Genoa 231
Georgia 34
  population *10*
German Democratic Republic (GDR) 123, 132, 249–55; *see also* Weimar
German language 205
Germany, see also *Lebensraum* and *Mitteleuropa* 299–300
  and idea of Europe 106–7
  changing boundaries 7–8, *111*
  colonization 64, 128–9, *301*
  EU and 147, 157–9
  GDP 170–1
  'lost' 243–9
  Nazism 107, 109–10, 240
  population *10*, 84
  unification 7, 38, 129, 225
  unification of West and East, 249–55
Gers 305
Gerschenkron, A. 77–80
Globalization 168–9
Goethe, Johann Wolfgang von 251–2
Gorbachev, Mikhail 138
GORs, see Government Offices in the Regions
Government Offices in the Regions, *see* United Kingdom
Great Britain 54, 71; *see also* United Kingdom
  Identity 2, 39–40
Greece 55, 83, 124
  EU and 147
  GDP 170–1
  population *10*
Groningen
  conservation area 279–80
  Jewish quarter 272–3
Gross Domestic Product (GDP), *see* European Union

Hanseatic League 240, 243, 246, 269
Havel, Václav 198
Herder, J. G. 136
Heritage 236–8, 242
  city as text 261–83
  dissonance of 43, 237, 275
  European 41–5, 258–9, 277–8
  German 249–55
  messages 279–83
  misplaced 243–9
  of war 113–14
  reconceptualization 249–55
  users 279–83
Hess, Rudolph 254
Hitler, Adolf 107, 246, 254
Holy Roman Empire 129, 191
Holocaust 42, 44, 110, 189, 240–1, 272; *see also* Jews
  death camps 241
  destruction of European Jewry *112*
  memorialization 241
Horthy, Mikós 272

Hungary 84, 123–4, 129, 132–3, 138–9, *147*, 200, 272
  population *10*
Iceland
  population *10*
Identity 1–2, 39–41
  and nationalism 213–16
  heritage and 41–5
  landscape and 213–16, 230–3
  war and 113–14
'Ioannina Compromise' 156–7
Ireland
  EU and 147, 158
  GDP 170–1
  language 204
  nationalism 40, 200, 315–16
  population *10*, 55, 84
  unemployment 171
Iron Curtain 7
Islam 23–8, 58, 127
Italy 54–5, 83; *see also* Rome, Tuscany
  Carbonari 220
  city-states 217–19
  EU and 147
  Fascism 107, 224, 227–8, 232
  GDP 170–1
  national landscapes 13, 219–33
  Macchiaoli 219–24
  population *10*, 84
  'Third Italy' 168–9
  unemployment 170–1, 174
  unification (Risorgimento) 38, 217–24
Ivan the Terrible 126–7

Jews 240–1; *see also* Holocaust
Joseph II 134, 136

Kaliningrad 246–9
Kallen, E. 195–6
Kant, Immanuel 246–7
Kemal, Mustafa (Atatürk) 267
Köningsberg, *see* Kaliningrad
Kraków 240, 259
  Kraków-Kazimierz 272
Kreuzberg, *see* Berlin
Kulikovo, Battle of 127

Landscape 213–16, 230–2, 242
  city as text 261–4
  cultural 19–21, 40
  Roman landscape ideal 214–32
  rural 287–91
  Tuscan landscape ideal 219–24
Language 128–9, 186–207
  and nation-building 188–91
  attributes of 199–200
  conflict 201–2
  hegemonic 202–6
  lesser 206
  policy and state-formation 199–200
Latvia 124, 135, *147*
  population *10*
Lebensraum 107, 110
Lega Lombarda, *see* Northern League
Lega, Silvestro *222*
Leiden 269
Lenin, Vladimir Ilyich 137

Liechtenstein
   population *10*
Liège 268
Lithuania 124, 135, *147*
   population *10*
Lombardy 35
London 269
Lukashenka, President 135
Luxembourg (city) 149
'Luxembourg Compromise' 155
Luxembourg (state)
   EU and 147, 158–9
   GDP 170–1
   population *10*
   unemployment 170
Lyons, F. S. L. 315–16

Maastricht (city) 269
   Maastricht Treaty, *see* Treaty of Maastricht
Macedonia (FYR of)
   population *10*, 138
Macchiaoli (painters), *see* Italy
Mackinder, Halford 129–30
Majdanek 241
Malta
   population *10*
Mansholt, Sicco 294
Manzoni, Alessandro 221
Maria Therese 134
Mérimée, Prosper 265
Migration 42, 138, 189, 205, 290
   misplaced peoples 243–9
   post-World War II 241
Milan 224, 231
Minorities 138, 192–6, 205–7, 239–40; *see also*
   Jews
Mitteleuropa 107, 110, 124, 205
Mitterand, François 258
Modernization 10–11, 53–60
   disparities in development 77–84
   Kondratieff cycles 61–5
   modern industrial development 72–6, *81*
   proto-industrialization 70–2
   twentieth-century 85
Moldova
   population *10*
Monaco
   population *10*
Mongol Empire *58*–9, 127
Monnet, Jean 148, 314
Morris, William 265
Multiculturalism 196–9, 255–8, 316
Mussolini, Benito 227–9

Naples 231
Napoleonic Wars, *see* France
Nationalism 20–1, 39–44
   and heritage 236–8
   and identity 214–18
   and nation-states 214–18, 277–8, 313–14
   in Central and Eastern Europe 128–9, 136–7
   ethnic and civic 136–7
NATO, *see* North Atlantic Treaty Organization
Nazism, *see* Germany
Netherlands 299
   cities 269, *274*
   Dutch United Provinces 71
   EU and 147, 158

GDP 170–1
   population *10*
   urban images 269
'New medievalism' 38–9
Normandy 305–6
North Atlantic Treaty Organization (NATO) 130,
   135
Northern League 35, 178, 231
Norway 55, 83, 200
   population *10*, 84
Norwich 276
Nuremberg 269

Operation Barbarossa 110
Ottoman Empire 55, *57*

Palermo 231
Papacy 225, 230
Paris 172, 269–70
Peace of Augsburg 191
Penn, William 95–6
Peter the Great 126, 134, 246
Pirenne, Henri 22
Poland 123–4, 129, 132–3, 138, *147*, 243–7, 296
   Jewish minority 240
   population *10*, 67
   religious history 191
   rural tourism *293*
Population 7, *10*, 67
   emigration 84
Portugal 23–4, 71, 83
   EU and 147, 158
   GDP 170–1
   population *10*, 84
Post-Communism 249–55
   marketization 135
   'return to Europe' 124, 131–2
Prague
   Jewish Museum 240–1
Pugin, A. W. N. 265

Regions 34–5, 169, 178–81, 313
   disparities 176–7
   unemployment 170
Reformation 22, 192
Religion 189–90
   religious plurality 191–2
Renaissance 22, 43, 219
Richardson, Lewis Fry 99
Risorgimento, *see* Italy
Roermond 269
Romania 123, 133, 139, *147*
   population *10*, 138
Roman Empire 6–7, 22, 125, 213
Romanticism 22, 219–24
Romany 205, 240
Rome
   and Fascism 227–8
   and Papacy 225–6, 229
   and unification of Italy 224–5
   city-plan 226–30
   landscape ideal 214–32
   Vittoriano 27
Rural tourism 303–4
   Poland *293*
Ruskin, John 265
Russia 3, 59, 84, 134; *see also* CIS and USSR
   Hitler's invasion of 110

Russia *cont.*
    Kievan 56, 66, 139
    Napoleon's invasion of (1812) *105*
    population *10*, 84
    Revolution 137
    rise of Muscovite state 58–9, 126–7

Saint-Pierre, Abbé de 97
Saint-Simon, Claude Henri de 97–8
San Francisco 276
San Marino
    population *10*
Santer, Jacques 149
Santiago de Compostela 23–8, 278
    Camino de Santiago 26–8, 278
Sarajevo 57
Savoy-Piedmont 217
Schiller, Friedrich von 250
Schindler's List 240
Schumpeter, Joseph 61
Schuman, Robert 314
Scotland 43, 84, 304
Scott, Walter 221
Scottish Nationalist Party 178
Sella, Quintano 225
SEM, *see* Single European Market
Separatist movements 43–4, 178–9
Sernesi, Raefello *223*
Seville 24
Signorini, Telemaco 220–2
Single European Act 169
Single European Market (SEM) 148–9, 164–5
Skolt Lapp 205
Slovakia 38, 124, *147*
    population *10*, 138
Slovenia 7, 57, *147*
    population *10*
Sobibor 241
Solidarity 246
South Africa 257–8
Spain 71, 83, 94, 278
    Civil War 107, *109*
    EU and 147
    GDP 170–1
    Moorish invasion 23–4
    pilgrimage in 23–8, 278
    population *10*, 84
    Reconquista 24, 26, 37, 64, 278
    unemployment 170, 174
Spanish language 205
Speer, Albert 254
Stalin, Joseph 110, 123
States
    formation 29–34, 36–9, 92, 129–30,
        199–200
    nation-states 39–41, 214–18
    sovereignty of 38–9
Sterne, Lawrence 221
Stewart, W. A. 199–200
Strasbourg 150
Structural Funds 165, 167, 181–2
Sully, Duc de (Maximilien de Béthune) 95
Sweden 55, 83
    EU and 147, 296
    GDP 170–1
    population *10*
Switzerland 200, 299
    population *10*

Tartars 58
Tatarstan 128
Thatcher, Margaret 258
Thirty Years War 92
Toulouse, County of 31, *33*
Treaty of Amsterdam 148, 312–13
Treaty of Paris 145, 147, 150–1
Treaty of Maastricht 148, 150–1, 158–9, 160, 164,
    169, 174, 177
Treaty of Rome 146–7, 150–1, 155
Treaty of Utrecht 92, 94
Treaty of Versailles 98, 107, 130
Treaty of Westphalia 92
Treblinka 241
Turin 224
Turkey
    creation of secular state 267–8
    population *10*
    Empire 127
Tuscany
    landscape ideal 219–24

UK, *see* United Kingdom
Ukraine 128, 135, 139
    population *10*
Unemployment, *see* European Union
United Kingdom (UK) 178, 299
    EU and 147, 157–8, 179
    GDP 170–1
    Government Offices in the Regions (GORs)
        178
    population *10*
United States of America 180, 232, 257
Urbanization
    city-states 31, 217–19
    Dark Age Europe 22
    medieval 30–1, 36, 69
USSR 124, 137–8
    break-up 135
    economy 129–35

Vichy, *see* France
Vidal de la Blache, Paul 32, 34
Vienna 127, 129
Viollett-le-Duc, Eugène 265
'Vital axis', *see* Europe

Wales 200, 204
Wallerstein, Immanuel 60–1
War 11, 43–4, 113–14; *see also* France, Germany
    and peace 95–100
    Europe built on 100–13
    geography and 89–91, *108*
    idea of Europe and 91–5
    mortality in 44, 91, 104–5, 110, *112*
Warsaw 43, 249
Warsaw Military Pact 123
Washington, DC 241
Weimar 250–2
Western Front 44, 114
Wolfsschanze 246
World War I
    roots of 106
    rural settlement and 290

Yeltsin, Boris 135
Yugoslavia 34, 42, 57, 123, 129, 132, 189
    population *10*